New Perspectives in Mathematical and Statistical Methods for Actuarial Sciences and Finance

Michele La Rocca · Massimiliano Menzietti ·
Cira Perna · Marilena Sibillo
Editors

New Perspectives in Mathematical and Statistical Methods for Actuarial Sciences and Finance

Springer

Editors
Michele La Rocca
Department of Economics and Statistics
University of Salerno
Fisciano, Salerno, Italy

Massimiliano Menzietti
Department of Economics and Statistics
University of Salerno
Fisciano, Salerno, Italy

Cira Perna
Department of Economics and Statistics
University of Salerno
Fisciano, Salerno, Italy

Marilena Sibillo
Department of Economics and Statistics
University of Salerno
Fisciano, Salerno, Italy

ISBN 978-3-032-05550-7 ISBN 978-3-032-05551-4 (eBook)
https://doi.org/10.1007/978-3-032-05551-4

This Springer imprint is published by the registered company Springer Nature Switzerland AG
The registered company address is: Gewerbestrasse 11, 6330 Cham, Switzerland

Preface

This volume is a collection peer-reviewed papers selected from those presented at the International Workshop on New Perspectives in Mathematical and Statistical Methods for Actuarial Sciences and Finance, held in Salerno on 27–28 June 2025.

This workshop, organized by the Department of Economics and Statistics of the University of Salerno, is part of the activities related to the MAF (Mathematical and Statistical methods for Actuarial Sciences and Finance) initiatives, whose main events are the biennial international conferences that have been held regularly since 2006 in Italy and abroad. The next edition is scheduled for 2026 and will be held in Barcelona.

In 2025, which is a transitional year between the two editions of the MAF conference, we proposed two study days aimed at presenting new ideas and innovative lines of research in the field of mathematical and statistical methods for insurance and finance, both from a theoretical and applied point of view. To this end, the workshop included both invited speakers and poster presentations. All the speakers, skilled in outlining new and exciting frontiers in research, presented relations aimed at encouraging interactions and discussions among participants around new ideas.

The main topics of interest were Artificial intelligence, Machine learning, Blockchain technology, Behavioral finance and insurance, Policy indications, Sustainability, Environmental, Social and Governance (ESG) criteria, Silver economy, Gender analysis, Climate-related Risks and Insurance, Insurtech and Fintech.

This workshop confirmed the growing interest of the international scientific community in these topics, with about 100 participants, about 30 posters and 8 prestigious plenary speakers, namely:

ROBERTO BALDONI (ACN- Agenzia per la Cybersicurezza Nazionale, Sapienza Università di Roma), who presented a talk titled: The AI Paradox: Between Global Power, Technology Sourcing, Errors and (De)regulation Shifts
ENRICO BIFFIS (Imperial College, London), who presented a talk titled: Short-lived Gases, Carbon Markets and Climate Risk Mitigation
MICHEL DACOROGNA (Partner at Prime Re Solutions, University of Zurich), who presented a talk titled: A Dynamic Framework for Correcting Procyclicality in Risk Estimation
GIAMPIERO GALLO (New York University in Florence), who presented a talk titled A Bird's Eye View on Volatility: from Measurement to Modeling
PAOLO GIUDICI (University of Pavia), who presented a talk titled: SAFE Artificial Intelligence
EDOARDO OTRANTO (Sapienza University of Rome) who presented a talk titled: Parsimonious Modeling of Covariance Matrices: Classical and High-Frequency Approaches
SALVATORE SCOGNAMIGLIO (Parthenope University of Naples), who presented a talk titled: Explainable Least Squares Monte Carlo for Solvency Capital Requirement Evaluation

DIMITRIS N. POLITIS (University of California, San Diego), who presented a talk titled: Model-free Bootstrap and Conformal Prediction in Regression: Conditionality, Conjecture Testing, and Pertinent Prediction Intervals

Also, this workshop, as has been the case for all MAF conferences, was accompanied by a book published by Springer, with the aim of offering a selection of scientific contributions in which the authors present their ideas and the methodology behind its development, providing, when possible, an illustrative application.

Finally, we would like to express our deep gratitude to the members of the Scientific and Organizing Committees and to all the people whose collaboration contributed to the success of the workshop. In particular, our heartfelt thanks go to Giulia Magni, who has worked unstintingly with great enthusiasm and efficiency, continually showing with her work the sharing of the aims of the initiative. We would also like to thank all the participants for their precious and indispensable contribution.

June 2025

Michele La Rocca
Massimiliano Menzietti
Cira Perna
Marilena Sibillo

Contents

Towards Fairer Sanction Systems: Income-Based Models with Aggregation Functions

Luca Anzilli[1(✉)], Marta Cardin[2], and Silvio Giove[2]

[1] Department of Economic Sciences, University of Salento, Lecce, Italy
luca.anzilli@unisalento.it

[2] Department of Economics, Ca' Foscari University of Venice, Venice, Italy
{mcardin,sgiove}@unive.it

Abstract. This paper presents a novel framework for designing income-based fines that integrate both the severity of the offence and the financial capacity of the offender. The model employs a grid-based interpolation approach, initially using bilinear interpolation and then extending to more expressive aggregation techniques such as the Choquet integral and t-norms. This generalization allows for customizable sanction functions that reflect diverse legal and ethical priorities. Numerical simulations illustrate the impact of different aggregation strategies on fine outcomes.

Keywords: Sanctions · Decision support · Aggregation functions · Multiple-criteria analysis · Interpolation

1 Introduction

One of the fundamental responsibilities of government is to deter individuals from engaging in actions that cause harm to others. Traffic accidents, for instance, represent the most prevalent unnatural cause of death globally. To mitigate this, governments employ measures such as speeding tickets to discourage dangerous driving behaviours. Economic theory suggests that increasing the severity of punishment, such as the size of speeding fines, should reduce criminal activity [2].

The imposition of fines is a widely used sanction in legal systems around the world [18]. Traditionally, fines are determined solely on the basis of the seriousness of the offence, without regard to the economic status of the offender. This conventional approach often leads to disproportionate consequences, where a fixed fine may be a minor inconvenience for a wealthy individual, but a significant financial burden for someone with limited resources. Many state courts recognize the importance of considering the defendant's wealth when determining punitive damages [17]. In a day fines system, sanctions are determined according to a person's financial circumstances first by assigning a number of units corresponding to the nature and severity of the offence ("units" or "days"). Subsequently, an

M. La Rocca et al. (Eds.): WMAF 2025, *New Perspectives in Mathematical and Statistical Methods for Actuarial Sciences and Finance*, pp. 1–13, 2025.
https://doi.org/10.1007/978-3-032-05551-4_1

amount is determined based on the offender's ability to pay per unit ("daily unit" or "daily rate"). An overview of the diverse legal frameworks of day fines systems across Europe is provided in [13].

Recent studies have explored the impact of punishment severity on crime, particularly in contexts where fines are scaled according to an individual's income [10]. Law-and-economics analysis suggests that the day-fine model, which adjusts fines based on the offender's income and the severity of the offence, is superior to other forms of fines. This model ensures that offenders with different socioeconomic statuses pay the same proportion of their wealth, thereby deterring both the rich and the poor equally [11]. Further analysis by [5] underscores the benefits of adjusting fine magnitudes based on the offender's wealth. This study highlights that deterrence is a function of the expected sanction, suggesting that day fines could be implemented by varying the probability of apprehension across different economic groups. However, making the probability of apprehension contingent on wealth proves more costly than adjusting fine magnitudes. The latter approach incurs verification costs only upon violation, whereas the former necessitates wealth verification for all individuals, irrespective of violations. Consequently, conditioning fines on wealth emerges as a more cost-effective strategy. Despite its potential, the day-fine system encounters significant obstacles to efficient implementation, primarily due to the high costs associated with collecting accurate financial information about offenders. While enforcement authorities can theoretically obtain this information, the process is often prohibitively expensive. To address this issue, [14] proposes a secondary day-fine system designed to induce truthful self-reporting of wealth. This system considers multiple factors, including the offender's wealth, the extent of misreporting, and the severity of the primary fine. By tailoring fines to individual circumstances, this approach enhances both individual and general deterrence.

Moreover, although there is some convergence, countries diverge on many aspects of the day fine model. These differences concern not only the design of the day fine system within specific legislation, but also its practical implementation [12]. Different jurisdictions vary considerably in how the amount per day is calculated, what information is used, and whether there are upper and lower limits on units, which affects the severity of offences that can be punished with daily fines. In addition, the amount per unit that can be charged to a person varies considerably.

From a theoretical perspective, there are strong justifications for introducing a fine system based on both retribution and deterrence arguments. However, despite these theoretical advantages, many countries remain reluctant to adopt day fines. The primary reasons for this reluctance include the negative perception of fines and the complexity of the system due to its reliance on the offender's financial information [12]. In particular, given the crucial role of judges in this system, it is essential to involve them in the design process. Before introducing daily fines, it is useful to inform judges in detail about the system and to address their concerns. By involving judges early in the process, their insights can be used to refine the system, thereby increasing its effectiveness and gaining their support [12].

Building on these considerations, in this paper we propose a novel decision-making based approach to establishing fines that addresses these challenges. Our methodology not only considers the gravity of the offence but also incorporates the wealth of the offender, ensuring a more balanced and just sanctioning process. The key feature of our approach is its flexibility, allowing policymakers to adjust sanctions based on their preferences by tuning suitable parameters. This adaptability ensures that the fine system can be tailored to different legal and social contexts, enhancing its applicability and effectiveness. Additionally, our approach produces smooth fines, avoiding abrupt tiered jumps. This method creates a monotonic increasing function of the violation, simplifying the process for legislators to define fines by setting key points and interpolating between them. We utilize aggregation functions within the interpolation mechanism to ensure a comprehensive and balanced fine determination process. This approach can serve as a useful tool for designing fine systems that are both equitable and effective. Some numerical applications are also provided. Furthermore, the proposed methodology, based on the use of aggregation functions [3,8,9,16], can be extended to address multidimensional problems.

The paper is structured as follows. Section 2 provides a brief overview of wealth-dependent fine systems. In Sect. 3, we introduce the core concepts of our proposed model. Section 4 extends this framework into a more flexible and generalizable system using advanced aggregation techniques. Finally, Sect. 5 offers concluding remarks and outlines directions for future research.

2 A Brief Discussion on Wealth-Dependent Fines

Wealth-dependent fines represent a significant advancement in the design of monetary sanctions, aiming to ensure proportionality and fairness in punishment. Traditional fixed-fine systems impose identical penalties regardless of the offender's financial situation, often resulting in disproportionate burdens: a minor inconvenience for the wealthy, but a severe hardship for those with limited means. In contrast, wealth-dependent systems adjust the magnitude of the fine based on the offender's economic capacity, thereby equalizing the relative impact of the sanction. We refer to [14] for a detailed discussion.

Among these systems, the day-fine model is one of the most prominent and widely studied. It operates through a two-step process. First, the court determines the number of penalty units—commonly referred to as "days"—based on the gravity of the offence. Second, the monetary value of each unit is calculated as a function of the offender's income, typically as a fixed proportion. The total fine is then the product of these two components. For instance, if an offender earns €100 per day and the daily unit is set at 50% of income, a 10-day fine results in a total sanction of €500. Conversely, an offender earning €20 per day would pay €100 for the same offence. This structure ensures that the financial burden is proportionally distributed across offenders, regardless of their income level. Formally, the sanction can be expressed as:

$$S = F(V, R) = n(V) \cdot d(R),$$

where $n(V)$ denotes the number of days determined by the severity of the offence V, and $d(R)$ represents the daily unit, typically defined as $d(R) = \alpha \cdot R$, with α being a legally prescribed constant.

Despite its theoretical appeal, the practical implementation of the day-fine system faces notable challenges—chief among them, the accurate assessment of the offender's financial status. Verifying income through official channels can be administratively burdensome and costly. As a result, courts often rely on self-reported income, which introduces the risk of strategic underreporting. According to deterrence theory, rational offenders will misrepresent their wealth if doing so reduces their expected sanction and carries no penalty.

To mitigate this issue, [14] propose an enhanced model that incorporates a secondary fine component to penalize misreporting. In this framework, the sanction becomes a function of three variables:

$$S = F(V, R_1, R_2),$$

where R_1 is the reported income and R_2 is the true income. The secondary fine increases with the discrepancy between R_1 and R_2, thereby incentivizing truthful declarations. This dual-layered structure—comprising a primary fine based on offence severity and reported wealth, and a secondary fine based on misrepresentation—strengthens both deterrence and fairness. Moreover, this framework opens the door to more sophisticated sanctioning systems that can incorporate additional dimensions, such as whether the offence is a first-time or repeat violation, or other contextual factors. As such, wealth-dependent fines, particularly when augmented with mechanisms to ensure truthful reporting, offer a promising path toward more equitable and effective penal policies.

3 The Income-Based Fine Design Model

In this section we introduce the proposed model for assigning sanctions on the basis of both gravity of the offence but also incorporates the wealth of the offender.

The amount of sanction S applied to an offence is determined by both the severity of the violation (e.g. speeding) V and the economic status of the offender R. Here, R represents the resources or financial capacity of the offender, i.e., the wealth (or income) indicator. The aim is to assign the amount of sanction $S = F(V, R)$ for every $(V, R) \in [V_{min}, V_{max}] \times [R_{min}, R_{max}]$. We divide the interval $[V_{min}, V_{max}]$ into n_V sub-intervals $[v_i, v_{i+1}]$, through the values $v_0, v_1, \ldots, v_{n_V}$, with $v_0 = V_{min}$ and $v_{n_V} = V_{max}$, and interval $[R_{min}, R_{max}]$ into n_R sub-intervals $[r_i, r_{i+1}]$, through the values $r_0, r_1, \ldots, r_{n_R}$, with $r_0 = R_{min}$ and $r_{n_R} = R_{max}$. Therefore, the rectangle $[V_{min}, V_{max}] \times [R_{min}, R_{max}]$ is now divided in a mesh of $n_v \cdot n_R$ sub-rectangles

$$[v_i, v_{i+1}] \times [r_j, r_{j+1}] \qquad i = 0, 1, \ldots, n_V - 1, \; j = 0, 1, \ldots, n_R - 1 .$$

The idea is to determine the amount of the penalty in each rectangle of the grid, knowing the values in the vertices. We denote

$$S_{i,j} = F(v_i, r_j) \qquad i = 0, 1, \dots, n_V - 1,\ j = 0, 1, \dots, n_R - 1$$

the values of sanction in the vertices (v_i, r_j). See Fig. 1.

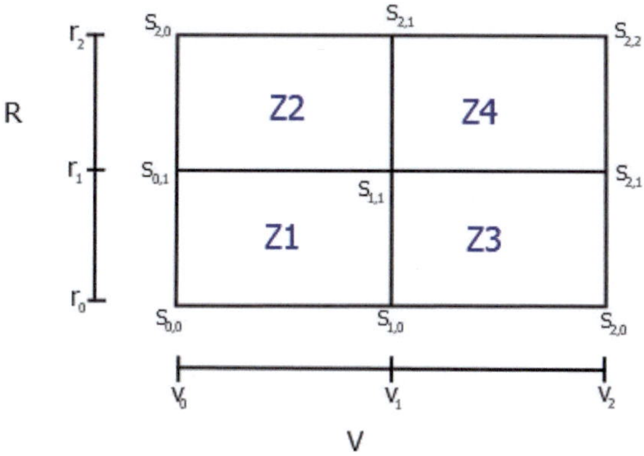

Fig. 1. The reference grid in the case where $n_V = n_R = 2$.

With reference to cell $[v_i, v_{i+1}] \times [r_j, r_{j+1}]$, note that

$$\begin{aligned} &S_{i,j} = F(v_i, r_j), \quad S_{i+1,j} = F(v_{i+1}, r_j), \quad S_{i,j+1} = F(v_i, r_{j+1}), \\ &S_{i+1,j+1} = F(v_{i+1}, r_{j+1}) \end{aligned} \tag{1}$$

with $S_{i,j} < \min(S_{i+1,j}, S_{i,j+1})$ and $S_{i+1,j+1} > \max(S_{i+1,j}, S_{i,j+1})$. For $v_i \leq V \leq v_{i+1}$ and $r_j \leq R \leq r_{j+1}$, it is convenient to express the sanction amount $F(V, R)$ as

$$F(V, R) = S_{i,j} + f(V, R) \cdot (S_{i+1,j+1} - S_{i,j}) \tag{2}$$

where $f(V, R)$ describes the *mark-up percentage* (o *mark-up rate*) applied (w.r.t. the maximum mark-up $S_{i+1,j+1} - S_{i,j}$), with

$$\begin{aligned} &f(v_i, r_j) = 0, & &f(v_{i+1}, r_j) = \frac{S_{i+1,j} - S_{i,j}}{S_{i+1,j+1} - S_{i,j}}, \\ &f(v_i, r_{j+1}) = \frac{S_{i,j+1} - S_{i,j}}{S_{i+1,j+1} - S_{i,j}}, & &f(v_{i+1}, r_{j+1}) = 1. \end{aligned} \tag{3}$$

Furthermore, it is convenient to introduce the following coefficients

$$\begin{aligned} \sigma_0 &= S_{i,j} \\ \sigma_1 &= S_{i+1,j} - S_{i,j} \\ \sigma_2 &= S_{i,j+1} - S_{i,j} \\ \sigma_3 &= S_{i+1,j+1} - S_{i+1,j} - S_{i,j+1} + S_{i,j} \\ \sigma &= S_{i+1,j+1} - S_{i,j}\,. \end{aligned} \tag{4}$$

We observe that the coefficient σ_3 can be negative. Since $\sigma = \sigma_1 + \sigma_2 + \sigma_3$, conditions (3) can be rewritten as

$$\begin{aligned} f(v_i, r_j) &= 0\,, & f(v_{i+1}, r_j) &= \frac{\sigma_1}{\sigma}\,, \\ f(v_i, r_{j+1}) &= \frac{\sigma_2}{\sigma}\,, & f(v_{i+1}, r_{j+1}) &= 1\,. \end{aligned} \tag{5}$$

We observe that σ_1/σ is the charge rate applied in (v_{i+1}, r_j), σ_2/σ is the charge rate applied in (v_i, r_{j+1}). Note that $\frac{\sigma_3}{\sigma} = 1 - \frac{\sigma_1}{\sigma} - \frac{\sigma_2}{\sigma}$. Also, we observe that (2) can be alternatively expressed as $F(V, R) = \sigma_0 + f(V, R) \cdot \sigma$.

We introduce the increasing value functions (or utility functions) $u_V = u_V(V)$ and $u_R = u_R(R)$ given by

$$u_V : [v_i, v_{i+1}] \to [0, 1]\,, \qquad u_R : [r_j, r_{j+1}] \to [0, 1]$$

with $u_V(v_i) = 0$, $u_V(v_{i+1}) = 1$, $u_R(r_j) = 0$ and $u_R(r_{j+1}) = 1$. For example, employing linear value functions $u_V(V)$ e $u_R(R)$ we have

$$u_V(V) = \frac{V - v_i}{v_{i+1} - v_i} \qquad v_i \le V \le v_{i+1} \tag{6}$$

and

$$u_R(R) = \frac{R - r_j}{r_{j+1} - r_j} \qquad r_j \le R \le r_{j+1}\,. \tag{7}$$

The idea is to formulate the sanction $F(V, R)$, as expressed in (2), using the representation

$$f(V, R) = g\left(u_V(V), u_R(R)\right) \tag{8}$$

for each $(V, R) \in [v_i, v_{i+1}] \times [r_j, r_{j+1}]$, where $g : [0, 1]^2 \to [0, 1]$ is an increasing function in each variable interpolating the value assigned in (5) to the vertices of the bidimensional unit square $Q = [0, 1]^2$, that is such that

$$g(0, 0) = 0\,, \quad g(1, 0) = \frac{\sigma_1}{\sigma} \quad g(0, 1) = \frac{\sigma_2}{\sigma}\,, \quad g(1, 1) = 1\,. \tag{9}$$

3.1 Bilinear Interpolation

First, we determine the function g using the bilinear interpolation. Therefore, we express $g(x, y)$ as

$$g(x, y) = g(0, 0)(1 - x)(1 - y) + g(0, 1)(1 - x)y + g(1, 0)x(1 - y) + g(1, 1)xy\,.$$

Substituting conditions (9), we get

$$g(x, y) = \frac{\sigma_2}{\sigma}(1 - x)y + \frac{\sigma_1}{\sigma}x(1 - y) + xy$$

that, since $\sigma_3 = \sigma - \sigma_1 - \sigma_2$, can be rewritten as

$$g(x,y) = \frac{\sigma_1}{\sigma}x + \frac{\sigma_2}{\sigma}y + \frac{\sigma_3}{\sigma}xy\,. \tag{10}$$

From (8) it follows

$$f(V,R) = \frac{\sigma_1}{\sigma}\cdot u_V(V) + \frac{\sigma_2}{\sigma}\cdot u_R(R) + \frac{\sigma_3}{\sigma}\cdot u_V(V)\cdot u_R(R)$$

from which, using (2), we obtain

$$S = F(V,R) = \sigma_0 + \sigma_1\cdot u_V(V) + \sigma_2\cdot u_R(R) + \sigma_3\cdot u_V(V)\cdot u_R(R)\,. \tag{11}$$

Expression (11) allows to determine the sanction when the bilinear interpolation is employed.

3.2 Numerical Simulations

To illustrate the features of our proposed methodology, we present a numerical example based on a discretized domain of offence severity and offender wealth. Specifically, we consider the intervals $[V_{\min}, V_{\max}] = [0, 80]$ for the severity of the violation, and $[R_{\min}, R_{\max}] = [800, 6000]$ for the offender's income. We partition these intervals as follows:

$$v_0 = 0\,, \qquad v_1 = 20\,, \qquad v_2 = 80$$

and

$$r_0 = 800\,, \qquad r_1 = 2500\,, \qquad r_2 = 6000$$

The sanction values are then specified at the vertices of the resulting grid:

$$\begin{array}{lll} S_{0,0} = 0 & S_{0,1} = 0 & S_{0,2} = 0 \\ S_{1,0} = 30 & S_{1,1} = 70 & S_{1,2} = 300 \\ S_{2,0} = 100 & S_{2,1} = 500 & S_{2,2} = 2000 \end{array} \tag{12}$$

Figure 2 shows the sanction $S = F(V,R)$ determined using Eq. 11 with linear utility functions (6) and (7).

We observe that the proposed income-based fine design model offers a flexible framework for determining sanctions. By discretizing the domain into a grid and applying bilinear interpolation, the model ensures a smooth and continuous transition of fines across different levels of income and violation severity. This avoids abrupt jumps or inconsistencies in sanctioning, which are common in tiered systems. Moreover, the use of value functions and aggregation techniques allows policymakers to tuning the model to reflect legal, ethical, or social preferences.

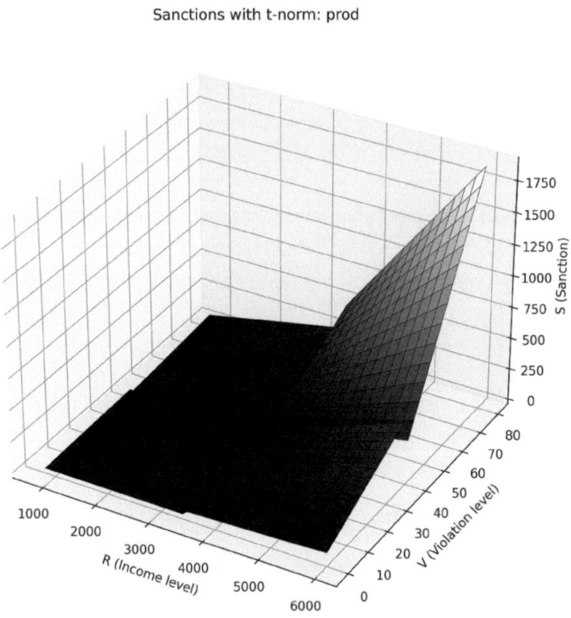

Fig. 2. Sanction determined using the bilinear interpolation (11) (corresponding to the employment of product t-norm in (15))

4 A More General Model Based on a Multi-criteria Approach

In this section, we extend the income-based fine model introduced earlier by incorporating a more flexible and expressive framework for sanction design. Specifically, by following a multi-criteria approach [4], we generalize the interpolation mechanism by first introducing the discrete Choquet integral and then proposing a broader formulation based on t-norms, which serve as generalized"AND" operators. This approach allows for a flexible aggregation of offence severity and offender wealth, enabling policymakers to tailor the sanctioning function to diverse legal and ethical contexts.

4.1 Interpolation Using the Choquet Integral

We begin by considering the Choquet integral as the interpolating function g:

$$g(x, y) = C_m(x, y)$$

where m is the non-additive measure (capacity) on the set of criteria $N = \{V, R\}$. The capacity values are determined by the sanction values at the vertices of the unit square:

$$m(\emptyset) = g(0,0) = 0, \qquad m(\{V\}) = g(1,0) = \frac{\sigma_1}{\sigma},$$

$$m(\{R\}) = g(0,1) = \frac{\sigma_2}{\sigma}, \qquad m(\{V,R\}) = g(1,1) = 1.$$

Using the identities $\sigma - \sigma_2 = \sigma_1 + \sigma_3$ and $\sigma - \sigma_1 = \sigma_2 + \sigma_3$, the Choquet integral can be expressed as follows: if $x \leq y$, $g(x,y) = C_m(x,y) = x + m(\{R\})\cdot(y-x) = (1 - m(\{R\}))\cdot x + m(\{R\})\cdot \mathrm{y} = \frac{\sigma - \sigma_2}{\sigma}x + \frac{\sigma_2}{\sigma}y = \frac{\sigma_1}{\sigma}x + \frac{\sigma_2}{\sigma}y + \frac{\sigma_3}{\sigma}x$; if $x \geq y$: $g(x,y) = C_m(x,y) = y + m(\{V\})\cdot(x-y) = (1 - m(\{V\}))\cdot y + m(\{V\})\cdot \mathrm{x} = \frac{\sigma - \sigma_1}{\sigma}y + \frac{\sigma_1}{\sigma}x = \frac{\sigma_1}{\sigma}x + \frac{\sigma_2}{\sigma}y + \frac{\sigma_3}{\sigma}y$. Thus, for all $(x,y) \in [0,1]^2$, we obtain a unified expression:

$$g(x,y) = \frac{\sigma_1}{\sigma}x + \frac{\sigma_2}{\sigma}y + \frac{\sigma_3}{\sigma}\min(x,y). \tag{13}$$

Substituting (13) into the sanction function, we get:

$$f(V,R) = \frac{\sigma_1}{\sigma}u_V(V) + \frac{\sigma_2}{\sigma}u_R(R) + \frac{\sigma_3}{\sigma}\min(u_V(V), u_R(R)),$$

and consequently, from (2), we have, for all $(V,R) \in [v_i, v_{i+1}] \times [r_j, r_{j+1}]$,

$$F(V,R) = \sigma_0 + \sigma_1 u_V(V) + \sigma_2 u_R(R) + \sigma_3 \min(u_V(V), u_R(R)). \tag{14}$$

As remark, we note that the Möbius transform m^* of the capacity m is given by: $m^*(\emptyset) = 0$, $m^*(\{V\}) = \sigma_1/\sigma$, $m^*(\{R\}) = \sigma_2/\sigma$ and $m^*(\{V,R\}) = \sigma_3/\sigma$. The Möbius transform decomposes a non-additive measure into marginal contributions of each subset of criteria. In our model, it clarifies how offence severity and offender wealth interact to influence sanctions.

4.2 A Generalized Aggregation Framework

The sanction obtained in (11) using the bilinear interpolation can be rewritten using the product t-norm $T_P(x,y) = xy$:

$$F(V,R) = \sigma_0 + \sigma_1\, u_V(V) + \sigma_2\, u_R(R) + \sigma_3\, T_P(u_V(V), u_R(R)).$$

Similarly, the sanction obtained in (14) using the Choquet-based interpolation can be expressed by employing the minimum t-norm $T_M(x,y) = \min(x,y)$:

$$F(V,R) = \sigma_0 + \sigma_1\, u_V(V) + \sigma_2\, u_R(R) + \sigma_3\, T_M(u_V(V), u_R(R)).$$

This motivates the introduction of a more general formulation of the sanction:

$$S = F(V,R) = \sigma_0 + \sigma_1\, u_V(V) + \sigma_2\, u_R(R) + \sigma_3\, T(u_V(V), u_R(R)), \tag{15}$$

where T is a continuous t-norm. Since $T_P(x,y) < T_M(x,y)$ for all $x, y \in (0,1)$, the sanction derived using T_P is generally lower than that obtained with T_M, reflecting different aggregation approaches.

Using the same setting as in Sect. 3.2, in Fig. 3 we have determined the sanction values $S = F(V,R)$ using Eq. (15) with the minimum t -norm $T = T_M$ (Fig. 3), corresponding to the Choquet integral based interpolation, and the Lukasiewicz t-norm $T = T_L$ (Fig. 4), defined by $T_L(x,y) = \max(x+y-1, 0)$.

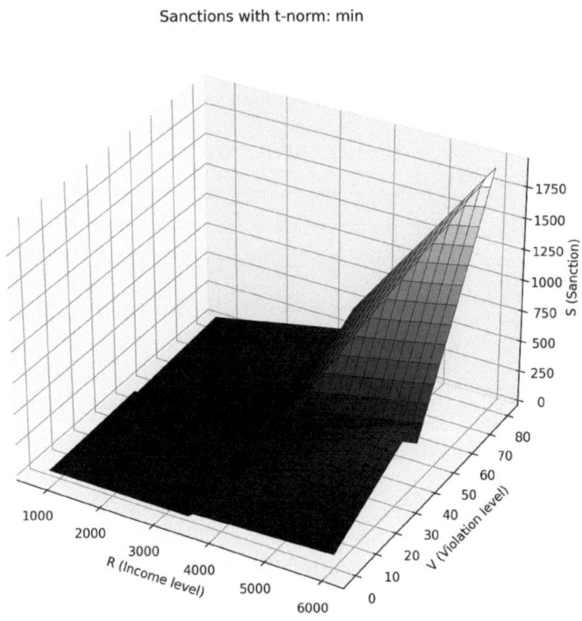

Fig. 3. Sanction determined using the minimum t-norm T_M

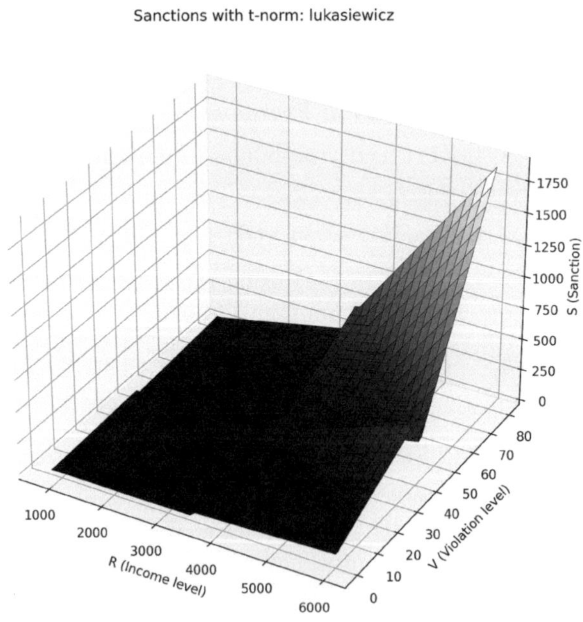

Fig. 4. Sanction determined using the Lukasiewicz t-norm T_L

4.3 Parametric T-Norms

To further enhance flexibility, we now consider in the expression (15) two well-known parametric families of t-norms:

- the SchweizerSklar family [1,15]:

$$T_\alpha^{(1)}(a,b) = 1 - \left[(1-a)^{1/\alpha} + (1-b)^{1/\alpha} - (1-a)^{1/\alpha}(1-b)^{1/\alpha}\right]^\alpha, \quad \alpha \in (0,1], \tag{16}$$

- the DuboisPrade family [6,7]:

$$T_\alpha^{(2)}(a,b) = \frac{ab}{\max(a,b,\alpha)}, \quad \alpha \in (0,1]. \tag{17}$$

We observe that these t-norms satisfy the following properties: $T_\alpha^{(1)} \to T_M$ and $T_\alpha^{(2)} \to T_M$ as $\alpha \to 0$; $T_\alpha^{(1)} = T_\alpha^{(2)} = T_P$ when $\alpha = 1$. Figure 5 illustrates how the sanction varies for a fixed offender profile ($v = 37$, $r = 4000$) under different t-norms and values of α. Numerical simulations show how different t-norms influence the resulting fines, offering a practical toolkit for designing equitable and context-sensitive sanction systems.

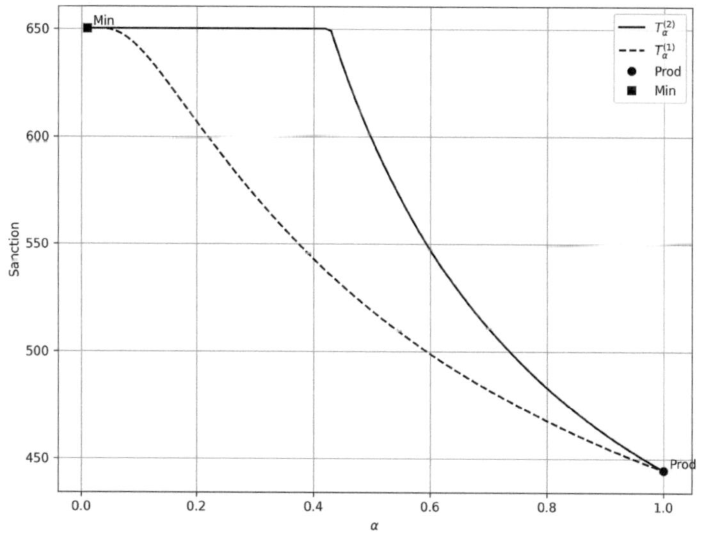

Fig. 5. Sanction for an offender with $v = 37$ and $r = 4000$, for different t-norms and values of α

5 Conclusions

In this study have proposed a novel framework for designing income-based fines that are both equitable and adaptable to diverse legal and social contexts. Building on the limitations of traditional fixed-fine systems and the theoretical advantages of day-fine models, we introduced a flexible methodology that integrates offence severity and offender wealth into a unified sanctioning function.

Our approach begins with a structured interpolation model based on a discretized grid of offence and income levels, using bilinear interpolation to ensure smooth transitions across the domain. We then extended this model by incorporating the Choquet integral, allowing for aggregation of evaluations through non-additive measures on criteria. Finally, we generalized the framework using t-norms, enabling policymakers to select from a wide range of aggregation operators that reflect different preferences regarding fairness, proportionality, and deterrence.

The proposed model offers several advantages. It ensures continuity and monotonicity in the sanction function, avoids abrupt jumps in penalties, and allows for the incorporation of additional criteria in future extensions. Moreover, the use of parametric t-norms provides a powerful tool for tuning the balance between offence gravity and offender wealth, making the system highly customizable. Numerical simulations highlight the potential of our approach as a decision-support tool for legislators and judicial authorities. In future research, we intend to enhance our multi-criteria approach by incorporating additional factors beyond the severity of the offence and the wealth of the offender, as suggested in [14]. In particular, we aim to explore the integration of additional dimensions—such as recidivism risk, social harm, or contextual aggravating factors—into the model. Future developments could also investigate the implicit optimisation of t-norm parameters through the use of tailored questionnaires, while addressing the issue of measuring and managing consensus in contexts involving multiple decision-makers. Moreover, we observe that while current access to personal financial data for judicial purposes may be limited, in the future developments in digital administration and legal frameworks will facilitate more systematic use of such information. This would enhance the practical applicability of our model in real-world sanctioning systems.

References

1. Alsina, C., Schweizer, B., Frank, M.: Associative Functions: Triangular Norms and Copulas. World Scientific (2006)
2. Becker, G.S.: Crime and punishment: an economic approach. J. Polit. Econ. **76**(2), 169–217 (1968)
3. Beliakov, G., Pradera, A., Calvo, T., et al.: Aggregation Functions: A Guide for Practitioners, vol. 221. Springer, Heidelberg (2007). https://doi.org/10.1007/978-3-540-73721-6
4. Bouyssou, D., Marchant, T., Pirlot, M., Tsoukias, A., Vincke, P.: Evaluation and Decision Models with Multiple Criteria: Stepping Stones for the Analyst. Springer, New York (2006). https://doi.org/10.1007/0-387-31099-1

5. Dari-Mattiacci, G., Immordino, G., Russo, F.F.: Fines for unequal societies. Eur. J. Polit. Econ. **86**, 102621 (2025)
6. Dubois, D., Prade, H.: New results about properties and semantics of fuzzy set-theoretic operators. In: Fuzzy sets: Theory and Applications to Policy Analysis and Information Systems, pp. 59–75. Plenum Press, New York (1980)
7. Dubois, D., Prade, H.: A class of fuzzy measures based on triangular norms a general framework for the combination of uncertain information. Int. J. General Syst. **8**(1), 43–61 (1982)
8. Grabisch, M., Labreuche, C.: A decade of application of the Choquet and Sugeno integrals in multi-criteria decision aid. Ann. Oper. Res. **175**(1), 247–286 (2010)
9. Grabisch, M., Marichal, J.L., Mesiar, R., Pap, E.: Aggregation functions. Encycl. Math. Appl., vol. 127. Cambridge University Press, Cambridge (2009)
10. Kaila, M.: How do people react to income-based fines? evidence from speeding tickets discontinuities. Technical report, CESifo Working Paper (2024)
11. Kantorowicz-Reznichenko, E.: Day-fines: should the rich pay more? Rev. Law Econ. **11**(3), 481–501 (2015)
12. Kantorowicz-Reznichenko, E., Faure, M.: Comparative law and economics perspective on day. Day Fines Eur. Assess. Income-Based Sanct. Crim. Justice Syst. **3**, 366 (2021)
13. Kantorowicz-Reznichenko, E., Faure, M.: Day Fines in Europe: Assessing Income-Based Sanctions in Criminal Justice Systems. Cambridge University Press, Cambridge (2021)
14. Kantorowicz-Reznichenko, E., Kerk, M.: Day fines: asymmetric information and the secondary enforcement system. Eur. J. Law Econ. **49**(3), 339–360 (2020)
15. Klir, G., Yuan, B.: Fuzzy Sets and Fuzzy Logic, vol. 4. Prentice hall New Jersey (1995)
16. Marichal, J.L.: Aggregation of interacting criteria by means of the discrete Choquet integral. In: Aggregation operators, pp. 224–244. Springer (2002). https://doi.org/10.1007/978-3-7908-1787-4_7
17. Perry, R., Kantorowicz-Rezinchenko, E.: Income-dependent punitive damages. Wash. UL Rev. **95**, 835 (2017)
18. Schokkaert, E., Overlaet, B.: Moral intuitions and economic models of distributive justice. Soc. Choice Welfare **6**(1), 19–31 (1989)

Multidimensional Inequality and Measurement of Social Well-Being

Luca Anzilli[1], Marta Cardin[2(✉)], and Silvio Giove[2]

[1] Department of Economic Sciences, University of Salento, Lecce, Italy
luca.anzilli@unisalento.it

[2] Department of Economics, Ca' Foscari University of Venice, Venice, Italy
{mcardin,sgiove}@unive.it

Abstract. In recent times, multidimensional measures of well-being are being utilized more widely within academic research and policy assessments. Understanding the multidimensional aspects of well-being is crucial for accurate measurement. This paper explores different types of societal well-being measures through an axiomatic framework while addressing the issue of inequality aversion. Different approaches to measuring inequality in a multidimensional framework are discussed.

Keywords: multidimensional · well-being · **OWA**

1 Introduction

This paper develops an axiomatic approach to measuring societal well-being in a multidimensional setting, where individual outcomes are represented by multidimensional variables, as proposed, among others, by [6,7].

It is widely recognized in both the literature and practical decision-making that well-being is a multidimensional concept and cannot be reduced to a single proxy, such as income (see [2]). The indicators used to assess individuals' achievements across various dimensions are numerous and expressed in different ways; some of them are continuous, others dichotomous or binary. Among the multiple techniques for constructing a multidimensional well-being index, we quote the column-first aggregation technique and the row-first method, depending on whether the aggregation proceeds first by column or by row of the deprivation matrix, whose i-th row collects the indicators of individual i. This paper explores the construction of multidimensional well-being measures that satisfy key properties. To this purpose, we consider several axioms and described properties along with their implications for potential inequality measures, focusing on a class of functions adhering to modified forms of the Pigou-Dalton Transfer principle. We prove that, under certain assumptions, the Ordered Weighted Averaging (**OWA**) algorithm, a specific instance of the Choquet integral, is applicable for row aggregation of the social matrix.

M. La Rocca et al. (Eds.): WMAF 2025, *New Perspectives in Mathematical and Statistical Methods for Actuarial Sciences and Finance*, pp. 14–25, 2025.
https://doi.org/10.1007/978-3-032-05551-4_2

The plan of the paper is as follows. In Sect. 2, we outline the framework we use. In Sect. 3, we characterize a class of multidimensional well-being evaluations. In Sect. 4, we introduce some classes of functions compatible with certain types of multidimensional transfers. In Sect. 5, we introduce the class of OWA functions, which are further considered in Sect. 6. Finally, Sect. 7 concludes.

2 Basic Notations and Motivation

We consider a society with n individuals and a set of m dimensions, which are used to evaluate both individual and societal well-being (e.g., income, education, health, etc.). Let $N = \{1, \ldots, n\}$ and $M = \{1, \ldots, m\}$ denote the sets of individuals and attributes, respectively, where m and n are positive integers, with $n \geq 2$. Thus, we assume that the society is given and fixed.

If X is a closed subset of the real line, let $\mathcal{X}^m = X^m$ be the set of row vectors of length m with entries in X, and let $\mathcal{X} = X^{n\times m}$ be the set of all $n \times m$ matrices with entries in X. Each individual i is represented by a row vector $\mathbf{x}_i \in \mathcal{X}^m$, which consists of m attributes and constitutes the achievement vector of individual i. The social well-being matrix $\mathbf{X} = (x_{ij}) \in \mathcal{X}$, of size $n \times m$, summarizes the well-being levels of all individuals across the m attributes.

For every matrix $\mathbf{X} \in \mathcal{X}$, we use the following notation:

- For each $i \in N$, the row vector $\mathbf{x}_{i\bullet} = (x_{i1}, \ldots, x_{im})$ represents the achievement vector of individual i across all dimensions.
- For each $j \in M$, the column vector $\mathbf{x}^{\bullet j} = (x_{1j}, \ldots, x_{nj})$ represents the achievement levels of all individuals in dimension j.

In this paper, we focus on measuring social well-being through a numerical function

$$F \colon \mathcal{X} \to \mathbb{R},$$

which aggregates well-being across dimensions for each individual and then across individuals, representing a social ordering $\succeq$.

3 Multidimensional Well-Being Evaluation Functions

In this section, multidimensional well-being evaluation functions are introduced and characterized. First, we consider the following standard axioms defined with respect to the set $\mathcal{X}$ and an order $\succeq$ in an set $\mathcal{X}$.

Anonimity. If $\mathbf{X}, \mathbf{Y} \in \mathcal{X}$, σ is a permutation of the set N and $\mathbf{y}_{ij} = \mathbf{x}_{\sigma(i),j}$ then $\mathbf{X} \sim \mathbf{Y}$.

Monotonicity. If $\mathbf{X}, \mathbf{Y} \in \mathcal{X}$ are such that $x_{ij} = y_{ij}$ for every $(ij) \neq (hk)$ and $x_{hk} > y_{hk}$ then $\mathbf{X} \succeq \mathbf{Y}$.

Our key axiom is related to the preferences defined in the set of rows and is similar to that proposed in [13]. It means that social preferences are increasing with respect to individual preferences.

ROW PREFERENCES. For all $i \in N$ there is an order $\succeq_i$ in $\mathcal{X}^m$ such that for every $\mathbf{X}, \mathbf{Y} \in \mathcal{X}$ if $\mathbf{x}_{k\bullet} = \mathbf{y}_{k\bullet}$ for every $k \in N, k \neq i$ then $\mathbf{X} \succ \mathbf{Y}$ if and only if $\mathbf{x}_{i\bullet} \succ_i \mathbf{y}_{i\bullet}$ and $\mathbf{X} \sim \mathbf{Y}$ if and only if $\mathbf{x}_{i\bullet} \sim_i \mathbf{y}_{i\bullet}$.

Note that Row Preferences is a separability property in our framework. We introduce another standard axiom for an order relation in the set $\mathcal{X}$.

CONTINUITY. The sets $\{\mathbf{X} \in \mathcal{X} : \mathbf{X} \succeq \mathbf{Y}\}$ and $\{\mathbf{X} \in \mathcal{X} : \mathbf{X} \preceq \mathbf{Y}\}$ are closed in the usual topology in $\mathbb{R}^{n \times m}$ for every $\mathbf{Y} \in \mathcal{X}$.

The implications of the above axioms are summarized in the following proposition.

Proposition 1. *An order relation $\succeq$ defined in $\mathcal{X}$ satisfies Anonymity, Monotonicity, Row Preferences, and Continuity if and only if there exists an increasing and continuous function $f \colon \mathbb{R}^m \to \mathbb{R}$ such that*

$$\mathbf{X} \succeq \mathbf{Y} \quad \textit{if and only if} \quad \sum_{i=1}^{n} f(\mathbf{x}_{i,\bullet}) \geq \sum_{i=1}^{n} f(\mathbf{y}_{i,\bullet}) \tag{1}$$

Proof. We consider only the sufficient part; the necessary part is immediate. We refer to Theorem 1 in [3] and we note that the order relation $\succeq$ satisfies property A1 (since it is a complete and transitive relation), A2(continuity), A3 (monotonicity), and A5 (anonymity) of [3].

Let us consider the following elements in $\mathcal{X}$

$$\begin{bmatrix} \mathbf{x}_{1\bullet} \\ \vdots \\ \mathbf{x}_{i\bullet} \\ \vdots \\ \mathbf{x}_{n\bullet} \end{bmatrix} \quad \begin{bmatrix} \mathbf{x}_{1\bullet} \\ \vdots \\ \mathbf{z}_{i\bullet} \\ \vdots \\ \mathbf{x}_{n\bullet} \end{bmatrix} \quad \begin{bmatrix} \mathbf{y}_{1\bullet} \\ \vdots \\ \mathbf{x}_{i\bullet} \\ \vdots \\ \mathbf{y}_{n\bullet} \end{bmatrix} \quad \begin{bmatrix} \mathbf{y}_{1\bullet} \\ \vdots \\ \mathbf{z}_{i\bullet} \\ \vdots \\ \mathbf{y}_{n\bullet} \end{bmatrix}$$

It can be stated that using Row Preferences that

$$\begin{bmatrix} \mathbf{x}_{1\bullet} \\ \vdots \\ \mathbf{x}_{i\bullet} \\ \vdots \\ \mathbf{x}_{n\bullet} \end{bmatrix} \succeq \begin{bmatrix} \mathbf{x}_{1\bullet} \\ \vdots \\ \mathbf{z}_{i\bullet} \\ \vdots \\ \mathbf{x}_{n\bullet} \end{bmatrix}$$

if and only if $\mathbf{x}_{i\bullet} \succ_j \mathbf{z}_{i\bullet}$. Moreover, we have that

$$\begin{bmatrix} \mathbf{y}_{1\bullet} \\ \vdots \\ \mathbf{x}_{i\bullet} \\ \vdots \\ \mathbf{y}_{n\bullet} \end{bmatrix} \succeq \begin{bmatrix} \mathbf{y}_{1\bullet} \\ \vdots \\ \mathbf{z}_{i\bullet} \\ \vdots \\ \mathbf{y}_{n\bullet} \end{bmatrix}$$

if and only if $\mathbf{x}_{i\bullet} \succ_j \mathbf{z}_{i\bullet}$. Thus we can say that

$$\begin{bmatrix} \mathbf{x}_{1\bullet} \\ \vdots \\ \mathbf{x}_{i\bullet} \\ \vdots \\ \mathbf{x}_{n\bullet} \end{bmatrix} \succeq \begin{bmatrix} \mathbf{x}_{1\bullet} \\ \vdots \\ \mathbf{z}_{i\bullet} \\ \vdots \\ \mathbf{x}_{n\bullet} \end{bmatrix}$$

if and only if

$$\begin{bmatrix} \mathbf{y}_{1\bullet} \\ \vdots \\ \mathbf{x}_{i\bullet} \\ \vdots \\ \mathbf{y}_{n\bullet} \end{bmatrix} \succeq \begin{bmatrix} \mathbf{y}_{1\bullet} \\ \vdots \\ \mathbf{z}_{i\bullet} \\ \vdots \\ \mathbf{y}_{n\bullet} \end{bmatrix}.$$

We proved the independence property as defined in [3] is satisfied in our framework and then, by Theorem 1 of [3] the thesis follows.

4 Multidimensional Inequality and Classes of Functions

One might naturally inquire if inequality can be effectively incorporated into a category of measures for well-being. We introduce various definitions of multidimensional inequality and examine the additive functionals that adhere to the proposed definitions.

4.1 Inframodular Functions

First, we propose a definition of inequality for a society of n individuals, where each individual's well-being is described by m attributes. This follows the approach considered in [1]. See also [3] and the definition of Clustered Deteriorations and well-being in [7].

We begin by defining a multidimensional transfer in $\mathcal{X}^m$. This definition will serve as a foundation for subsequent definitions and axioms imposed on the function that represents a given order.

Let $\mathbf{x}, \mathbf{y} \in \mathcal{X}^m$ with $\mathbf{x} \leq \mathbf{y}$. We say that the pair $(\mathbf{x}', \mathbf{y}')$ is derived from $(\mathbf{x}, \mathbf{y})$ through a Strong PigouDalton Transfer (SPDT) if there exists a vector $\mathbf{h} \in \mathbb{R}^m$ such that

$$0 \leq \mathbf{h} \leq \mathbf{y} - \mathbf{x}, \quad \mathbf{x}' = \mathbf{x} + \mathbf{h}, \quad \mathbf{y}' = \mathbf{y} - \mathbf{h}.$$

This definition allows the transfer $\mathbf{h}$ to be large enough to reverse the relative positions of the two individuals who initially held $\mathbf{x}$ and $\mathbf{y}$. However, it does not permit $\mathbf{h}$ to be so large that the originally wealthier individual becomes poorer than the initially less wealthy individual after the transfer.

If $\mathbf{X}, \mathbf{Y} \in \mathcal{X}$ such that $\mathbf{y}_{i_1,\bullet} \leq \mathbf{y}_{i_2,\bullet}$ and $(\mathbf{x}_{i_1,\bullet}, \mathbf{x}_{i_2,\bullet})$ is obtained from $(\mathbf{y}_{i_1,\bullet}, \mathbf{y}_{i_2,\bullet})$ through a SPDT and $\mathbf{x}_{i,\bullet} = \mathbf{y}_{i,\bullet}$ if $i \notin \{i_1, i_2\}$, we say that $\mathbf{X}$ is less unequal than $\mathbf{Y}$. More generally, if $X, Y \in \mathcal{X}$ we say that $\mathbf{X}$ is less unequal than $\mathbf{Y}$ denoted $X \succeq_{\mathrm{I}} Y$, if $\mathbf{X}$ is obtained from $\mathbf{Y}$ through a sequence of SPDTs applied to the rows. Our interest lies in relations $\succeq$ within $\mathcal{X}$ such that, whenever $\mathbf{X} \succeq_{\mathrm{I}} \mathbf{Y}$, it follows that $\mathbf{X} \succeq \mathbf{Y}$.

An *inframodular function* is a mathematical concept relevant in fields such as economics and game theory. It generalizes the notion of concavity to functions of multiple variables. In essence, an inframodular function is one in which the "difference" between the function's values at different points decreases as the points become closer together. This property is significant because it connects to important economic ideas, such as diminishing marginal utility and the *Pigou-Dalton principle of transfer.*

A function $f : \mathbb{R}^m \to \mathbb{R}$ is said to be inframodular if, for all $x, y, h \in \mathbb{R}^m$,

$$x \leq y, \quad h \geq 0 \quad \text{implies that} \quad f(x+h) - f(x) \geq f(y+h) - f(y). \tag{2}$$

By Proposition 1 in [1] we can prove the following result that characterizes well-being measures with additive representation that respect the inequality order.

Proposition 2. *Let $F \colon \mathcal{X} \to \mathbb{R}$ a function such that*

$$F(\mathbf{X}) = \sum_{i=1}^{n} f(\mathbf{x}_{i,\bullet})$$

where f is a function $f \colon \mathcal{X}^m \to \mathbb{R}$. Then

$$\mathbf{X} \succeq_I \mathbf{Y} \quad \textit{implies that} \quad F(\mathbf{X}) \geq F(\mathbf{Y}) \tag{3}$$

if and only if the function f is an inframodular function.

According to Theorem 5.5 in [11], a function $f : \mathbb{R}^2 \to \mathbb{R}$ that is twice continuously differentiable will be inframodular if and only if its pure and mixed partial derivatives fulfill the condition:

$$\frac{\partial^2 f}{\partial x_i \, \partial y_j}(x, y) \leq 0, \quad \text{for every } i, j\,.$$

4.2 Submodular Functions

Here, we introduce an alternative definition of inequality applicable to a society consisting of n individuals, each characterized by m well-being attributes. To start, we define a multidimensional transfer within $\mathcal{X}^m$ and this initial definition will act as a basis for the forthcoming definitions and axioms applied to the function that represents a particular order.
Let $\mathbf{x}, \mathbf{y} \in \mathcal{X}^m$with $\mathbf{x} \leq \mathbf{y}$ we say that the pair $(\mathbf{x}', \mathbf{y}')$ is derived from $(\mathbf{x}, \mathbf{y})$ through a Strong Submodular Transfer (SST) if there exists $i, j, 1 \leq i, j \leq m$ such that $x_i < y_i$ and $x_j < y_j$, $x'_i = x_i$, $x'_j = y_j$, $y'_i = y_i$, $y'_j = x_j$ and $x_k = y_k = x'_k = y'_k$ if $1 \leq k \leq m, k \neq i, j$. Observe that this means $\mathbf{x}' \wedge \mathbf{y}' = \mathbf{x}$ and $\mathbf{x}' \vee \mathbf{y}' = \mathbf{y}$. If $\mathbf{X}, \mathbf{Y} \in \mathcal{X}$ such that $\mathbf{y}_{i_1,\bullet} \leq \mathbf{y}_{i_2,\bullet}$ and $(\mathbf{x}_{i_1,\bullet}, \mathbf{x}_{i_2,\bullet})$ is obtained from $(\mathbf{y}_{i_1,\bullet}, \mathbf{y}_{i_2,\bullet})$ through a SST and $\mathbf{x}_{i,\bullet} = \mathbf{y}_{i,\bullet}$ if $i \notin \{i_1, i_2\}$, we say that $\mathbf{X}$ is less s-unequal than $\mathbf{Y}$. Then if $X, Y \in \mathcal{X}$ we say that $\mathbf{X}$ is less unequal than $\mathbf{Y}$ denoted $X \succeq_{\mathrm{SI}} Y$, if $\mathbf{X}$ is obtained from $\mathbf{Y}$ through a sequence of SSTs applied to the rows. We focus on the relations $\succeq$ over the set $\mathcal{X}$ where, if $\mathbf{X} \succeq_{\mathrm{SI}} \mathbf{Y}$ is true, then it implies that $\mathbf{X} \succeq \mathbf{Y}$. Let $f : \mathbb{R}^m \to \mathbb{R}$ be a function defined on a lattice $(\mathbb{R}^m, \leq)$ where the order is component-wise. We say that f is *submodular* if for all $\mathbf{x}, \mathbf{y} \in \mathbb{R}^m$,

$$f(\mathbf{x}) + f(\mathbf{y}) \geq f(\mathbf{x} \wedge \mathbf{y}) + f(\mathbf{x} \vee \mathbf{y})$$

where the operations $\wedge$ (infimum) and $\vee$ (supremum) are taken component-wise: $(\mathbf{x} \wedge \mathbf{y})_i = \min(x_i, y_i)$ and $(\mathbf{x} \vee \mathbf{y})_i = \max(x_i, y_i)$.

This form is particularly useful when dealing with functions over lattices and extends naturally both to discrete and continuous settings. It can also be shown to be equivalent to the diminishing returns property in the appropriate contexts. In particular, it can be proved that inframodular functions are submodular but the converse is not true (see, for instance, [1] and the references cited within).

Moreover, a twice-differentiable function is submodular if and only if (see [10])

$$\frac{\partial^2 f}{\partial x_i \, \partial y_j}(x, y) \leq 0, \quad \forall i \neq j \, .$$

Proposition 3. *If* $F \colon \mathcal{X} \to \mathbb{R}$ *a function such that*

$$F(\mathbf{X}) = \sum_{j=1}^{n} f(\mathbf{x}_{i,\bullet})$$

where f *is a function* $f \colon \mathcal{X}^m \to \mathbb{R}$ *such that* $f(\mathbf{0}_m) = 0$. *Then*

$$X \succeq_{SI} Y \quad \text{implies that} \quad F(\mathbf{X}) \geq F(\mathbf{Y}) \tag{4}$$

if and only if the function f *is a submodular function.*

Proof. Let $\mathbf{X}, \mathbf{Y} \in \mathcal{X}$ such that $\mathbf{X}$ is a transfer of $\mathbf{Y}$ then $F(\mathbf{X}) - F(\mathbf{Y}) =$
$f(x_{i_1 1}, \ldots, x_{i_1 j_1}, \ldots, x_{i_1, j_2}, \ldots, x_{i_1 m}) + f(x_{i_2 1}, \ldots, x_{i_2 j_1}, \ldots, x_{i_2, j_2}, \ldots, x_{i_2 m})$
$-f(x_{i_1 1}, \ldots, y_{i_1 j_1}, \ldots, y_{i_1, j_2}, \ldots, x_{i_1 m}) - f(x_{i_2 1}, \ldots, y_{i_2 j_1}, \ldots, y_{i_2, j_2}, \ldots, x_{i_1 m}) =$
$f(x_{i_1 1}, \ldots, x_{i_1 j_1}, \ldots, x_{i_1, j_2}, \ldots, x_{i_1 m}) + f(x_{i_2 1}, \ldots, x_{i_2 j_1}, \ldots, x_{i_2, j_2}, x_{i_2, m})$
$-f(x_{i_1 1}, \ldots, x_{i_1 j_1} \vee, x_{i_2 j_1}, \ldots, x_{i_1 j_1} \vee x_{i_2 j_1}, \ldots x_{i_1 m})$
$-f(x_{i_1 1}, \ldots, x_{i_1 j_1} \wedge, x_{i_2 j_1}, \ldots, x_{i_1 j_1} \wedge x_{i_2 j_1}, \ldots x_{i_1 m})$
Hence if f is submodular $F(\mathbf{X}) - F(\mathbf{Y}) \geq 0$ and so we can say that if $X \succeq_{SI} Y$ then $F(\mathbf{X}) \geq F(\mathbf{Y})$.
Now let $F \colon \mathcal{X} \to \mathbb{R}$ a function, $F(\mathbf{X}) = \sum_{i=1}^{n} f(\mathbf{x}_{i,\bullet})$ where f is a function $f \colon \mathcal{X}^m \to \mathbb{R}$, $f(\mathbf{0}_m) = 0$ and such that Eq. 4 is satisfied.

If $\mathbf{x}, \mathbf{y} \in \mathcal{X}^m$ we define the two elements of $\mathcal{X}$, $\mathbf{W}, \mathbf{Z}$,

$$\mathbf{Z} = \begin{pmatrix} \mathbf{x} \\ \mathbf{y} \\ \mathbf{0}_m \\ \vdots \\ \mathbf{0}_m \end{pmatrix} \qquad \mathbf{W} = \begin{pmatrix} \mathbf{x} \vee \mathbf{y} \\ \mathbf{x} \wedge \mathbf{y} \\ \mathbf{0}_m \\ \vdots \\ \mathbf{0}_m \end{pmatrix}.$$

It can be easily proved that $\mathbf{Z} \succeq_{SI} \mathbf{W}$ and so $F(\mathbf{Z}) \geq F(\mathbf{W})$. Since $F(\mathbf{Z}) = f(\mathbf{x}) + f(\mathbf{y})$ and $F(\mathbf{W}) = f(x \vee \mathbf{y}) + f(\mathbf{x} \wedge \mathbf{y})$ we have proved that $f(\mathbf{x}) + f(\mathbf{y}) \geq f(\mathbf{x} \vee \mathbf{y},) + f(\mathbf{x} \wedge \mathbf{y})$ and then we have that f is submodular.

4.3 S-Concave Functions

In this section, we explore a column-first aggregation strategy. Specifically, we examine the ordering derived from column majorization, which involves comparing multivariate distributions by sequentially evaluating the distributions of their individual attributes. For the following definition of majorization, refer to e.g. [12].

Let $\mathbf{x}, \mathbf{y} \in \mathbb{R}^n$. We say that $\mathbf{x}$ is majorized by $\mathbf{y}$, written as $\mathbf{x} \preceq \mathbf{y}$, if the following two conditions hold:

- $$\sum_{i=1}^{k} x_{(i)} \leq \sum_{i=1}^{k} y_{(i)} \quad \text{for all } k = 1, 2, \ldots, n-1,$$
- and the total sums are equal:
$$\sum_{i=1}^{n} x_{(i)} = \sum_{i=1}^{n} y_{(i)}.$$

when the components of both vectors are rearranged in decreasing order that is $x_{(1)} \geq \cdots \geq x_{(n)}$ and $y_{(1)} \geq \cdots \geq y_{(n)}$.

Intuitively, majorization provides a way to compare the "spread" or "inequality" of two vectors. If $\mathbf{x} \prec \mathbf{y}$, then the elements of $\mathbf{x}$ are more "evenly distributed" than those of $\mathbf{y}$. This concept is widely used in areas such as economics (to compare income distributions), quantum information theory, and matrix analysis.
A real-valued function $f : \mathbb{R}^n \to \mathbb{R}$ is called S-concave (or Schur-concave) if for

all $\mathbf{x}, \mathbf{y} \in \mathbb{R}^n$,

$$\mathbf{x} \preceq \mathbf{y} \Rightarrow f(\mathbf{x}) \geq f(\mathbf{y})$$

That is, if $\mathbf{x}$ is majorized by $\mathbf{y}$, then f assigns a larger (or equal) value to $\mathbf{x}$ than to $\mathbf{y}$. S-concave functions assign higher values to vectors whose components are more evenly distributed. In other words, they decrease as the inequality (dispersion) among the components increases.
We define an inequality order in $\mathcal{X}$ where if $\mathbf{X}, \mathbf{Y} \in \mathcal{X}$, $\mathbf{X} \succeq_{\mathrm{M}} \mathbf{Y}$ if and only if for each $j \in M$, the column vector $\mathbf{x}^{\bullet j} \preceq \mathbf{y}^{\bullet j}$ that is to say that the distribution of every attribute in $\mathbf{X}$ is less unequal than that in $\mathbf{Y}$. The following proposition follows directly from the definition of S-concave function.

Proposition 4. *If $F\colon \mathcal{X} \to \mathbb{R}$ a function such that*

$$F(\mathbf{X}) = \sum_{j=1}^{m} f(\mathbf{x}^{\bullet j})$$

where f is a function $f\colon \mathcal{X}^n \to \mathbb{R}$. Then

$$X \succeq_M Y \quad \textit{implies that} \quad F(\mathbf{X}) \geq F(\mathbf{Y}) \tag{5}$$

if and only if the function f is an S-concave function.

5 OWA Functions

The Ordered Weighted Averaging (**OWA**) functions, introduced in [14] represent an important family of aggregation operators that generalize classical means. **OWA** functions satisfying specific properties constitute notable examples of both submodular and S-convex functions, thereby providing an important class of functions for the formulation of inequality measures within the framework introduced in the preceding sections. The **OWA** function related to the weighting vector $(w_1, \ldots, w_m) \in [0,1]^m$, which satisfies $\sum_{i=1}^{m} w_i = 1$, is specified by the function $f\colon \mathbb{R}^m \to \mathbb{R}$, given by

$$f_{\mathbf{w}}(\mathbf{x}) = \sum_{i=1}^{m} w_i x_{(i)} \quad \text{where} \quad x_{(1)} \geq \cdots \geq x_{(m)}. \tag{6}$$

This definition makes *OWA* operators order-dependent, assigning weights based on the rank of values rather than their source.

A significant connection exists between **OWA** functions and the Choquet integral within the framework of non-additive integrals (see [8]). It can be verified that any Choquet integral with respect to a *symmetric* non-additive measure - a measure that depends only on the cardinality of the subset - can be represented with an *OWA* operator. Thus **OWA** functions offer a simpler yet flexible alternative by focusing on the ordered position of inputs, permitting the flexible modeling of different aggregation behaviors through the adjustment of weight vectors.

The following result plays an important role in our framework.

Proposition 5. *Let* $f\colon \mathcal{X}^m \to \mathbb{R}$ *a function defined as*

$$f_{\mathbf{w}}(\mathbf{x}) = \sum_{i=1}^{m} w_i x_{(i)}$$

where $x_{(1)} \geq \cdots \geq x_{(m)}$. *Then*

- f *is submodular if and only if* $w_1 \leq \cdots \leq w_m$.
- f *is S-concave if and only if* $w_1 \leq \cdots \leq w_m$.

Proof. The Choquet integral with respect to a submodular capacity is a submodular function of its inputs (see for example [9]). Since **OWA** functions can be seen as Choquet integrals with respect to symmetric capacities, if the underlying symmetric capacity is submodular, then the resulting **OWA** aggregation would inherit submodularity properties. This statement is valid if and only if the weighting vector $(w_1, \ldots, w_m)$ satisfies the condition $w_1 \leq \cdots \leq w_m$, as proved in [9]. The second statement is proved in [4].

6 OWA Elicitation and a Numerical Example

It is well known that suitably tuning the weight values of the **OWA** operator (6), can obtain different aggregation operators, see [14], as the MAX operator if $w_1 = 1, w_i = 0, \forall i = 2, ..., n$, the MIN operator if $w_n = 1, w_i = 0, \forall i = 1, ..., n-1$, and the simple averaging if $w_i = \frac{1}{n}, \forall i = 1, ..., n$, together with many others[1]. Moreover, the more or less tendency to represent a MAX-type operator (MIN-type), can be represented by its *Andness*-degree (*Orness*-degree) where $Orness(OWA^w) = \frac{1}{n-1}\sum_{i=1}^{n}(n-i)w_i$, and $Andess(OWA^w) = 1 - Orness(OWA^w)$. An *Orness*-degree close to zero (one) considers only the best (worst) value of the criteria[2]. Given that the **OWA** weights characterize the behavior of the aggregation function, their elicitation represents a strategic phase, and many algorithms were proposed. For what it concerns the elicitation of the weights, among other algorithm, we consider the *Monotonic Quantifier* approach $Q : [0, 1] \to [0, 1]$, a monotonically increasing and continuous function, with $Q(0) = 0, Q(1) = 1$, see [14], that represents natural language term-sets like "it exists," "most," "at least half," "for all," and so on, [14]. Given a Quantifier, the **OWA** weights can be computed as follows:

$$w_i = Q\left(\frac{i}{n}\right) - Q\left(\frac{i-1}{n}\right) \tag{7}$$

A commonly used Quantifier is the following one, see [5]:

$$Q(z) = z^{\alpha} \tag{8}$$

[1] Like the median, the k-th order statistic, the Hurwizc operator and so on.

[2] *Orness*-degree equal to one corresponds to the logic quantifier *At least one*, so called *optimistic* behaviour, while *Orness*-degree equal to zero, *pessimistic* behaviour, corresponds to the quantifier *All*.

If $\alpha = 0$ ($\alpha \to +\infty$) the obtained **OWA** weights characterize an *Orness*-type (*Andness*-type) behavior.

We define, for $\mathbf{X} \in \mathcal{X}$, $Val(\mathbf{X}) = \sum_{i=1}^{n} OWA^{\mathbf{w}}(\mathbf{x}_{i,\bullet})$. Observe that, from Proposition 5, if $w_1 \leq w_2 \leq \cdots \leq w_m$ then $OWA^{\mathbf{w}}$ is a S-concave function, and consequently, from Proposition 4, if $\mathbf{X}$ is less unequal than $\mathbf{Y}$, i.e. $\mathbf{X} \succeq_{\mathrm{SI}} \mathbf{Y}$, then $Val(\mathbf{X}) \geq Val(\mathbf{Y})$.

Unfortunately, *OWA* does not satisfy the inframodularity property, and thus the Prop. 2 cannot be applied. But in particular case, the Pigou-Dalton transfer principle remains satisfied, as verified by the following example. Let us consider a society formed by 6 individuals with 5 dimensions, characterized by the following Achievement matrix A:

$$\mathbf{A} = \begin{bmatrix} 4\,7\,8\,1\,2 \\ 3\,4\,5\,9\,4 \\ 7\,2\,3\,6\,4 \\ 5\,4\,4\,8\,5 \\ 3\,2\,2\,7\,6 \\ 7\,4\,8\,1\,2 \end{bmatrix}$$

Suppose now that the second individual transfer 5 unit of the fourth dimension and one unit of the fifth dimension to the first individual. Moreover, let us suppose that the third individual transfer 2 units of the first dimension to the fourth individual, thus obtaining the following matrix B:

$$\mathbf{B} = \begin{bmatrix} 4\,7\,8\,6\,3 \\ 3\,4\,5\,4\,3 \\ 5\,2\,3\,6\,4 \\ 7\,4\,4\,8\,5 \\ 3\,2\,2\,7\,6 \\ 7\,4\,8\,1\,2 \end{bmatrix}$$

To respect the Pigou-Dalton transfer principle, the evaluation of the society represented by matrix B, denoted as Val(B), should be greater than the evaluation of society A, denoted as Val(A). We consider an **OWA**, where the weights are derived from the *Monotonic Quantifier* $Q(z) = z^{\alpha}$. We note that: if $\alpha < 1$, the operator exhibits a more *orness*-oriented behavior (favoring higher values); if $\alpha > 1$, the operator behaves more *and*-like (favoring lower values).

We consider the following cases: 1) $\alpha = 0$ (corresponding to the MAX operator), 2) $\alpha = 0.5$, 3) $\alpha = 0.8$, 4) $\alpha = 1.5$, 5) $\alpha = 3$. The **OWA** weights and the evaluation function for both A and B are respectively:

1) $OWA = \{1, 0, 0, 0, 0\}$, $Val(A) = 47$, $Val(B) = 42$,
2) $OWA = \{0.44, 0.18, 0.14, 0.12, 0.10\}$, $Val(A) = 34.5$, $Val(B) = 32.89$,
3) $OWA = \{0.28, 0.20, 0.18, 0.17, 0.16\}$,$Val(A) = 29.79$, $Val(B) = 29.29$.
4) $OWA = \{0.09, 0.16, 0.21, 0.25, 0.29\}$,$Val(A) = 23.13$, $Val(B) = 23.93$.
5) $OWA = \{0.01, 0.06, 0.15, 0.30, 0.48\}$,$Val(A) = 17.4$, $Val(B) = 18.9$.

Let us observe that increasing the value of α, the **OWA** weights becomes smoother, and the differences between the evaluations of A and B decrease[3]. These results are in line with what is reported in Proposition 5.

As a counterexample, observe that for the cases $\alpha < 1$ (OR-type) and $\alpha > 1$ (AND-type), the preferences between A and B are reversed.

In Fig. 1 we reported the graphs of $Val(B)$ and $Val(B)$ in function of $\alpha \geq 0$. In line with what has been stated, the inversion of the ranking between A and B is observed at the point $\alpha = 1$.

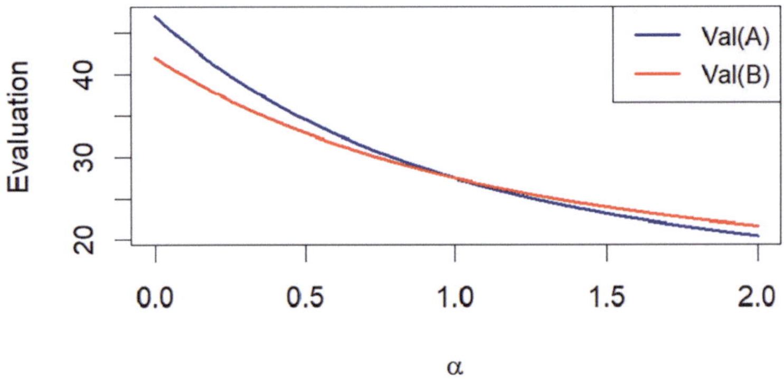

Fig. 1. Comparison among $Val(A)$ and $Val(B)$ in function of the parameter α

But consider now the society described by the following matrix:

$$\mathbf{C} = \begin{bmatrix} 3\,7\,4 \\ 2\,1\,3 \end{bmatrix}, \qquad \mathbf{D} = \begin{bmatrix} 3\,5\,4 \\ 2\,3\,3 \end{bmatrix}$$

Even here, D is obtained from C by a Pigou-Dalton transfer, removing from the first row of C 2 units by the second criterion and allocating them in the second element of the second row. If we proceed as above and for example select the MAX operator, it is straightforward to verify that $Val(C) > Val(D)$; in this case the Pigou-Dalton transfer is not satisfied.

7 Conclusions and Future Work

In multidimensional assessment of well-being, a crucial aspect is the consideration of inequalities, the recognition of which can guide public policies towards a different allocation of resources. We proposed an evaluating approach to the social well-being, considering a class of functions which satisfy some modified

[3] If $\alpha = 1$ all the **OWA** weights are equal to $\frac{1}{5}$, thus for each row we obtain the simple mean, and given that the sum of the elements of both the two matrices is the same (because A and B contain the same amount of total richness), we obtain $V(A) = V(B)$.

versions of the Pigou-Dalton Transfer principle, using suitable mathematical characterizations. As a possible example of aggregation function of the social matrix, we proposed (**OWA**), putting in evidence both potentialities and limits. A future research will go deeper into this topic, with particular attention to the parameters optimization phase, possibly including Experts and/or Stakeholders.

References

1. Aouani, Z., Chateauneuf, A.: Multidimensional inequality and inframodular order. J. Math. Econ. **90**, 74–79 (2020)
2. Atkinson, A.B.: Multidimensional deprivation: contrasting social welfare and counting approaches. J. Econ. Inequality **1**, 51–65 (2003)
3. Basili, M., Casaca, P., Chateauneuf, A., Franzini, M.: Multidimensional Pigou-Dalton transfers and social evaluation functions. Theor. Decis. **83**, 573–590 (2017)
4. Bortot, S., Pereira, R.A.M.: The binomial gini inequality indices and the binomial decomposition of welfare functions. Fuzzy Sets Syst. **255**, 92–114 (2014)
5. Cardin, M., Giove, S.: A grid-based optimization algorithm for parameters elicitation in WOWA operators: an application to risk assesment. Advances in Neural Networks: Computational and Theoretical Issues, pp. 207–215 (2015)
6. Dhongde, S., Li, Y., Pattanaik, P.K., Xu, Y.: Binary data, hierarchy of attributes, and multidimensional deprivation. J. Econ. Inequality **14**, 363–378 (2016)
7. Dhongde, S., Pattanaik, P.K., Xu, Y.: Well-being, deprivation, and the great recession in the us: a study in a multidimensional framework. Rev. Income Wealth **65**, S281–S306 (2019)
8. Grabisch, M.: Fuzzy integral in multicriteria decision making. Fuzzy Sets Syst. **69**(3), 279–298 (1995)
9. Grabisch, M.: The application of fuzzy integrals in multicriteria decision making. Europ. J. Operat. Res. **89**(3), 445–456 (1996)
10. Lovász, L.: Submodular functions and convexity. Mathematical Programming The State of the Art: Bonn **1982**, 235–257 (1983)
11. Marinacci, M., Montrucchio, L.: Ultramodular functions. Math. Oper. Res. **30**(2), 311–332 (2005)
12. Marshall, A.W., Olkin, I., Arnold, B.C.: Inequalities: theory of majorization and its applications. Springer (1979)
13. Mongin, P., Pivato, M.: Ranking multidimensional alternatives and uncertain prospects. J. Econ. Theory **157**, 146–171 (2015)
14. Yager, R.R.: On Ordered weighted averaging aggregation operators in multicriteria decisionmaking. IEEE Trans. Syst. Man Cybern. **18**(1), 183–190 (1988)

A Neural Network Model Approach to Longevity Risk Management

Giovanna Apicella[1(✉)], Michele La Rocca[2], Cira Perna[2], and Marilena Sibillo[2]

[1] Department of Economics and Statistics, University of Udine, 33100 Udine, Italy
giovanna.apicella@uniud.it

[2] Department of Economics and Statistics, University of Salerno, 84084 Fisciano, SA, Italy
{larocca,perna,msibillo}@unisa.it

Abstract. Insurance activities intrinsically deal with the management of risks. According to the mark-to-model approach to the assessment of the insurer's debt position, the models used to forecast the probability structures involved in the computation of the fair value of the liabilities are central. We use artificial neural networks to the purpose of forecasting in this context. We focus on a portfolio of endowment insurance policies and on the ex ante estimation of the related mathematical reserve. We consider the Lee-Carter model for the forecasting of the random number of the policies in-force at each future policy anniversary, with either linear time-series model or autoregressive neural network models (ARX-NN models)for the time index. We find out that ARX-NN models allow to reduce the bias that linear time series models may imply. In particular, narrower prediction intervals around central forecasts translates into a much milder upward shift of the portfolio reserve in the case where systematic reduced mortality than expected should occur along the insurance contracts' duration.

Keywords: neural network models · mortality forecasting · longevity risk management · portfolio reserve

1 Introduction

The management of risks is instrinsically part of insurance activities. Nowadays, insurance companies exploit the actuarial framework, along with notions and methods from other disciplines, for the purpose of being resilient against known and emerging risks while creating value.

Among the various sources of risk (e.g., financial risks), the risks related to the random human lifetime are of high concern for life insurance companies, that have to care about their solvency position and their social sustainability.

The assumptions about how policyholders' mortality will evolve along the duration of the contract enter several actuarial evaluations aimed at assessing assets and liabilities, such as, for instance, premium arrangements and mathematical reserves. In particular, for a life insurance company, the systematic

M. La Rocca et al. (Eds.): WMAF 2025, *New Perspectives in Mathematical and Statistical Methods for Actuarial Sciences and Finance*, pp. 26–37, 2025.
https://doi.org/10.1007/978-3-032-05551-4_3

deviations from the expected age-pattern of mortality are highly relevant [1]; the potential for these upward or downward deviations is known as either mortality risk (higher mortality than expected), that impacts particularly on insurance policies providing benefits in case of death, or aggregate longevity risk (lower mortality than expected), being a downside risk for insurance products that provide benefits in case of survival.

In our paper, we focus on endowment insurance policies that originate a cashflow paid to the beneficiaries if the insured dies by the end of the contract or a cashflow paid to the insured if the insured himself/herself is alive at the policy term. In this framework, both mortality and longevity risk play a role in shaping the insurer's assets and liabilities. In particular, the longevity risk component's impact is due to the long term over which the outflow in case of survival could be paid by the insurer and to the uncertainty around the expected number of in-force policies over time. The life tables used for predicting future mortality and the stochastic modeling choices underlying the construction of these tables can make a remarkable difference, in that the higher the accuracy in the estimation of the probability distribution of mortality rates over long and very long time horizons, the more reliable the assessment of the losses that may arise from deviations of mortality rates from central forecasts.

Among the most widely adopted approaches to forecast future mortality rates there are extrapolative models, that describe a measure of mortality as a function of time with parameters calibrated on past data and develop a model to obtain projections of mortality trends and age patterns, along with some measure of the uncertainty involved. As stressed by [3], the Lee-Carter (LC) method [2] seems to reconcile the plausibility of forecasts and the ease of measuring uncertainty; furthermore, the LC model has been used and remains widely used for mortality projections for actuarial and demographic purposes [4].

The LC model exploits a single stochastic time index to capture the general evolution of mortality; the modeling of this time index through time-series methods allows to obtain point and interval forecasts of mortality rates. Autoregressive integrated moving average (ARIMA) models, that have been widely adopted for a long time, are linear time-series models that can describe properly the dynamics of the time index only if this shows, indeed, a linear trend over the fitting period. In all those cases when the time index shows, instead, deviations from linearity, ARIMA methods tend to increase the variance of the estimated time index, thus producing biased forecasts and wide prediction intervals. As demonstrated in the literature, the case of nonlinear trend of the time index is empirically verified in relation to historical data relative to various countries [5,6], especially when the fitting period extends over a long past horizon [7]. For actuarial evaluations of life insurance policies made in the framework of the LC model, adopting a time-series model that is not able to capture the possible non-linear dynamics of the time index could lead to inflated forecast levels of uncertainty around central mortality projections and, thus, to a biased assessment of the mortality and of the downside risk that prospectively affect the portfolio of policies. Recent contributions in the literature have addressed the issue of improving the ability of the LC model to process historical mortality information, by using neural network models for the time index [8–12].

In our paper, we contextualize the LC structures with neural network models in the framework of the risk management in life insurance, with a specific focus on the assessment of portfolio liabilities, according to a 'mark-to-model' approach. According to this approach, the role of the models used to forecast the probability structures involved in the computation of the fair value of the liabilities is central. We compute the safe-side individual reserve associated with an endowment insurance contract at each policy anniversary. We then shift to the assessment of the time profile of the portfolio reserve (for a pool of homogeneous contracts) and, in this specific context, we measure the implications of two alternative modeling choices for the forecasting of the random number of the policies in-force at each future policy anniversary: a standard LC model with a linear time-series model for the time index and the neural network LC model (ARX-NN model) combined with the bootstrap scheme proposed in [12]. As shown in [12], the ARX-NN models, even those with a few number of neurons in the hidden level, outperform time-series linear models and smooth the uncertainty around central forecasts over time. In the context of our study, this advantageous feature in statistical terms translates into an actuarial fact, that becomes more and more relevant as we move forward along the policies' duration, in relation to the ex ante measurement of the effects that longevity risk may have on the estimation of the portfolio reserve. The goal of the paper is to verify if the use of the ARX-NN models for the LC time index implies a reduction in the gap between the portfolio reserve assessed under a realistic, central, mortality scenario and a stressed scenario (systematic reduced mortality according to the lowest quantiles of the forecast distribution of mortality rates), and to measure this reduction. A lower gap means that the life insurer, at the time of policies' issue, should set aside less resources to provide for the deviations from the best-estimate mortality and thus to would have available a larger amount of free capital to invest in another activities that create value for the insurance company (e.g., product innovation, strategic improvements).

The paper is organized as follows. In Sect. 2, we introduce notation and discuss the actuarial measures we focus on. In Sect. 3, we describe the main features of the Lee-Carter mortality model, while in Sect. 4, we explain how neural network models are integrated into the LC framework. In Sect. 5, we describe the data and the methodology underpinning the numerical application and illustrate the results. Section 6 concludes the paper.

2 Actuarial Notation and Measures

As it is conventional in life insurance mathematics [13], we consider measures of mortality that vary with time t and age x. In what follows, the right subscript of mortality-related measures has to be interpreted as the specific age and time the mortality measure refers to; for instance $T_{x,t}$ denotes the random lifetime of a person aged x at time t. Within our study, we consider the following probabilities, that are directly involved in mortality modeling and actuarial computations:

- probability of death within h years: ${}_hq_{x,t} = P(T_{x,t} \leq h)$;
- one-year probability of death (or mortality rate): $q_{x,t}$;
- probability to survive at least further h years: ${}_hp_{x,t} = 1 - {}_hq_{x,t} = P(T_{x,t} > h)$;
- deferred probability of dying between ages $x+h$ and $x+h+k$:

$$ {}_{h|k}q_{x,t} = P(h < T_{x,t} \leq h+k | T_x > h) = {}_hp_{x,t}\ {}_kq_{x+h,t+h}; $$

- one-year death rate $m_{x,t}$ whose relationship with the mortality rate can be described as follows [14]:

$$ m_{x,t} = -\log(1-q_{x,t}) = -\log(p_{x,t}). $$

Note that the empirical estimate of $m_{x,t}$ is obtained as follows:

$$ m_{x,t} = \frac{d_{x,t}}{E^c_{x,t}}, $$

where $d_{x,t}$ is the expected number of deaths, in t, between age x and age $x+1$, while $E^c_{x,t}$ is the average number of people alive in the same age interval, the so-called exposed-to-risk.

The standard endowment insurance combines a pure endowment with a term insurance having common insured capital and duration. Accordingly, the standard endownment insurance issued on an insured aged x provides: on the one hand, the opportunity to save, since the insured capital is paid to the policyholder at maturity in case of the insured's survival, and, on the other hand, protection from the event of death, as the insured capital is paid to the beneficiaries if the insured's death occurs at any time before the maturity. We denote by Y the random present value of the benefit provided by the endowment insurance product.

The expected present value (or actuarial value), $E(Y)$, of the endowment insurance contract with insured capital C, issued in $t=0$ on a person aged x and having duration m years, is as follows:

$$ E(Y) = C\left[\sum_{h=1}^{m}(1+i)^{-h}\ {}_{h-1|1}q_{x,0} + (1+i)^{-m}\ {}_mp_{x,0}\right]. \quad (1) $$

As shown in Eq. 1, the computation of $E(Y)$ involves discounting the outflows and requires the adoption of a *technical basis*, that is constituted by the interest rate i and a proper life table (i.e., probabilities of death and survival). The death probability ${}_{h-1|1}q_{x,0}$ refers to the event that the benefit is paid at the given time h within the policy duration following the policyholder's death between ages $x+h-1$ and $x+h$, while ${}_mp_{x,0}$ refers to the event that the benefit is paid at maturity the policyholder being alive at age $x+m$.

We assume that a single premium Π is paid by the policyholder, calculated according to the equivalence principle, that states the equality between the actuarial value of premiums and the actuarial value of benefits, so that $\Pi = E(Y)$.

The policy prospective mathematical reserve, at a given time t, with $t = 0, 1, 2, \ldots, m$, allows to assess the prospective insurer's debt from time t onwards, until the maturity, in relation to each insurance policy in the portfolio. The mathematical reserve at time t is thus given by the difference between the actuarial value in t of the prospective outflows (benefits) and the actuarial value, in t, of the prospective inflows (level premiums). In the case where a single premium is due by the policyholder, the mathematical reserve coincides with the actuarial value of future benefits.

For the endowment insurance we are dealing with, the mathematical reserve V_t at time t, with $1 \leq t < m$, is defined as follows:

$$V_t = C\left[\sum_{h=0}^{m-t-1} (1+i)^{-(h+1)} \; {}_{h|1}q_{x+t,t} + (1+i)^{-(m-t)} \; {}_{m-t}p_{x+t,t}.\right] \tag{2}$$

If $t = m^-$, $V_{m^-} = C$.

The traditional approach adopts the first-order, prudential, basis to discount benefits and premiums. For our application, we use A62I mortality tables for immediate life annuity contracts, from ANIA (Italian Association of Insurance Companies); such tables account for specific self-selection factors [15,16]. For the insurer's making valuations at the time of the policy issue, computing the policy reserve at all future anniversaries leads to an analysis of the time profile of the expected insurer's liabilities. In our paper, we limit ourselves to consider only integer durations.

Our perspective is forward-looking, since we assume to know information until time 0; thus all the evaluations relative to subsequent times are made at time 0. When considering a pool of identical policies, issued in $t = 0$, that is closed with respect to new policies, the related portfolio reserve at any future time t can be determined starting from the estimated policy reserve for year t and from the expected size of the portfolio at that time. In particular, the number of policies in-force at the future time $t (0 < t \leq m)$ is random when the computation is performed ($t = 0$). It is thus needed to make probabilistic assumptions on the decumulation over time of the initial number of policies, that involves building a life table describing the age-pattern of mortality for the policyholders. For this purpose, it is generally adopted a second-order, realistic, basis. Let us denote by N_0 the initial number of policies in the portfolio and by N_t the random number of policies still in-force at time t. The portfolio reserve at time t is the random amount $V_t^{[P]}$ given by [13]:

$$V_t^{[P]} = N_t \, V_t; \quad t = 1, 2, \ldots, m \tag{3}$$

and its expected value, $\hat{V}_t^{[P]} = E(V_t^{[P]})$, depends on the expectations that can derived on N_t, conditional on the information available at the time $t = 0$ of the policies' issue:

$$\hat{V}_t^{[P]} = E(N_t)\, V_t; \tag{4}$$

where:

$$E(N_t) = N_0\ {}_tp_{x,0}; \tag{5}$$

under the assumption that only the policyholders' death can trigger the policy termination. ${}_tp_{x,0}$ represents the probability that a person aged x at time 0 is still alive at age $x+t$, according to a second-order or realistic basis.

We obtain estimates of the survival probabilities involved in Eq. 5, by means of stochastic mortality modeling. In particular, we resort to the Lee-Carter (LC) model's framework and select two different types of time-series models to describe the dynamics of the time index of the LC model, that, as explained in the next sections, drives mortality evolution:

- autoregressive integrated moving average (ARIMA) models, that we can regard as standard linear models, as in the original contribution of Lee and Carter [2];
- autoregressive exogeneous neural network models, that account for possible non-linearities, as in [12].

3 Lee-Carter Model

The Lee-Carter (LC) model was developed by [2] to project the level and the age pattern of mortality in the long run, based on parsimonious stochastic representation of the log of a time series of death rates $m_{x,t}$ [18].

Under the LC predictor structure, a single stochastic time-varying index k_t expresses the level of mortality and thus drives its evolution, as follows:

$$log\ m_{x,t} = a_x + b_x k_t + e_{x,t}. \tag{6}$$

a_x and b_x are age-specific parameters, denoting, respectively, the general shape of mortality across ages and the sensitivity of the logged death rate for age x to variations in k_t. $e_{x,t}$, with mean 0 and variance σ_e^2, explains residual age-specific historical effects namely the variance that k_t parameter is not able to capture.

According to the LC method, the b_x's are normalized to sum to unity and the k_t's to sum to 0. Furthermore, a second-step adjustment of the fitted time-varying index $\hat{k}_t$ is needed to realize the matching between the fitted and the observed death rates in each year of the fitting period.

Standard statistical methods are traditionally used to model and forecast the estimates of $\hat{k}_t$ and to build probabilistic predictions of mortality rates. Linear time-models are conventionally used for this purpose. For instance, [2] selected a Random Walk with drift based on the U.S. mortality data from 1933 to 1989.

4 Forecast Intervals with Neural Network Bootstrap

Following [12], we assume that the mortality rates can be modelled as:

$$\hat{k}_t = g(\mathbf{z}_t) + \varepsilon_t \tag{7}$$

where $\mathbf{z}_t$ is a vector of $r = m + 1$ explicative variables which, consistently with the structure of mortality rates, include m lagged values of $\hat{k}_t$, to capture the possible non linear dynamics of $\hat{k}_t$, and the exogenous trend variable, to capture its long-run pattern. That is:

$$\mathbf{z}_t' = \left(\hat{k}_{t-1}, \ldots, \hat{k}_{t-m}, t\right) \tag{8}$$

.

The regression function $g(\cdot)$ can be approximated by using neural networks with a single output and additive nodes, clarifying what is actually learned by the artificial neural networks in a nonlinear regression framework. The class of used ARX-NN models is:

$$f_s(\mathbf{z}_t; \mathbf{w}) = \boldsymbol{\beta}'\mathbf{z}_t + \sum_{j=1}^{s} c_j \psi\left(\boldsymbol{\alpha}_j'\mathbf{z}_t + \alpha_{j0}\right) + \beta_0 \tag{9}$$

where $\mathbf{w}' = (\boldsymbol{\beta}', c_1, \ldots, c_s, \boldsymbol{\alpha}_1', \ldots, \boldsymbol{\alpha}_s', \alpha_{10}, \ldots, \alpha_{s0}, \beta_0)$ is the parameter vector of dimension $s(r+2)+r+1$; s is the hidden layer size; $\{\boldsymbol{\alpha}_j' = (\alpha_{j1}, \ldots, \alpha_{jr}), j = 1, \ldots, s\}$ are the weight vectors of the connections between the input layer and the hidden layer; $\{c_j, j = 1, \ldots, s\}$ are the weights of the link between the hidden layer and the output neuron; $\boldsymbol{\beta}' = (\beta_1, \ldots, \beta_r)$ is the vector of skip-layer connections; $\alpha_{10}, \ldots, \alpha_{s0}$, and β_0 are bias terms; $\psi(\cdot)$ is a properly chosen activation function for the hidden neurons. As usual in the neural network literature, a sigmoidal function, such as the logistic or hyperbolic tangent, is assumed for the hidden layer. In this case, a single-layer neural network can arbitrarily closely approximate the unknown function as well as its derivatives, up to a given order (provided that they exist), as measured by a proper norm.

Once the topology of the neural network is fixed, the estimate of the weights can be obtained by minimizing a loss function compromising both the primary squared loss function and a penalty on the L_2 norm of the weights. That is:

$$\hat{\mathbf{w}} = \arg\min_{\mathbf{w}\in\mathbf{W}} \sum_{t=1}^{T-m} \left(\hat{k}_t - f_s(\mathbf{z}_t, \mathbf{w})\right)^2 + \frac{\lambda}{2}\|\mathbf{w}\|^2 \tag{10}$$

where $\|\cdot\|$ is the L_2-norm, and λ is a regularization parameter which forces weights to decay towards zero. It is usually fixed by cross-validation.

Let I_T be the conditioning set including the information on $\hat{k}_t$ available up to time T. The one-step ahead point forecast of $\hat{k}_t$ is:

$$\hat{k}_{T+1\,T} = E(\hat{k}_{T+1}|I_T) = f_s(\mathbf{z}_{T+1}, \hat{\mathbf{w}}) \tag{11}$$

The bootstrap procedure, to construct the forecasting interval for the future observation $\hat{k}_{T+1}$, runs as follows.

1. Select the order of autoregression m in the explicative variables, the hidden layer size s and the regularization parameter λ.
2. Estimate the neural network weights by (10).
3. Compute the residuals $\hat{\epsilon}_t = \hat{k}_t - f_s(\mathbf{z}_t, \hat{\mathbf{w}})$
4. Define the empirical distribution function of the centered residuals $\tilde{\epsilon}_t$ as:

$$\hat{F}(x) = \frac{1}{T} \sum_{t=m+1}^{T} \mathbb{I}(\tilde{\epsilon}_t \leq x)$$

   where $\tilde{\epsilon}_t = \hat{\epsilon}_t - \bar{\epsilon}$ with $\bar{\epsilon} = (T-m)^{-1} \sum_{t=m+1}^{T} \hat{\epsilon}_t$ and $\mathbb{I}$ is the indicator function.
5. Fix the bootstrap runs B
6. For $b = 1, \ldots B$
   (a) Let $\epsilon_{T+1}^{(b)}$ a random draw from $\hat{F}(.)$
   (b) Compute the future bootstrap observations as $\hat{k}_{T+1}^{(b)} = f(\hat{\mathbf{z}}_{T+1}^{(b)}, \hat{\mathbf{w}}) + \epsilon_{T+1}^{(b)}$ where $\hat{\mathbf{z}}_{T+1}^{(b)} = (\hat{k}_T^{(b)}, \ldots, \hat{k}_{T+1-p}^{(b)}, T+1)$ with $\hat{k}_t^{(b)} = k_t$ if $t \leq T$.
7. Given the set of B forecasts $\left\{\hat{k}_{T+1}^{(b)}, b = 1, \ldots, B\right\}$, compute the ECDF $\hat{F}_{T+1}^{*}$, which can be used to estimate the unknown distribution $F_{T+1|T}(.)$ of $\hat{k}_{T+1}$
8. The $(1-\alpha)$ forecast interval for $\hat{k}_{T+1}$ is given by $\left[\hat{Q}^*(\alpha/2), \quad \hat{Q}^*(1-\alpha/2)\right]$ where $\hat{Q}^*(.)$ is the quantile function associated to the estimated bootstrap distribution $\hat{F}_{T+1}^{*}$.

5 Numerical Application

We consider a portfolio of 1000 identical endowment insurance products, issued in 1997 on a pool of female and male individuals aged 65, with duration 23 years and insured capital, in case of death and survival, equal to 1000 monetary units. The policyholder pays a single premium at the time of policy issue.

We assess the time profile of the policy reserve according to Eq. 2, for t moving forward from the first up to the 22nd policy anniversary; $V_0 = 0$ and $V_{23} = 1000$, considering $V_t = V_{t^-}$. The computations rely on a safe-side technical basis, that is constituted by a constant annual interest rate equal to 2% and A62I mortality tables for immediate life annuity contracts, from ANIA (Italian Association of Insurance Companies) [15,16], according to a composition made by 50% men and, accordingly, 50% women. Year by year along the contract duration, the policy reserve increases by an amount ranging from approximately 17% to 19%, similarly to the case of a pure endowment insurance contract, due to the mild impact of the term insurance component.

The computation of the expected portfolio reserve as in Eq. 4 requires the estimation of the expected number of in-force policies $E(N_t), t = 1, 2, \ldots, 23$, out of the initial 1000 insurance contracts, according to Eq. 5. This implies to obtain the forecast distributions of the mortality rates that will characterize the cohort of policyholders for the next 22 years; this is crucial to estimate the realistic sequence of probabilities ${}_tp_{65,0}$, having posed 1997 as our initial time

$t = 0$, with $t = 1, 2, \ldots, 23$, according to the standard relations described in Sect. 2. To this purpose, we use the predictor structure for logged mortality rates provided by the Lee-Carter model, according to Eq. 6. From the empirical point of view, we extrapolate and project the mortality pattern of the Italian unisex population. Our case study is based on the mortality data that are available in the Human Mortality Database [17], for the total Italian population. The time frame spanning years from 1908 to 1997 is selected as the training period. The Lee-Carter model provides the predictor structure for logged mortality rates. The forecast distributions of mortality rates over the policy duration is obtained within this common predictor structure, but under two alternative type of models for the time index $\hat{k}_t$: a nonlinear autoregressive neural network model (ARX-NN) and standard ARIMA models. In relation to the same population and the same historical horizon considered in our study, [12] found that the best ARIMA model coincides with the Random Walk (RW) model, while the autoregressive neural network model with two hidden neurons (ARX-NN(2)) outperforms all competitive models, both linear and nonlinear, and regardless of the hidden layer size. Accordingly, for our empirical application, we select an ARX-NN(2) model and use, as a benchmark linear model, the RW model with drift. Estimation and forecasting procedures are consistent with those described in [12].

In Panels a) and b) of Fig. 1, we display the amount of the expected portfolio reserve, $\hat{V}_t^{[P]}$, against policy anniversaries, respectively in the case of a RW or a ARX-NN(2) model for the LC time index. The single dots represent the value of $\hat{V}_{23}^{[P]}$, that coincides with the expected number of in-force policies multiplied by the insured capital: this is due to the fact that for all the insured people alive at time 22, both surviving or not at time $t = 23$, the benefit payment has to be performed by the insurer. Apart from the realistic mortality scenario, that we obtain from central forecasts of $\hat{k}_t$, we also consider a stress scenario derived by the 2.5% quantile of the forecast distribution for $\hat{k}_t$, to give graphical insights into the model-implied exposure to longevity risk associated with the portfolio of endowment insurance contracts. In both graphs, the red curves relate to the stress mortality scenario, while the green ones to the realistic mortality scenario. The gap between the two curves quantifies the extent to which the life insurer risks in underestimating its portfolio reserve, as the time goes forward, and thus in understating its debt position. One of the main issues with linear time-series models for the LC time index, as stressed in the literature, is the tendency to inflate the variance of $\hat{k}_t$ and thus to produce amplified forecast levels of uncertainty in mortality projections ([4]). Figure 1 gives sense of some of the actuarial implications of this limitation. Indeed, in Panel a), we can notice a much sharper gap between the two curves than in Panel b). For instance, based on the probabilistic predictions of the LC model with RW for the time index, if the insurer relied on the realistic scenario for the estimation of the expected portfolio reserve but a lower mortality scenario realized, the variation in its debt position would be highly impactful; for instance, the realistically predicted portfolio reserve would be 18% lower than needed under the stress scenario at the 12th policy anniversary, and 43% lower than needed at one-year to maturity.

This implies that the life insurer, at the time of policies'issue, should set aside more resources to provide for the deviations from the best-estimate mortality. As shown in Panel b) of Fig. 1, instead, relying on the realistic scenario would lead to underestimate the expected portfolio reserve to a lower extent; for instance, by 4% at the 12th anniversary and by 15% at one-year to maturity.

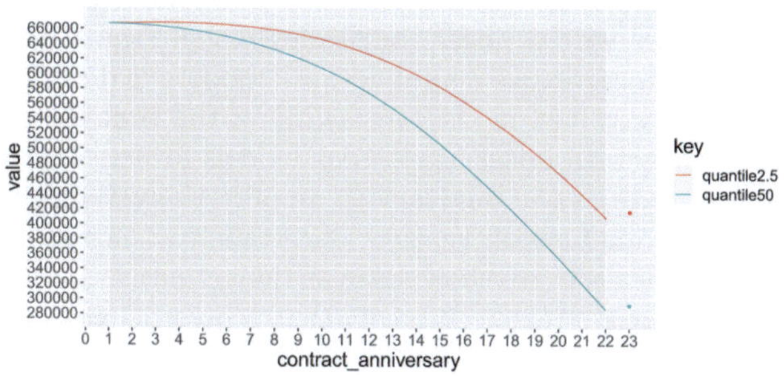

(a) RW model for the LC time index.

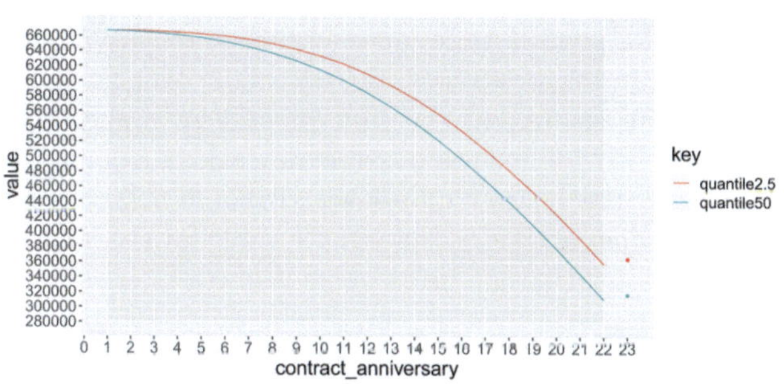

(b) ARX-NN(2) model for the LC time index.

Fig. 1. Time profile of the expected portfolio reserve

6 Conclusions

Insurance activities intrinsically deal with the management of risks, that more and more often integrates into the actuarial framework notions and methods from various disciplines. Artificial neural networks have become effective tools to approximate nonlinear functions and dynamics, for instance to the purpose of forecasting. According to the mark-to-model approach to the assessment of the insurer's debt position, the role of the models used to forecast the probability structures involved in the computation of the fair value of the liabilities is central.

The benefits of artificial neural networks for forecasting is thus worth to be analysed in this context.

In our paper, we focus on a portfolio of endowment insurance policies and on the ex-ante measurement of the effects that longevity risk may have on the estimation of the mathematical reserve for this portfolio. We measure the implications of two alternative modeling choices for the forecasting of the random number of the policies in-force at each future policy anniversary: a standard Lee-Carter (LC) model with a linear time-series model for the time index and the LC model with autoregressive neural network models (ARX-NN models) and bootstrap procedures involved in the forecasting of the time index (as in [12]). One of the main issues with linear time-series models for the LC time index, as stressed in the literature, is the tendency to amplify the variance of time index and thus to produce wide forecast levels of uncertainty in mortality projections. In our numerical application, we show some of the actuarial implications of this limitation. We consider a stressed scenario, that represents the case where systematic reduced mortality with respect to the central, best-estimate mortality occurred along the duration of the insurance contracts. From a practical point of view, this implies considering the lowest quantiles of the forecast distribution of mortality rates. In a framework analogous to a what-if analysis, we explore ex-ante the impact, on the value of the portfolio reserve, of experiencing low mortality than expected along the policy duration. We find out that the use of the ARX-NN models for the LC time index implies a much milder upward shift of the portfolio reserve in the stress scenario, especially over the long run. A lower gap means that the life insurer, at the time of policies' issue, may reliably set aside less resources to provide for the possible deviations from the best-estimate mortality; accordingly, the insurer may have available a larger amount of free capital to invest in another activities that create value for the insurance company (e.g., product innovation, strategic improvements).

References

1. Pitacco, E.: ERM and QRM in Life Insurance. An Actuarial Primer. Springer, Cham (2020). https://doi.org/10.1007/978-3-030-49852-8
2. Lee, R. D., Carter, L. R.: Modeling and forecasting U.S. mortality. Journal of the American Statistical Association 87(419), 659–671
3. Wong-Fupuy, C., Haberman, S.: Projecting Mortality Trends. Recent Developments in the United Kingdom and the United States. North Am. Actuarial J. **8**(2), 56–83 (2013). https://doi.org/10.1080/10920277.2004.10596137
4. Basellini, U., Camarda, C.G., Booth, H.: Thirty years on: a review of the Lee-Carter method for forecasting mortality. Int. J. Forecast. **39**(3), 1033–1049 (2023). https://doi.org/10.1016/j.ijforecast.2022.11.002
5. Booth, H., Tickle, L.: Mortality modeling and forecasting: a review of methods. Ann. Actuarial Sci. **2**(1–2), 3–43 (2008). https://doi.org/10.1017/S1748499500000440
6. Bergeron-Boucher, M.-P., Kjærgaard, S.: Mortality forecasting at age 65 and above: An age-specific evaluation of the Lee-Carter model. Scand. Actuar. J. **2022**(1), 64–79 (2022). https://doi.org/10.1080/03461238.2021.1928542

7. Li, H., Li, J.S.: Optimizing the Lee-Carter approach in the presence of structural changes in time and age patterns of mortality improvements. Demography **54**, 1073–1095 (2017). https://doi.org/10.1007/s13524-017-0579-x
8. Levantesi, S., Pizzorusso, V.: Application of machine learning to mortality modeling and forecasting. Risks **7**(1), 26 (2019). https://doi.org/10.3390/risks7010026
9. Nigri, A., Levantesi, S., Marino, M., Scognamiglo, S., Perla, F.: A deep learning integrated Lee-Carter model. Risks **7**(1), 33 (2019). https://doi.org/10.3390/risks7010033
10. Lindholm, M., Palmborg, L.: Efficient use of data for LSTM mortality forecasting. Eur. Actuar. J. **12**, 749–778 (2022). https://doi.org/10.1007/s13385-022-00307-3
11. Marino, M., Levantesi, S., Nigri, A.: A neural approach to improve the Lee-Carter mortality density forecasts. North Am. Actuarial J. **27**(1), 148–165 (2023). https://doi.org/10.1080/10920277.2022.2050260
12. Apicella, G., La Rocca, M., Perna, C., Sibillo, M.: Neural network Lee-Carter model and the actuarial relevance of longevity risk assessment. Scand. Actuar. J. (2024). https://doi.org/10.1080/03461238.2024.2443833
13. Olivieri, A., Pitacco, E.: Introduction to insurance mathematics. Technical and financial features of risk transfers. Springer (2011). https://doi.org/10.1007/978-3-642-16029-5
14. Pitacco, E., Denuit, M., Haberman, S., Olivieri, A.: Modelling longevity dynamics for pensions and annuity business. Oxford University Press (2009). ISBN:9780199547272
15. ANIA.: Le basi demografiche per rendite vitalizie A1900-2020 e A62. Relazione tecnico-metodologica (2014)
16. ANIA.: Le basi demografiche per assicurazioni di rendita A1900-2020 e A62 (2015). https://www.ania.it/studi-e-rapporti-demografici
17. Human Mortality Database. Max Planck Institute for Demographic Research (Germany), University of California, Berkeley (USA), and French Institute for Demographic Studies (France) (2025). http://www.mortality.org
18. Lee, R.D.: The lee-carter method for forecasting mortality, with various extensions and applications. North Am. Actuarial J. **4**(1), 80–91 (2000)

Climate-Related Extensions of the Lee-Carter Model

Imma Lory Aprea[1], Francesca Perla[2], Raffaele Clemente Petrella[3(✉)], Mariafortuna Pietroluongo[4], and Salvatore Scognamiglio[2]

[1] Department of Economics, Law, Cybersecurity, and Sports Sciences, University of Naples "Parthenope", 80035 Nola, Italy
[2] Department of Management and Quantitative Studies, University of Naples "Parthenope", 80132 Naples, Italy
[3] Department of Economic and Legal Studies, University of Naples "Parthenope", 80132 Naples, Italy
raffaeleclemente.petrella001@studenti.uniparthenope.it
[4] Department of Business and Economic Studies, University of Naples "Parthenope", 80132 Naples, Italy

Abstract. This study examines the impact of climate risk-related variables on mortality patterns across diverse population groups. An innovative version of the Lee-Carter model is proposed, incorporating a climate risk-related variable - specifically, CO2 emissions - to enhance the accuracy of mortality prediction. A three-step estimation process using Ordinary Least Squares is employed to calibrate the model parameters. Experiments conducted on data from 42 populations indicate that integrating climate risk information leads to more accurate mortality predictions. Especially, combining climate variables with macroeconomic indicators such as GDP yields further improvements in forecast performance.

Keywords: Mortality Modelling · Stochastic Models · Climate risk

1 Introduction

The impact of climate change on global health has become a central focus in recent actuarial and demographic research due to its significant influence on mortality trends both directly and indirectly [12]. The frequency and intensity of extreme weather events, such as heatwaves, have been linked to increased cardiovascular and respiratory diseases [1,4,5]. Additionally, greenhouse gas emissions contribute to air pollution, exacerbating respiratory conditions and increasing lung cancer risk [11,14].

Addressing these challenges necessitates advanced predictive models capable of integrating climate risk variables to improve mortality forecasts. In the actuarial science literature, the Lee-Carter (LC) model [8] is a widely utilized framework for stochastic mortality projections, leveraging historical data to generate

M. La Rocca et al. (Eds.): WMAF 2025, *New Perspectives in Mathematical and Statistical Methods for Actuarial Sciences and Finance*, pp. 38–49, 2025.
https://doi.org/10.1007/978-3-032-05551-4_4

probabilistic forecasts. Although primarily extrapolative, the LC model has been extended to incorporate macro-economic variables, demonstrating improved predictive accuracy [2,9,10].

On the other side, recent studies suggest that climate-related factors, particularly CO2 exposure, may serve as critical explanatory variables in mortality forecasting [13,16]. In this vein, this study proposes extending the classical LC model by incorporating CO2 exposure as a climate change proxy, aiming to enhance mortality forecast accuracy.

Numerical experiments using *Human Mortality Database*[1] data for 21 countries reveal a strong relationship between the LC model residuals and CO2 trends, underscoring the variable's explanatory potential in mortality analysis.

2 Lee-Carter Model

The LC model [8] was a milestone in the field of mortality forecasting by introducing a straightforward but powerful stochastic framework based only on historical data. By relying on a single time-varying index with a linear trend, the model effectively captures long-term mortality patterns. Unlike previous approaches, it is entirely extrapolative, deliberately excluding external variables such as medical advancements or social factors.

Let X denote the set of ages under consideration and T the set of years analyzed. The LC model expresses the logarithm of the central death rate $\log(m_{x,t}) \in \mathbb{R}$, for age $x \in X$ in year $t \in T$, using the following formulation for a given single population:

$$\log(m_{x,t}) = a_x + b_x k_t + \epsilon_{x,t}, \quad \text{with} \quad \epsilon_{x,t} \sim N(0, \sigma_\epsilon^2). \tag{1}$$

In this framework, $a_x \in \mathbb{R}$ represents the average mortality pattern across different ages; $b_x \in \mathbb{R}$ reflects how responsive the mortality rate at age x is to changes over time, essentially measuring the age-specific impact of the temporal component k_t; $k_t \in \mathbb{R}$ accounts for the general time evolution trend of mortality; and $\epsilon_{x,t}$ denotes the residual term, representing the difference between the observed and the modeled logarithmic mortality rates.

In order to address identifiability concerns, the authors recommend the imposition of two constraints:

$$\sum_{x \in X} b_x = 1 \quad \text{and} \quad \sum_{t \in T} \frac{k_t}{|T|} = 0. \tag{2}$$

The Ordinary Least Squares (OLS) estimator is used to calibrate the LC model. In particular, parameter estimates are derived by addressing the following optimization problem:

$$\underset{(a_x)_x,(b_x)_x,(k_t)_t}{\operatorname{argmin}} \sum_{x \in X} \sum_{t \in T} \left(\log(m_{x,t}) - a_x - b_x k_t\right)^2. \tag{3}$$

[1] See [7] for further details.

The estimate of a_x, denoted by $\hat{a}_x$, is calculated as the logarithm of the geometric mean of the mortality rates over time, for each age $x \in X$:

$$\hat{a}_x = \log(\prod_{t \in T} (m_{x,\,t})^{1/|T|}). \tag{4}$$

The parameters $(b_x)_x$ and $(k_t)_t$ are estimated via the first-order Singular Value Decomposition (SVD) of the centered log-mortality matrix $H = \{h_{x,\,t}\}_{x \in X,\, t \in T} \in \mathbb{R}^{|X| \times |T|}$, where:

$$h_{x,\,t} = (\log(m_{x,\,t}) - \hat{a}_x). \tag{5}$$

For future forecasts, the model assumes that the parameters a_x and b_x are constant over time, while the forecasts of k_t are made using the modern time series approach. Specifically, despite the multiple formulations of the ARIMA(p, d, q) models, in practice, the random walk with drift model is used almost exclusively:

$$k_t = k_{t-1} + \gamma + \xi_t \quad \text{with i.i.d.} \quad \xi_t \sim N(0, \sigma_\xi^2), \tag{6}$$

where $\gamma \in \mathbb{R}$ represents the drift. Assuming multiple populations, for each country $r \in \mathcal{R}$ and each gender $g \in \mathcal{G} = \{\texttt{males}, \texttt{females}\}$, it would be appropriate to apply an LC model to each subgroup considered. In this way, the mortality dynamics of each group is modeled in a consistent and comparable manner:

$$\log(m_{x,\,t}^{(r,g)}) = a_x^{(r,g)} + b_x^{(r,g)} k_t^{(r,g)} + \epsilon_{x,\,t}^{(r,g)} \quad \forall r \in \mathcal{R},\ \forall g \in \mathcal{G}. \tag{7}$$

The model is estimated separately for each $r \in \mathcal{R}$, while the time indices $k_t^{(r,g)}$, specific to each population group, are forecasted independently using ARIMA(0,1,0) processes.

3 Mortality and Climate Risk Modelling

This section is focused on the integration of the climate risk feature into the LC model, specifically with CO2 emissions per capita as a proxy for climate risk. This greenhouse gas is central to global warming [6], with human activities altering the earth's temperature [15]. Along with methane and nitrous oxide, carbon dioxide contributes to the atmospheric greenhouse effect, amplified by fossil fuel emissions. Land use changes also have shown a key role, increasing greenhouse gas emissions and global temperatures. These events have led to warming of the climate system and global climate change, affecting weather patterns and climate balance. CO2 concentrations may reach critical levels indoors, affecting air quality.

Inclusion of climate risk variables in the model could improve mortality predictions. The model that we propose will be called LC_CO2. In this context, the designation $\text{CO2}_t^{(r)}$ indicates the emissions related to the region r that is the same for both gender populations within a given region; as a consequence, $\text{CO2}_t^{(r,\,male)} \equiv \text{CO2}_t^{(r,female)}$.

It is assumed that the residuals of the LC model for each population (r, g) are linearly dependent on the emission level $CO2_t^{(r)}$ observed in country r. This relationship is formally represented as:

$$\epsilon_{x,t}^{(r,g)} = w_x^{(r,g)} CO2_t^{(r)} + u_{x,t}^{(r,g)}, \tag{8}$$

where $u_{x,t}^{(r,g)}$ denotes the unexplained component of the residuals that is not captured by the climate risk variable. By incorporating this relationship into the LC residual structure, the extended model can be expressed as:

$$\log(m_{x,t}^{(r,g)}) = a_x^{(r,g)} + b_x^{(r,g)} k_t^{(r,g)} + w_x^{(r,g)} CO2_t^{(r)} + u_{x,t}^{(r,g)}, \tag{9}$$

where $w_x^{(r,g)} \in \mathbb{R}$ represents the coefficient associated with CO2 emissions for country r specific for each age and population. This parameter captures how sensitive the logarithmic mortality rate is to variations of CO2 emissions, and it is important to note that it differs between male and female populations.

Model calibration is carried out through an OLS estimator, following a three-step procedure. Initially, the classical LC parameters are estimated via the two-step SVD method previously described. Subsequently, the coefficient $w_x^{(r,g)}$, which links mortality to CO2 exposure, is estimated using the following expression:

$$w_x^{(r,g)} = \frac{\sum \left(CO2_t^{(r)} - \overline{CO2_t^{(r)}}\right)\left(\epsilon_{x,t}^{(r,g)} - \overline{\epsilon}_x^{(r,g)}\right)}{\sum \left(CO2_t^{(r)} - \overline{CO2_t^{(r)}}\right)^2} = \frac{\mathrm{Cov}\left(CO2_t^{(r)}, \epsilon_{x,t}^{(r,g)}\right)}{\mathrm{Var}\left(CO2_t^{(r)}\right)}. \tag{10}$$

Here, $\overline{CO2_t^{(r)}}$ and $\overline{\epsilon}_x^{(r,g)}$ represent the mean values of CO2 over time and $\epsilon_{x,t}^{(r,g)}$, respectively.

To produce forecasts, it is necessary to define a dynamic process governing the evolution of CO2 emissions. In this framework, an ARIMA(p, d, q) specification [3] can be employed, modeling the progression of $CO2_t^{(r)}$ over time using the following general expression:

$$\left(1 - \sum_{k=1}^{p} \alpha_k L^k\right)(1 - L)^d CO2_t = \left(1 + \sum_{k=1}^{q} \beta_k L^k\right)\eta_t. \tag{11}$$

In this formulation, $CO2_t$ denotes the observed time series at time t, as previously introduced, while η_t is the stochastic error term - assumed to be white noise with zero mean, constant variance, and no autocorrelation ($\eta_t \sim N(0, \sigma^2)$). The lag operator L is defined such that $LCO2_t = CO2_{t-1}$ and $L^2 CO2_t = CO2_{t-2}$, and so forth. The parameter p denotes the order of the autoregressive (AR) component, referring to the number of past values $CO2_{t-k}$, with $k = 1, \ldots, p$, included in the model. The parameter q corresponds to the order of the moving average (MA) component, incorporating q past error terms η_{t-k}, with $k = 1, \ldots, q$, into the current prediction. The differencing order d indicates how many times the series must be differentiated to achieve stationarity. The coefficients α_k and β_k represent the weights associated with the AR and MA terms, respectively. The model can be refined by incorporating additional external variables that are expected to influence mortality dynamics. Expanding the set of explanatory factors may improve both predictive accuracy and interpretability. These additional variables can be estimated using the same OLS method previously discussed. For example, adding GDP to the LC_CO2 model allows the inclusion of both economic and climatic influences, highlighting their combined impact on mortality trends. The resulting augmented model can be represented as follows:

$$\log(m_{x,t}^{(r,g)}) = a_x^{(r,g)} + b_x^{(r,g)} k_t^{(r,g)} + \delta 2_x^{(r,g)} \mathrm{GDP}_t^{(r)} + w 2_x^{(r,g)} \mathrm{CO2}_t^{(r)} + v_{x,t}^{(r,g)}, \quad (12)$$

where $\delta 2_x^{(r,g)}$ and $w 2_x^{(r,g)}$ denote the age- and population-specific coefficients linked to GDP and CO2, respectively, and $v_{x,t}^{(r,g)}$ denotes the unexplained component of the model. In this context, the temporal evolution of $\mathrm{GDP}_t^{(r)}$ is also modeled using an ARIMA process, following the same approach adopted for $\mathrm{CO2}_t^{(r)}$.

4 Numerical Experiments

This section presents the outcomes of the numerical experiments performed on the set of population groups. It includes a description of the data sources employed, the estimation of parameters for the stochastic mortality models, and an assessment of their forecasting accuracy. Furthermore, a comparative analysis is carried out to evaluate the relative performance of the proposed modeling approaches.

4.1 Data Description

The analysis is based on data from the Human Mortality Database [7] to study mortality, GDP data from the World Bank Open Data platform[2], and CO2 emissions data from the OECD website[3].

The study includes data on male and female populations for all ages from 0 to 99 years, covering the period from 1960 to 2021. The dataset was divided into a training and a test sets. Specifically, it will be used to evaluate the models forecast performance. The training set covers the years prior to 1999, while the test set includes the subsequent years to assess predictive performance. The Table 1 reports the list of countries analyzed.

[2] https://data.worldbank.org.
[3] https://www.oecd.org.

Table 1. List of countries considered.

Country	Initialism
Australia	AUS
Austria	AUT
Belgium	BEL
Canada	CAN
Denmark	DNK
Spain	ESP
Finland	FIN
France (Total Population)	FRATNP
United Kingdom (Total Population)	GBR_NP
Greece	GRC
Ireland	IRL
Iceland	ISL
Italy	ITA
Japan	JPN
Luxembourg	LUX
Norway	NOR
Netherlands	NLD
Poland	POL
Portugal	PRT
Sweden	SWE
United States of America	USA

4.2 Parameter Estimates

The starting point of the analysis is given by the observation of parameter estimates of the LC model. Specifically, for the two populations, we observe the estimates of $a_x^{(r,g)}$, $b_x^{(r,g)}$ and $(k_x^{(r,g)})_{x \in X}$. Specifically, in Fig. 1 the estimates of $a_x^{(r,g)}$ are shown. There is a minimal difference between the distributions among the various nations. This result was predictable given that the curve is calculated as the average of the observed logarithmic mortality rates.

Similarly, Fig. 2 illustrates the consistent trend of the $b_x^{(r,g)}$ curves across countries, showing higher values at younger ages that decrease with age, except for a slight increase in adulthood. This pattern confirms a recurring dynamic in mortality response by age, aligning with general trends observed over time. In particular, in Iceland (ISL), women exhibit more pronounced fluctuations than men and other populations, whereas in Poland (POL), the coefficient for men shows greater variability than for women. Despite these distinctions, the overall trend underscores the general dynamic that mortality variations are associated with common underlying factors.

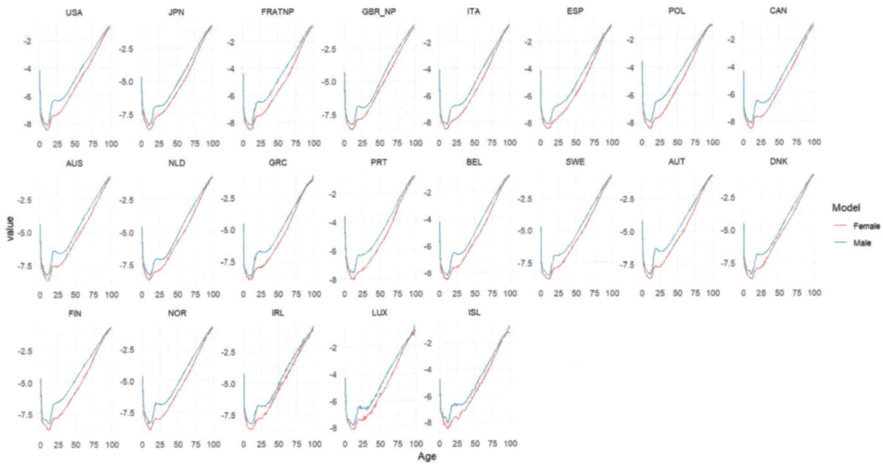

Fig. 1. Estimates of the parameter $(a_x^{(r,g)})_{x \in X}$ for different populations.

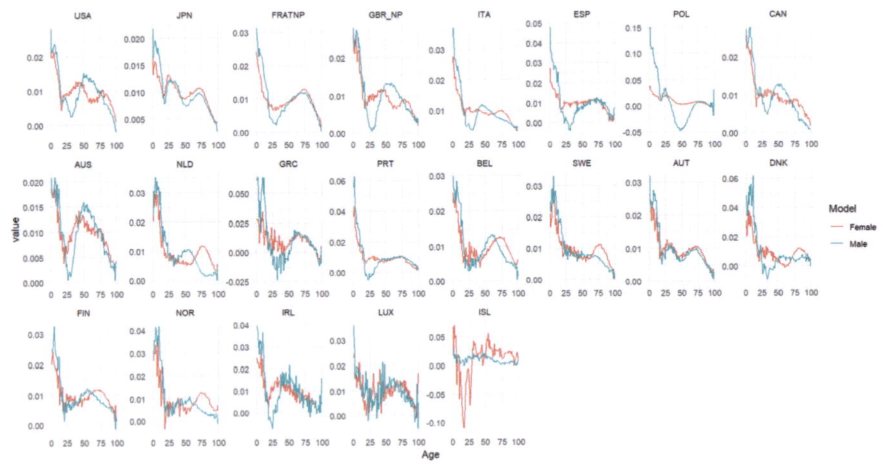

Fig. 2. Estimates of the parameter $(b_x^{(r,g)})_{x \in X}$ for different populations.

Estimates of $k_t^{(r,g)}$ for various regions are presented in Fig. 3. It shows a dotted line at the height of the year 1999. Beyond it the values are generated by forecasts made by the ARIMA(0,1,0) model. A decreasing trend is evident for all nations and all populations although with a considerable difference between the trend of k_t for men compared to women for some nations such as Poland (POL), Greece (GRC), and Iceland (ISL).

To evaluate the possibility that external variables improve the LC model, the correlation between the model residuals and CO2 emissions is preliminarily examined. This exploratory analysis provides a preliminary indication of the possible dependence between mortality dynamics and climate risks, justifying

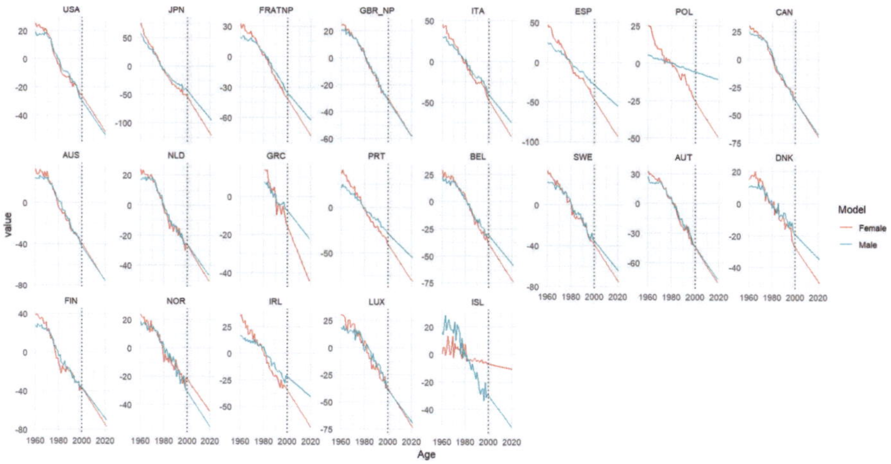

Fig. 3. Estimates of the parameter $(k_x^{(r,g)})_{x \in X}$ for different populations.

the inclusion of this climate variable to explain the error component not captured by the model.

The relationship between residuals and CO2 is assessed using the Pearson correlation test, which measures the linear association between two quantitative variables. The null hypothesis ($\rho = 0$) assumes the absence of linear correlation, while the alternative hypothesis ($\rho \neq 0$) postulates its presence. The correlation coefficient ρ and the p value allow us to evaluate the statistical significance of the association. Tests are conducted for specific ages $x \in \{5, 15, 25, \dots, 95\}$, in all countries $r \in \mathcal{R}$ and genders $g \in \mathcal{G}$. The results of the correlation test between the LC model residuals and CO2 are reported in Fig. 4. It shows a bar plot with the percentage of tests that showed a significant correlation for each age group. The figure highlights how this percentage is particularly significant for the ages $35 \leq x \leq 85$. The evidence that emerges in the figure allows us to use CO2 as a proxy variable of climate change for our analysis.

At this stage, it is both reasonable and feasible to estimate the coefficients $w_x^{(r,g)}$ of the LC_CO2 model using the OLS estimator previously discussed. The results are shown in Fig. 5. They appear to be very heterogeneous and different from population to population and from nation to nation. Overall, they show an oscillation above and below the value of zero, thus pointing to a first positive and then negative relationship of CO2 on mortality, or in general, a relationship that is in any case not constant. This oscillation could indicate that the influence of climate change depends on demographic and environmental particularities that result in this difference.

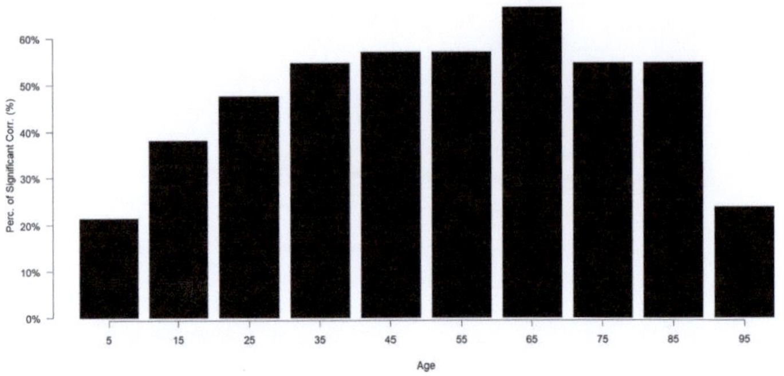

Fig. 4. Percentage of Significant Correlations by Age between CO2 and the residual of the Lee-Carter model.

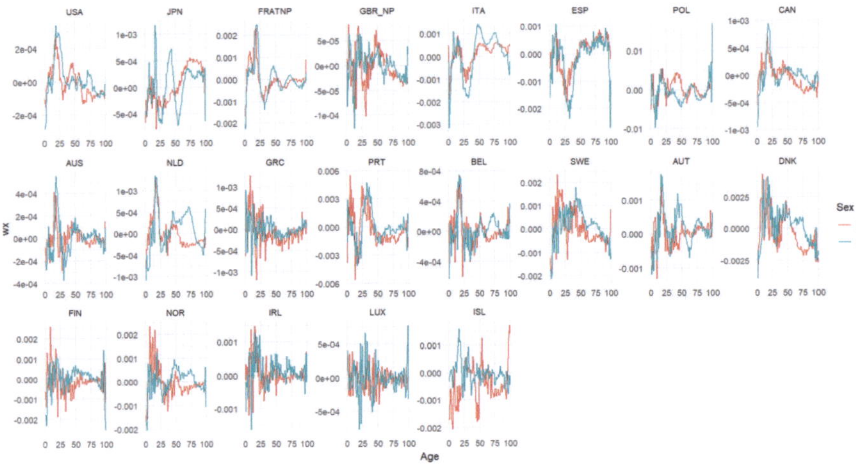

Fig. 5. Value of the coefficient $w_x^{(r,g)}$ for each country and each sex.

Following the above discussion, the next step in the analysis is to forecast CO2 for the years 2000 to 2021. As outlined, the forecasts will be produced after considering the appropriate ARIMA model for each nation, so that each historical series for each nation can be used and capture its own particularities. Figure 6 shows, in this sense, both the historical CO2 series for each nation - up to 1999 - and the forecasts generated with the ARIMA model identified for each nation. Two clusters emerge: the first identifies nations that observe an increase in CO2 emissions, among them we find, for example, Australia (AUS), Austria (AUT), Japan (JPN), Italy (ITA), and Ireland (IRL). In contrast, the second cluster consists of nations showing a reduction, including such as, Luxembourg (LUX) and Poland (POL). This suggests that some nations, more than others, are making strides in reducing their carbon footprint.

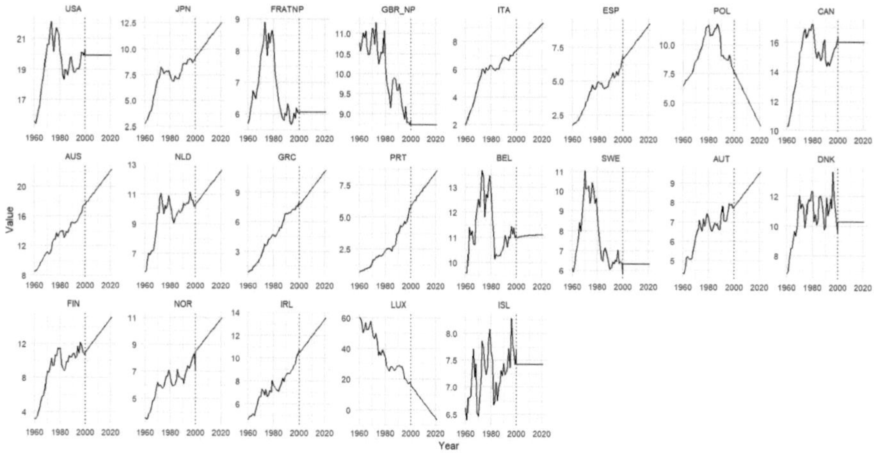

Fig. 6. CO2 time series for each country, including the forecasted values from 2000 to 2021.

4.3 Predictive Performance

This section is used to present the main results related to the comparison of the forecast performance of the various models. We compare the baseline LC model with three alternative versions: (i) the LC_GDP model, which incorporates merely GDP-related economic data [2]; (ii) the LC_CO2 model, previously discussed and reported in (9), which integrates climate risk information; and (iii) the LC_CO2_GDP model, introduced in (12), which combines both economic and climate risk data within the LC framework. This comparison aims to evaluate the contribution of each information source to the models performance. To assess predictive capabilities, we employ two standard metrics: Mean Squared Error (MSE) and Mean Absolute Error (MAE).

Table 2 presents the evaluation metrics for the models in both the training and the test sets, with the results multiplied by one hundred for an improved interpretability. The analysis of models with a single additional variable highlights marginal improvements in predictive performance, confirming that GDP and CO2 emissions are relevant, but their individual contributions to forecasting are limited. The LC_GDP_CO2 model shows the best performance on both sets, with lower MSE and MAE, indicating greater generalization and predictive accuracy compared to the other models. The interplay between GDP and CO2 significantly enhances the model's ability to forecast mortality, suggesting higher effectiveness in capturing the influencing factors.

Table 2. Evaluation metrics (MSE and MAE) for all models on training and test sets.

Model	In-sample		Out-of-sample	
	MSE	MAE	MSE	MAE
LC	2.95	10.04	11.57	23.41
LC_GDP	2.93	10.01	11.31	23.02
LC_CO2	2.95	10.03	11.58	23.44
LC_GDP_CO2	**2.89**	**9.85**	**11.05**	**22.44**

5 Conclusion

This study explores the impact of climate risk on mortality forecasting for a set of countries. Specifically, it employs models extending the well-known Lee-Carter model. These extended models incorporate a climate risk variable in order to provide more accurate mortality forecasts. The variable used is the level of per capita CO2 emissions. Throughout the study, it is shown that the significant correlation between the Lee-Carter's error terms and CO2 levels supports the hypothesis of a potential relationship between climate risk and mortality dynamics.

The proposed strategy is based on a three-stage Ordinary Least Squares estimation procedure. The dataset used for the analysis and the series of numerical experiments consists of a sample of 42 populations. The results illustrate that integrating climate risk data improves forecast accuracy for some populations, and that the combined inclusion of economic indicators, such as GDP, further enhances the model's predictive power.

Possible future developments of this work include several directions. First, it would be appropriate to consider additional variables, such as temperature. In addition, future works may explore alternative more flexible modeling assumptions concerning the relationship between climate indicators and mortality trends. These formulations could include non-linear regression models, machine learning approaches, or interaction effects between variables. Finally, it would be interesting to extend the analysis to stochastic mortality models beyond the Lee-Carter family, such as those belonging to the Generalized Age-Period-Cohort (GAPC) framework.

References

1. Bell, J., et al.: Changes in extreme events and the potential impacts on human health. J. Air Waste Manage. Assoc. **68**, 265–287 (2018)
2. Boonen, T., Li, H.: Modeling and forecasting mortality with economic growth: a multipopulation approach. Demography **54**, 1921–1946 (2017)
3. Box, G., Jenkins, G., Reinsel, G., Ljung, G.: Time series analysis: forecasting and control. (John Wiley & Sons 2015)

4. Chang, A., Tan, A., Nadeau, K., Odden, M.: Aging hearts in a hotter, more turbulent world: The impacts of climate change on the cardiovascular health of older adults. Curr. Cardiol. Rep. **24**, 749–760 (2022)
5. Ebi, K., et al.: Extreme weather and climate change population health and system implications. Annu. Rev. Public Health **42**, 293–315 (2021)
6. Florides, G., Christodoulides, P.: Global warming and carbon dioxide through sciences. Environ. Int. **35**, 390–401 (2009)
7. Jdanov, D., Jasilionis, D., Shkolnikov, V., Barbieri, M.: Human mortality database. Encyclopedia of Gerontology and Population Aging, pp. 2495–2503 (2022)
8. Lee, R., Carter, L.: Modeling and forecasting US mortality. J. Am. Stat. Assoc. **87**, 659–671 (1992)
9. Li, N., Lee, R.: Coherent mortality forecasts for a group of populations: an extension of the Lee-Carter method. Demography **42**, 575–594 (2005)
10. Ma, Q., Boonen, T.: Longevity risk modeling with the consumer price index. North Am. Actuarial J. **28**, 593–610 (2024)
11. Manisalidis, I., Stavropoulou, E., Stavropoulos, A., Bezirtzoglou, E.: Environmental and health impacts of air pollution: a review. Front. Public Health **8**, 14 (2020)
12. Patz, J., Campbell-Lendrum, D., Holloway, T., Foley, J.: Impact of regional climate change on human health. Nature **438**, 310–317 (2005)
13. Richards, C., Gauch, H., Allwood, J.: International risk of food insecurity and mass mortality in a runaway global warming scenario. Futures **150**, 103173 (2023)
14. Tran, H., et al.: The impact of air pollution on respiratory diseases in an era of climate change: a review of the current evidence. Sci. Total Environ., 166340 (2023)
15. Yoro, K., Daramola, M.: CO2 emission sources, greenhouse gases, and the global warming effect. Advances In Carbon Capture, pp. 3–28 (2020)
16. Zhao, Q., et al.: Global climate change and human health: Pathways and possible solutions. Eco-Environment Health **1**, 53–62 (2022)

Quantification of Operational Flexibility in Wastewater Treatment Projects

Marta Biancardi[1], Antonio Di Bari[2(✉)], and Giovanni Villani[1]

[1] Department of Economics and Finance, University of Bari, Largo Abbazia S. Scolastica, 53, Bari, 70124 Bari, Italy
{marta.biancardi,giovanni.villani}@uniba.it

[2] Department of Economics, University of Campania, "L. Vanvitelli", Corso Gran Priorato di Malta, 81043 Capua, Italy
antonio.dibari@unicampania.it

Abstract. Today, wastewater treatment projects are very important to preserve human health and the terrestrial and aquatic ecosystems. The valuation of these projects presents some difficulties related to the uncertainty of wastewater flows over time and the operational flexibilities that influences their performance.

This paper aims to provide a valuation methodology of wastewater treatment projects that can consider the stochastic nature of the wastewater volume, the operational flexibility to expand the project, and the possibility of turning off the system when the wastewater volume is too low to avoid financial losses. To pursue this goal, we apply a Real Options Approach (ROA) fitted for the characteristics of the wastewater volume in an agricultural firm, such as its evolution through a standard Brownian motion to which it is applied a seasonality component referred to the availability of fruits and vegetables in certain months. We also propose a case study related to a wastewater treatment project carried out by the Rosso Gargano Company in southern Italy.

Keywords: Wastewater treatment plants · Operational flexibility · Real Options · Monte Carlo Simulations

1 Introduction

There is no doubt that water is a fundamental renewable resource to support human life ([22]). The increase in population and the expansion of agricultural and industrial areas have caused an increase in water demand. However, this increasing water demand has found difficulties in being met as humans, with their activity, pollute the existing conventional water sources ([8] and [7]). Thus, considering that fresh and potable water is not an unlimited resource and that its demand is increasing, it is necessary to fulfill this by implementing wastewater treatments ([7]). Today, wastewater management represents a relevant issue as it influences human health and the terrestrial and aquatic ecosystem. The

M. La Rocca et al. (Eds.): WMAF 2025, *New Perspectives in Mathematical and Statistical Methods for Actuarial Sciences and Finance*, pp. 50–60, 2025.
https://doi.org/10.1007/978-3-032-05551-4_5

wastewater is all water which, after its use in the household, agricultural, or industrial context, requires treatment (purification) before it can be reintroduced into the environment or reused ([11,17] and [23]). The sewage derived from these sectors contains inside organic carbon, nitrogenous organics, inorganic, suspended/dissolved solids, and heavy metals ([17,20,25]). For these reasons, it is important to purify this water and the wastewater treatment plant represents a good opportunity to purify water and, possibly, to recycle and reuse it.

Evaluation of these plants requires specific mathematical models that can consider the unpredictability of the wastewater volume ([13]). This unpredictably can change the project performance over time. In these cases, the classical discounted cash flow approaches, such as the NPV, could give an incorrect result ([19]).

The Real Options Approach (ROA) overcomes the limitations of the NPV by providing a reliable valuation methodology of uncertain projects ([24]). The ROA includes uncertainty in the project valuation through pricing of the managerial flexibility to change investment decision according to the changeable project performance. [24] explained that an investor could have various options during the life of the project. In his work some examples are represented by: the option to defer the investment, the option to default during the construction, the option to expand or contract the project, the option to shut down operations, the option to abandon for salvage value, the option to switch use, and the corporate growth options. In this sense, different from the classical discounted cash flow approaches, the ROA can include this "optionality" in the valuation model. ROA can be applied in various fields, such as renewable energy ([1,2,12], etc.), technological projects ([6]), R&D investments ([5] and [21]), other projects to pursue sustainability goals (for example [4] developed a ROA to evaluate a corn ethanol plant).

Previous works applied ROA in the field of wastewater treatment. For example, [15] and [13] applied the ROA to the case of a wastewater treatment plant in a PPP context under risk mitigation strategies. [10] provided a model to optimize the expansion of a wastewater treatment plant through Monte Carlo simulation using Real Options. They create a framework for finding the optimal timing and size of a two-stage wastewater treatment plant expansion.

These studies have faced ROA as a valuation approach of wastewater treatment plants built to provide social benefits to a large community. In fact, [13] and [15] considered a wastewater project developed through the Public-Private Partnership (PPP), and [10] considered a wastewater treatment plant for a municipality whose growth is uncertain. However, none of them analyzes a private wastewater investment for a company that needs to purify industrial or agricultural water. In this work, we focus on developing a valuation model of a wastewater treatment plant built into an agricultural firm to purify wastewater deriving from tomato production. Unlike a public plant, the choice to pursue a private wastewater treatment project depends on the financial benefits it could generate. In this work, we focus on the valuation model of a wastewater treatment plant built in an agricultural firm to purify the wastewater derived from tomato production. The mathematical model is built starting from the ROA, including two operational flexibilities: the first one related to activating or deactivating the plant following the changeable wastewater volume; and the second one related to the possibility to expand or not the project.

The paper is organized as follows. Section 2 provides the mathematical approach based on simulations, stochastic process, and real options; Sect. 3 provides a case study; Sect. 4 describes the conclusive remarks.

2 Methodology

In this section, we provide a mathematical model for evaluating private wastewater treatment plants. In the first subsection, firstly we simulate the wastewater flows that follow stocasticity and seasonality characteristics, and then, we calculate a new version of the NPV that contains the operational flexibility to activate or deactivate the plant following the increasing or decreasing wastewater flows. In the second subsection, we set the real options approach to evaluate the possibility to expand the project.

2.1 Discounted Cash Flows Analysis Through Monte Carlo Simulations

We consider an investment in a wastewater plant for an agricultural firm that needs to purify wastewater resulting from the washing of fruits and vegetables. In fact, fruits and vegetables are harvested first and then washed to eliminate mud and other impurities. The wastewater flows produced by agricultural firms are affected by uncertainty and seasonality, as fruits and vegetables are not available for all days of the year. Taking these characteristics into account, we model wastewater flows through a stochastic process that includes a seasonality factor. The inclusion of seasonality component in stochastic evolution is quite common in renewable energy investment, its performance being dependent on climate and meteorological conditions ([9] and [16]). Using the same logic, we include seasonality in the most appropriate stochastic process to model wastewater flows. Thus, to evaluate a wastewater treatment plant we need to simulate the wastewater flows starting from a daily expected value W_0 that would represent the mean of the previous periods. We assume that W_0 evolves over time $t \in [0, T]$ using the stochastic differential equation:

$$\frac{dW_t}{W_t} = (\alpha + \theta(t))dt + \sigma dZ_t \tag{1}$$

where $\alpha + \theta(t)$ is the drift, σ is the standard deviation of the changes in wastewater, Z is a standard Brownian motion and $\theta(t) = \beta \sin\left(2\pi t\right)$ is the seasonality function with β is a positive parameter of seasonality considering one business year (360 days).

Using a logarithm transformation and applying the Itô's Lemma, we write discrete-time format for a geometric process in Eq. 1 with seasonal drift as:

$$W^i_{t_j} = W^i_{t_{j-1}} \exp\left[\left(\alpha + \theta(t_j) - \frac{\sigma^2}{2}\right)\Delta t + \sigma\sqrt{\Delta t}\,\varepsilon^i_j\right] \tag{2}$$

where t_j are the periods expressed in days, $j = 0, 1, ..., J$ are the days considered in our simulations, $\theta(t_j) = \beta \sin\left(\frac{2\pi t_j}{360}\right)$, $\Delta t = \frac{1}{360}$ and ε_i represent the standard normal random variables given for $i = 1, ..., m$ simulations. To simulate expected daily water flows, we find the mean of all simulations to obtain $\overline{W}_{t_j} = \frac{\sum_{i=1}^{m} W^i_{t_j}}{m}$.

The company that uses the plant expects that, by purifying the water internally, it would reduce the total disposal costs, which would otherwise have paid if the public authority had done the purification. Thus, the benefits are represented by the difference between what the company would have spent without the plant and what the company spends with the plant:

$$b_{t_j} = e \cdot \overline{W}_{t_j} - \left[w + s \cdot q \cdot \overline{W}_{t_j} + e \cdot (1 - q) \cdot \overline{W}_{t_j}\right] \tag{3}$$

where w represents the fixed costs to keep the plant turned on; s represents the cost to dispose of wastewater that has been purified by the system (cleaner water); q represents the percentage of water that can be purified by the plant; $1 - q$ represents the percentage of water that cannot be purified by the plant; e represents the cost to purify and dispose of water through public authority service.

The savings b_{t_j} are given by the difference between what the company would have spent without the plant and what the company spends with the plant. The Eq. 3 assumes that the manager activates the plant even if the volumes are too low. This could generate negative savings as a result of the fixed costs needed to activate the plant. We need to consider the operational flexibility to activate the plant only if the volume is quite high to make the savings positive.

To capture this operational flexibility, we substitute the second part of the Eq. 3 with:

$$\omega_{t_j} = \begin{cases} w + s \cdot q \cdot \overline{W}_{t_j} + e \cdot (1 - q) \cdot \overline{W}_{t_j} \quad \text{if} \quad W_{t_j} \geq U \\ \min\{[w + s \cdot q \cdot \overline{W}_{t_j} + e \cdot (1 - q) \cdot \overline{W}_{t_j}]; e \cdot \overline{W}_{t_j}\} \quad \text{if} \quad L < W_{t_j} < U \\ e \cdot \overline{W}_{t_j} \quad \text{if} \quad W_{t_j} \leq L \end{cases} \tag{4}$$

where U represents the threshold above which to keep the system turned on; L represents the threshold below which to keep the system turned off.

The new savings values ($bflex$) consider the operational flexibility to use the plant only if the water volume allows it, and are calculated as:

$$bflex_{t_j} = e \cdot \overline{W}_{t_j} - \omega_{t_j} \tag{5}$$

In this way ω_{t_j} represents a kind of minimized costs. The value of $bflex_{t_j}$ represents the savings resulting from having the wastewater treatment plant available, without implying its use a priori. Thus, The Eq. 4 implies that $bflex_j \geq 0$. We calculate the present value (PV) under operational flexibility :

$$PV = \sum_{t_j=1}^{t_J} \frac{bflex_{t_j}}{(1+r)^{t_j+\eta}} \tag{6}$$

where t_j represents the operation days, t_J represents the final day of the project lifetime, η is the time period to realize the plant in which there are no savings, and r is the daily discount rate.

At this point, the NPV will be equal to the present value of annual savings minus the cost to create the wastewater treatment plant (I):

$$NPV = PV - I \tag{7}$$

2.2 The Real Options to Expand the Project

During the lifetime of the project, the manager could decide to expand the project as a consequence of an increase in wastewater flows. In this case, the NPV cannot capture the potential financial benefits derived from the possible expansion project. To include the expansion option in the project valuation, we need to apply the ROA. The expansion real options have been already valuated through the Real Options Approach in various fields. For example, the expansion real options can be related to the renewable energy field ([14]), R&D projects ([21]), TFT manufacturing industry ([26])). In this section, we set an expansion real option model for a wastewater plant inserted into a private agricultural firm.

We assume that the project value PV evolves from year 0 to year H ($h = 0, ..., H$) along the binomial tree through up ($PV \cdot u$) or down ($PV \cdot d$) movements where $u = e^{\sigma \cdot \sqrt{\Delta h}}$ and $d = e^{-\sigma \cdot \sqrt{\Delta h}}$ following the CRR ([3]) model:

$$PV_H = PV \cdot u^n \cdot d^{z-n} \tag{8}$$

where z is the number of discrete intervals into which we divide H, $\Delta h = \frac{H}{z} = 1$ year, and $n = 0, ..., z$ represents the number of up movements and $z - n$ represents the number of down movements. At time H we compute this real option (e_H):

$$e_H = \max[PV_H; PV_H \cdot (1+\gamma) - K] \tag{9}$$

where K represents the expansion investment, γ represents the expansion factor.

The Eq. 9 explains that the expansion investment will be pursued only if: $PV_H \cdot (1+\gamma) - K > PV_H$. At this point, we proceed backward up to time $t = 0$:

$$e_0 = \frac{\sum_{n=0}^{z} \binom{z}{n} \cdot \pi^n \cdot (1-\pi)^{z-n} \cdot e_H}{(1+r_f)^z} \tag{10}$$

where $\pi = \frac{(1+r_f)^{\Delta h}-d}{(u-d)}$. By subtracting the first investment to realize project I from e_0, we obtain the final Real Option value (ROV) at time 0:

$$ROV = -I + e_0 \tag{11}$$

The ROV value represents the wastewater plant valuation at the current time $h = 0$ that considers the operational flexibility to expand the project if the expanded project value is greater than the project value without expansion. This operational flexibility is added to that to activate the system only if the water volume is enough to allow the fixed costs of the plant to be borne.

3 Case Study

In this section, we apply our methodology to evaluate a wastewater treatment plant project implemented by an agricultural firm. We used data from an Italian agricultural firm named Rosso Gargano, located in southern Italy. The company was founded in 2008 in the industrial area of Foggia, a city in southern Italy. The company produces and sells 120 000 tons of tomatoes per year. Once harvested, the tomatoes must be washed eliminating mud and impurities. Considering the huge amount of tomato production and, consequently, tomato wastewater, the company in 2019 decided to create a private wastewater treatment plant within the company.[1]

3.1 Data Description

We implement our model using the likely data of the wastewater produced by the Rosso Gargano firm. We start from $W_0 = 2\,000$ m^3/day, which has been chosen as the average value between what the firm consumes and purifies in the full production and what the firm consumes and purifies in the low production. The threshold below which it is advisable to turn off the system is $L = 700$ m^3/day, while that above which it is advisable to turn on the system is $U = 1\,200$ m^3/day. The cost to purify and dispose the wastewater externally is $e = 1.20$ €$/m^3$. In contrast, if the company decides to purify the water internally, it would have to incur fixed costs $w =$ € 1 000, to which the cost of removing the purified wastewater externally must be added $s = 0.20$ €/m^3. Although the plant allows the purification of most wastewater $q = 95\%$, there is a part that cannot be purified internally $(1 - q) = 5\%$. This last part will be purified and disposed of externally by paying $e = 1.20$ €$/m^3$. The investment cost to carry out the wastewater treatment project is $I =$ € 3 million and the possible expansion

[1] Please note that our methodology represents a generic framework for the evaluation of these projects. Therefore, to have a more accurate evaluation of the company's wastewater treatment plant, a more specific analysis would be needed that should take into account parameters that we do not consider in this work. For example, possible total shutdown of the plant due to internal causes of the company, problems with the harvest, possible public incentives, etc.

cost is $K = € \ 50\,000$. We developed our analysis over 10 years $z = 1, 2, ..., 10$ and therefore over 3600 days $t_j = 1, 2, ..., 3\,600$. Moreover, there is an additional year in which no savings are generated and that represents the time period to realize the plant. This time period is assumed to be 1 year, and so $\eta = 360$ days. Regarding the financial parameters, we make our analysis assuming $\alpha = 0$, $\sigma = 0.7$, $r = 7\%$,[2] $r_f = 3\%$ and $m = 10\,000$ simulations. For the seasonality component we use a $\beta = 1.05$ that fits well the evolution of the wastewater flows during the time.[3] The data are summarized in the Table 1.

Table 1. Summary and description of parameters

Parameters	Description	Values
W_0	Initial expected wastewater flows	$2\,000$ m^3/day
L	Threshold needed to turn off the system	700 m^3/day
U	Threshold needed to turn on the system	$1\,200$ m^3/day
e	Cost to purify and dispose the wastewater externally	$e = 1.20$ €$/m^3$
w	Fixed cost to activate the system	€ 1 000
s	Cost to dispose the purified wastewater	0.20 €/m^3
q	Proportion of the wastewater that can be purified internally	95%
$1 - q$	Proportion of the wastewater that can not be purified internally	5%
I	Investment cost to acquire the wastewater treatment plant	€ 3 million
K	Investment cost to expand the project	$K =$ € 50 000
z	Time period in years for the ROA	$0, 1, ..., 10$
t_j	Time period in days for wastewater simulations	$0, 1, ..., 3\,600$
α	Drift	0
σ	Standard deviation of the wastewater changes	0.7
r	Daily discount rate for the NPV calculation	0.01879%
r_f	Risk-free rate for the ROA	3%
β	Positive parameter of seasonality	1.05
m	Simulations	10 000

3.2 Discussion of Results

Applying the data described in the previous section, we obtain $PV =$ € 3.1642 million, $NPV =$ € 164 200, and $ROV =$ € 448 872. The PV has been obtained by discretizing the stochastic differential equation and making a mean of all the simulations. We plotted these average values $\overline{W}_{t_j}$ for 10 years in Fig. 1. This

[2] The yearly value of the discount rate, for our analysis, is converted into a daily value $r = 0.01879$.

[3] The choice of the $\beta = 1.05$ parameter has been done by considering adequate fluctuations of the wastewater flows evolution that fits the firm water consumption.

figure also makes a comparison between the case in which $\sigma = 0.7$ (blue line) and the case in which $\sigma = 0$ (red line). In the latter case, only the seasonal component influences the evolution of the wastewater flows. The results of PV, NPV, and ROV obtained so far allow us to consider the operational flexibility to activate/deactivate the plant and the operational flexibility to expand the project following financial profitability. Specifically, we apply Eq. 6 to calculate PV which is used for NPV and ROA. In addition to this, we can compare these PV, NPV and ROV values under operational flexibility to switch ON/OFF the plant with the one without this operational flexibility. In particular, it involves replacing Eq. 5 with Eq. 3 for the PV calculation. The comparison between these values is given by Table 2.

This comparison suggests that the possibility of turning the system ON/OFF according to economic convenience increases the project valuation compared to keeping the wastewater treatment plant always active. Moreover, we can state that the ROV is always higher than NPV, which means that the pricing of the operational flexibility to expand the project increases the project valuation. Moreover, in the absence of operational flexibility, the NPV method would suggest rejecting the investment, while ROA would indicate its acceptance. This highlights the importance of incorporating operational flexibility into the evaluation process, as ignoring this dimension can lead to suboptimal or incorrect investment decisions. These result is in line with the majority of the Real Options literature.

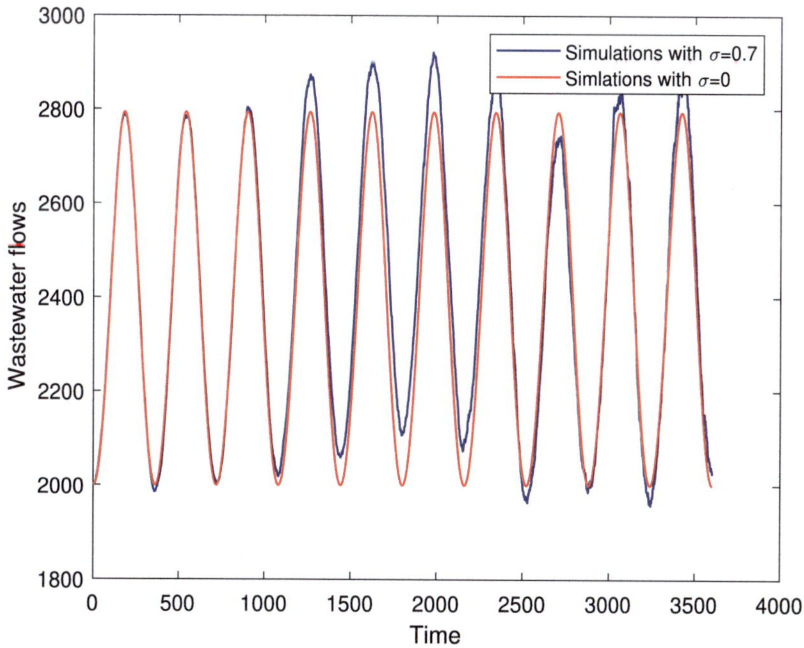

Fig. 1. Wastewater flows simulations

Table 2. Results

System always ON	ON/OFF depending on economic convenience
$PV =$ € 2.9007 million	$PV =$ € 3.1642 million
$NPV =$ € −99 300	$NPV =$ € 164 200
$ROV =$ € 159 703	$ROV =$ € 448 872

4 Conclusions

This paper aims to develop a mathematical model to support a private company that wants to invest in a wastewater treatment plant to save money allocated to purify and dispose of wastewater. The savings are represented by the difference between what the company would have spent to purify and dispose of the wastewater without the plant and what the company spends with the plant. When a firm decides to make these investments, it should consider that the wastewater flows can change during the time evolving as a seasonal stochastic process. The evolution of wastewater depends on the productive activity of the company.

It is also important to include in the valuation the possibility, also called optionality, to switch ON/OFF the system according to the financial benefits that the plant could generate. In fact, since the activation of the plant implies that the company should bear the fixed costs, it is not always best to keep it. In addition to this, the firm should include in the valuation methodology the operational flexibility to expand the project if the wastewater volume increases over time.

The goal of this paper is to provide a methodology that can consider the stochastic nature of the wastewater flows and the various operational flexibilities thanks to which an investor could mitigate the possible financial losses. Thus, the mathematical framework is based on the ROA. The valuation methodology includes the operational flexibility to expand the project and the possibility of turning off the system when the wastewater volume is too low.

We also propose a case study referring to Rosso Gargano company. We apply our mathematical model to assess the performance of this plant considering all the characteristics described above. The results show that the operational flexibility of turning the system ON/OFF increases the project valuation compared to the case without this flexibility. Moreover, we compare the static NPV analysis with the Real Options Approach, and results show that the ROA is always higher than the NPV.

Statements and Declarations

Acknowledge. The authors Marta Biancardi and Antonio Di Bari acknowledge the financial support and data from the italian regional program RIPARTI program (assegni di Ricerca per riPARTire con le Imprese) POC PUGLIA FESR-FSE 2014 / 2020 - CUP: H94C22001090002) n. 13.48 entitled “Rossogargano: una filiera sostenibile dal seme allo scaffale”.

Funding. The authors declare that no funds, grants, or other support were received during the preparation of this manuscript.

Competing Interests. The authors have no relevant financial or non-financial interests to disclose.

Authorship contribution. Authors whose names appear on the submission have contributed sufficiently to the scientific work and therefore share collective responsibility and accountability for the results.

Compliance with Ethical Standards. This article does not contain any studies with human participants or animals performed by any of the authors. The authors declare that they have no conflict of interest.

Data Availability. Manuscript has no associate data.

References

1. Biancardi, M., Bufalo, M., Di Bari, A., Villani, G.: Flexibility to switch project size: a real option application for photovoltaic investment valuation. Commun. Nonlinear Sci. Numer. Simul. **116**, 106869 (2023)
2. Bufalo, M., Di Bari, A., Villani, G.: Multi-stage real option evaluation with double barrier under stochastic volatility and interest rate. Ann. Finance **18**, 247–66 (2022)
3. Cox, J.C., Ross, S.A., Rubinstein, M.: Option pricing: a simplified approach. J. Financ. Econ. **7**, 229–263 (1979)
4. Biancardi M., Bufalo M., Di Bari A., Villani G. A valuation of a corn ethanol plant through a compound options model under skew-Brownian motions. *Annals of Operations Research* (2023)
5. Cassimon, D., Engelen, P.J., Thomassen, L., Van Wouwe, M.: The valuation of a NDA using a 6-fold compound option. Res. Policy **33**, 41–51 (2004)
6. Chen, T., Zhang, J., Lai, K.K.: An integrated real options evaluating model for information technology projects under multiple risks. Int. J. Project Manage. **27**, 776–786 (2009)
7. Djukic, M., Jovanoski, I., Ivanovic, O.M., Lazic, M., Bodroza, D.: Cost-benefit analysis of an infrastructure project and a cost-reflective tariff: a case study for investment in wastewater treatment plant in Serbia. Renew. Sustain. Energy Rev. **59**, 1419–1425 (2016)
8. Fessehaye, M., Abdul-Wahab, S., Savage, M., Kohler, T., Gherezghiher, T., Hurni, H.: Fog-water collection for community use. Renew. Sustain. Energy Rev. **29**, 52–62 (2014)
9. Kanamura, T., Homann, L., Prokopczuk, M.: Pricing analysis of wind power derivatives for renewable energy risk management. Appl. Energy **304**, 117827 (2021)
10. Lawryshyn, Y., Zhang, B., Davison, M.: Optimizing two-stage modular wastewater plant expansions using numerical methods and simulation in a real options context. J. Environ. Inf. Lett. **4**, 64–72 (2020)

11. Siddharth, L.S.: Sustainable and eco-friently approach for controlling industrial wastewater quality imparting succour in water-energy nexus system. Energy Nexus **3**, 100020 (2021)
12. Lee, S.C.: Using real option analysis for highly uncertain technology investments: The case of wind energy technology. Renew. Sustain. Energy Rev. **15**, 4443–50 (2011)
13. Liu, J., Cheah, C.Y.J.: Real option application in PPP/PFI project negotiation. Constr. Manag. Econ. **27**, 331–342 (2009)
14. Loncar, D., Milovanovic, I., Rakic, B., Radjenovic, T.: Compound real options valuation of renewable energy projects: the case of a wind farm in Serbia. Renew. Sustain. Energy Rev. **75**, 354–367 (2017)
15. Marzouk, M., Ali, M.: Mitigating risks in wastewater treatment plant PPPs using minimum revenue guarantee and real options. Utilities Policy **53**, 121–133 (2018)
16. Nibbering, D., van Buuren, C. and Wei, W. Real options valuation of wind energy based on the empirical production uncertainty. *Working Paper, Department of Econometrics and Business Statistics, Monash Business School* (2021)
17. Obaideen, K., Shehata, N., Sayed, E.T., Abdelkareem, M.A., Mahmoud, M.S., Olabi, A.G.: The role of wastewater treatment in achieving sustainabe development goals (SDGs) and sustainability guideline. Energy Nexus **7**, 100112 (2022)
18. Or, B., Bilgin, G., Akcay, E.C., Dikmen, I., Birgonul, M.T.: Real options valuation of photovoltaic investments: a case from Turkey. Renew. Sustain. Energy Rev. **192**, 114200 (2024)
19. Ross, A.S.: Uses, abuses and alternatives to the net-present-value rule. Financ. Manag. **24**(3), 96–102 (1995)
20. Senfter, T., et al.: Sludge thickening in a wastewater treatment plant using a modified hydrocyclone. Carbon Resour. Convers. **4**, 132–141 (2021)
21. Schneider, M., Tejeda, M., Dondi, G., Herzog, F., Keel, S., Geering, H.: Making real options work for practitioners: a generic model for valuing R&D projects. R&D Manag. **38**, 85–106 (2008)
22. Siegel, F.: Demands of Expanding Populations and Development Planning: Clean Air, Safe Water, Fertil Soils. Springer, Heidelberg (2008)
23. Tom, A.P., Jayakumar, J.S., Biju, M., Somarajan, J., Ibrahim, M.A.: Aquaculture wastewater treatment technologies and their sustainability: a review. Energy Nexus **4**, 100022 (2021)
24. Trigeorgis, L.: Real options and interactions with financial flexibility. Financ. Manage. **22**, 202–224 (1993)
25. Wang, J., Ji, Y., Zhang, F., Wang, D., He, X., Wang, C.: Treatment of coking wastewater using oxic-anoxic-oxic process followed by coagulation and ozonation. Carbon Resour. Convers. **2**, 151–156 (2019)
26. Yuan, F.C.: Simulation-optimization mechanism for expansion strategy using real option theory. Expert Syst. Appl. **36**, 829–837 (2009)

Roughness in VIX Index and in Realized Volatility: Rolling Window Estimation by Randomized Kolmogorov-Smirnov Distribution

Sergio Bianchi and Daniele Angelini(✉)

MEMOTEF, Sapienza University of Rome, Rome, Italy
daniele.angelini@uniroma1.it

Abstract. The modeling and forecasting of financial market volatility constitute fundamental components of effective risk management and optimal asset allocation. Traditional models like GARCH and SV often fail to capture the long memory and roughness empirically observed in volatility, prompting the adoption of fractional processes. Accurate estimation of the log-volatility roughness parameter is thus key to validating rough volatility models, with several methodologies proposed, including spectral, wavelet, and machine learning techniques. In contrast to approaches focused on moment behavior, we adopt a novel method based on the self-similarity of fractional processes, examining how the entire log-volatility distribution scales across time horizons. We deduce the variance of the estimator and study the roughness of both CBOE VIX and realized volatility.

Keywords: VIX · Realized volatility · Hurst exponent · Kolmogorov-Smirnov test

1 Introduction

Modeling and forecasting volatility in financial markets is a fundamental task in risk management, asset pricing, and portfolio optimization. Traditional approaches, such as GARCH models and stochastic volatility (SV) models, typically assume short-range dependence, which constrains their ability to capture the persistent and irregular behavior observed in empirical volatility data. However, substantial empirical evidence indicates that volatility displays both long memory and roughness, motivating the use of fractional processes for a more accurate representation of its dynamics.

Recent advances in the literature have introduced fractional Brownian motion (fBm) and its rough volatility extensions as powerful tools for modeling such features. In particular, rough fractional stochastic volatility models [11] posit that volatility evolves according to a fractional Ornstein–Uhlenbeck process characterized by a Hurst parameter $H < 0.5$, which implies short memory and high-frequency roughness. These models offer significant improvements in volatility

M. La Rocca et al. (Eds.): WMAF 2025, *New Perspectives in Mathematical and Statistical Methods for Actuarial Sciences and Finance*, pp. 61–73, 2025.
https://doi.org/10.1007/978-3-032-05551-4_6

forecasting by effectively capturing fine-scale fluctuations, which are especially relevant in the context of high-frequency trading and risk assessment.

Beyond single-asset frameworks, the modeling and forecasting of volatility spillovers across financial markets has become a prominent area of research. Fractional processes provide a natural framework for capturing cross-market volatility transmission, wherein long-memory effects in one asset's volatility can influence the volatility dynamics of another. Approaches such as fractional cointegration, particularly those based on the fractional vector autoregressive (FIVAR) framework, have been employed to study volatility interdependencies across asset classes, industry sectors, and international markets. Compared to classical vector autoregression (VAR) models, these methods exhibit superior predictive performance by better accommodating persistent shocks and heterogeneous memory structures in volatility.

Consequently, estimating the roughness parameter H of log-volatility is essential for the empirical validation of rough volatility models. A variety of estimation methodologies have been proposed, including Power Spectral Density (PSD) estimation [11], wavelet-based techniques such as Wavelet Leaders and Multiresolution Analysis [1], variational methods based on Quadratic Variation [14], as well as Bayesian and maximum likelihood estimation procedures [4]. More recently, deep learning and machine learning approaches have also been explored as promising tools for estimating the Hurst parameter [3].

In this study, we focus on the estimation of the Hurst parameter H by exploiting a methodology that markedly departs from conventional approaches based on the behavior of individual moments of the log-volatility distribution. Leveraging the self-similarity property inherent to fractional processes, we investigate the scaling behavior of the entire distribution of log-return across multiple time horizons. To this end, we employ the Kolmogorov–Smirnov (KS) statistic, extending its implementation to accommodate data that exhibit dependence structures, including long-range dependence. From a methodological standpoint, we enhance the applicability of the KS statistic by incorporating a random permutation framework, which effectively eliminates autocorrelation effects while preserving the self-similar nature of the underlying process. Through a conceptually straightforward yet not trivial procedure involving some methodological technicalities, we dynamically estimate the self-similarity parameter of the log-volatility process. The resulting estimates exhibit financial consistency, since they fluctuate around a value that can reflects a market in equilibrium, in the sense of its informational efficiency [8].

The paper is organized as follows: Sect. 2 recalls some preliminaries and describes the methodology; Sect. 3 provides the estimates, while discussion and conclusions are presented in Sect. 4.

2 Methodology

2.1 Preliminaries

This section reviews the fundamental concepts underlying the Hurst parameter estimation method proposed by [5]. Since these are largely established concepts, only the definitions will be provided, without delving into their respective justifications.

Definition 1. *A nontrivial stochastic process $X = \{X_t\}_{t\geq 0}$, stochastically continuous at $t = 0$, is self-similar with $H_0 \geq 0$ (H_0-ss), if for any $a \in \mathbb{R}^+$*

$$\{X_{at}\}_{t\geq 0} \stackrel{d}{=} \{a^{H_0} X_t\}_{t\geq 0} \tag{1}$$

in the sense of finite-dimensional distributions.

Remark 1. If a H_0-*ss* stochastic process $\{X_t\}_{t\geq 0}$, with $X_0 = 0$ a.s., has stationary increments, the increment process is self-similar with the same parameter (H_0-*sssi*):

$$\{X_{t+a} - X_t\}_{t\geq 0} \stackrel{d}{=} \{a^{H_0}(X_{t+1} - X_t)\}_{t\geq 0}, \ \forall a \in \mathbb{R}^+. \tag{2}$$

Building upon the approach presented in [5], the self-similarity exponent H_0 can be estimated by testing the equality of the distributions in Eq. (1) (or Eq. (2)). To accomplish this, the Kolmogorov-Smirnov test may be employed.

In fact, given a compact timescale set $\mathcal{A} = [\underline{a}, \overline{a}] \subset \mathbb{R}^+$ and denoted by $\Phi_{X_t}(x)$ the cumulative distribution function of the stationary increment process $Z_{t,a} := X_{t+a} - X_t$, for any $a \in \mathcal{A}$, equality (2) can be written as

$$\Phi_{Z_{t,a}}(x) := \mathbb{P}\left(Z_{t,a} < x\right) \stackrel{\text{using eq. (2)}}{=} \mathbb{P}\left(a^{H_0} Z_{t,1} < x\right) = \Phi_{Z_{t,1}}\left(a^{-H_0} x\right),$$

$$\Phi_{a^{-H} Z_{t,a}}(x) := \mathbb{P}\left(a^{-H} Z_{t,a} < x\right) \stackrel{\text{using eq. (2)}}{=} \mathbb{P}\left(a^{H_0 - H} Z_{t,1} < x\right) = \Phi_{Z_{t,1}}\left(a^{H - H_0} x\right),$$

and

$$\delta_{Z_t}\left(\Psi_H\right) = \sup_{x\in\mathbb{R}} \left|\Phi_{Z_{t,1}}\left(\underline{a}^{H-H_0} x\right) - \Phi_{Z_{t,1}}\left(\overline{a}^{H-H_0} x\right)\right|, \tag{3}$$

where $\Psi_H := \{\Phi_{a^{-H} Z_{t,a}}(x), a \in \mathcal{A}, x \in \mathbb{R}\}$ is the set of the distribution functions of $\{a^{-H} Z_{t,a}\}_{t\geq 0}$ and δ_{Z_t} is the diameter of Ψ_H on the metric space (Ψ_H, ϱ), where ϱ is the distance function induced by the sup-norm $||\cdot||_\infty$ with respect to $\mathcal{A}$. It is easy to recognize in (3) the Kolmogorov-Smirnov statistic.

The diameter $\delta\left(\Psi_H\right)$ satisfies the following propositions (see [5] for proofs):

Proposition 1. *$\{Z_{t,a}\}_{t\geq 0}$ is H_0-ss if and only if, for any $\mathcal{A} \subset \mathbb{R}^+$, $\delta_{Z_t}\left(\Psi_{H_0}\right) = 0$.*

Proposition 2. *Let $\{Z_{t,a}\}_{t\geq 0}$ be H_0-ss. Then $\delta_{Z_t}\left(\Psi_H\right)$ is non-increasing for $H \leq H_0$ and non-decreasing for $H \geq H_0$.*

Proposition 3. *Let $\{Z_{t,a}\}_{t\geq 0}$, $\boldsymbol{x} \geqq 0$ or $\boldsymbol{x} \leqq 0$, $\{\mathcal{A}_i\}_{i=1,\ldots,n}$ be a sequence of timescale sets such that, denoted by $\underline{a}_i = \min\left(\mathcal{A}_i\right)$ and by $\overline{a}_i = \max\left(\mathcal{A}_i\right)$, it is $\underline{a}_i \leq \underline{a}_j$ and $\overline{a}_i \geq \overline{a}_j$ for $i > j$. Then, with respect to the sequence $\{\mathcal{A}_i\}$, $\delta_{Z_t}\left(\Psi_H\right)$ is: non-decreasing, if $H \neq H_0$, or zero, if $H = H_0$.*

Denoted by X_t the log-volatility process and by Ψ_H the set of rescaled distribution of its increments, the idea is to estimate the self-similarity parameter H_0 by seeking the value of $H \in (0,1]$ that minimizes the Kolmogorov-Smirnov statistic $\delta_{Z_t}(\Psi_H)$ of any pair of rescaled distributions of $Z_{t,a}$. In fact, Proposition 1 ensures that if $Z_{t,a}$ is H_0-ss then $\delta_{Z_t}(\Psi_H)$ has a unique minimum with respect to $H \in (0,1]$, in correspondence of the value H_0. Thus,

$$\hat{H}_0 = \underset{H \in (0,1]}{\arg\min}\ \hat{\delta}_{Z_t}(\Psi_H) \tag{4}$$

The methodology above described poses two challenges:

- assessing the statistical significance of the minimum, as the distribution of the Kolmogorov-Smirnov (KS) test holds only when the samples are independent and identically distributed. Therefore, it is necessary to modify the input processes of Eq. (3) to eliminate their correlation and force the statistic to follow the KS distribution. This task can be achieved by employing some results from random permutations theory which are shortly summarized below.
- evaluate the variance of estimator (4). Clearly, determining the variance is important for constructing confidence intervals and assessing the extent to which the parameter—potentially modeled as time-varying—deviates from the equilibrium value consistent with the market efficiency hypothesis.

2.2 Random Permutations

Since the power spectrum $S_Z(\omega)$ of a stationary process $\{Z_n\}_{n\in\mathbb{Z}}$ is defined as the Fourier transform of its covariance function, the autocovariance function can be obtained as the inverse Fourier transform of its power spectrum, i.e.

$$K_Z(q) = \mathbb{E}[Z_n Z_{n-q}] = \int_{-\pi}^{\pi} e^{iq\omega} S_Z(\omega) d\omega. \tag{5}$$

Applying this to the power spectrum of the fractional Gaussian noise $Z^H_{t,a}$, $a > 0$, one has

$$S_{Z^H}(\omega) = 2c_H(1-\cos\omega)\sum_{j\in\mathbb{Z}} |2\pi j + \omega|^{-1-2H}, \qquad \forall \omega \in [-\pi, \pi], \tag{6}$$

with $c_H = \frac{C^2}{2\pi}\sin(\pi H)\Gamma(2H+1)$ [7]. The structure of (6) can be disrupted by random permutations: given the random sequence $Z = \{Z_\ell, \ell \in \mathbb{Z}\}$, a new sequence $U = \{U_\ell, \ell \in \mathbb{Z}\}$ is defined such that

$$U_\ell = Z_{\overline{\ell}L + b_{\underline{\ell}}}, \ \ \ell = \overline{\ell}L + \underline{\ell}, \ \ \overline{\ell} \in \mathbb{Z}, \ \ 0 \le \underline{\ell} \le L-1, \tag{7}$$

where $b = (b_0, b_1, \ldots, b_{L-1})$ is a random permutation of the vector $(0, 1, \ldots, L-1)$, uniformly distributed such that $\mathbb{P}(b) = \frac{1}{L!}$, with $\overline{\ell}$ and $\underline{\ell}$ the quotient and the remainder, respectively, of $\frac{\ell}{L}$. Based on Sect. 4 in [2,12] prove the following.

Proposition 4. *Let* $Z^H = \{Z^H_{\ell,a}, \ell \in \mathbb{Z}, a > 0\}$ *be a fractional Gaussian noise. Then, the new sequence* $\tilde{Z}^H = \{Z^H_{\ell L + b_{\underline{l}} + \phi, a}, \ell \in \mathbb{Z}, a > 0\}$, *with an independent random phase* ϕ *uniformly distributed on the set of integers* $(0, 1, \ldots, L-1)$, *has factorizable covariance in the limit of large permutation length* L

$$\lim_{L\to\infty} K_{\tilde{Z}^H}(q) = \mathbb{E}\left[Z^H_{\ell L + b_{\underline{l}} + \phi, a}\right] \mathbb{E}\left[Z^H_{\ell + qL + b_{\underline{l+q}} + \phi, a}\right] = 0 \tag{8}$$

for $q \neq 0$.

Remark 2. Given the stationarity of a fractional Gaussian noise (fGn), aside from a multiplicative parameter, we can state that $S_{Z^H}(\omega)$ is the power spectrum of its autocorrelation function; since random permutation acts only on the order of the elements of a sequence and not on their distribution, the structure of self-similarity is maintained, i.e.

$$\{Z^{H_0}_{i,a}\} \stackrel{d}{=} \{a^{H_0} Z^{H_0}_{i,1}\} \quad \Rightarrow \quad \{Z^{H_0}_{\bar{i}L + b_{\underline{i}} + \phi, a}\} \stackrel{d}{=} \{a^{H_0} Z^{H_0}_{\bar{i}L + b_{\underline{i}} + \phi, 1}\}.$$

Proposition 4 asserts that, for sufficiently large values of L, the randomly permuted sequence $\tilde{Z}^H$ can be regarded as a white noise process. For an fGn this occurs starting from $L \sim 100$. The rate at which the power spectrum $K_{\tilde{Z}^H}(q)$ converges to that of white noise depends on the regularity of the original power spectral density $S_{Z^H}(\omega)$. For example, the randomly permuted versions of NRZ-type (Non-Return-to-Zero) and Biphase processes approximate white noise behavior as $L \geq 500$ [12].

2.3 Variance of the Estimator

Concerning the variance of estimator (4), we prove the following

Proposition 5. *When the process is a fractional Brownian motion decorrelated by random permutation, the variance of estimator* (4) *is*

$$Var(\hat{H}_0) = \frac{2\pi c}{(\ln a)^2} \left(\frac{1}{\sqrt{n}} + \frac{1}{\sqrt{m}} \right)^2. \tag{9}$$

Proof. We derive the variance of the estimator $\hat{H}_0$ using the Fisher information approach, which quantifies the amount of information an observable sample contains about a parameter. The Cramér-Rao bound provides a fundamental lower bound on the variance of any unbiased estimator.

To this aim, let us observe that when the observed process is an fBm, the estimator $\hat{H}_0$ can be obtained from an observable variance function $V(H)$, which follows the approximate model:

$$V(H) \approx a^{2(H_0 - H)}, \tag{10}$$

that is

$$H_0 \approx H + \frac{\ln V(H)}{2 \ln a}.$$

The log-likelihood function for a sample of size n is given by

$$\ell(H) = \sum_{i=1}^{n} \ln f(X_i; H), \tag{11}$$

where $f(X_i; H)$ is the probability density function of the observed sample, which depends on H.
The Fisher information $I(H)$ is defined as:

$$I(H) = -\mathbb{E}\left[\frac{\partial^2 \ell(H)}{\partial H^2}\right]. \tag{12}$$

For large samples, the Cramér-Rao bound states that the variance of any unbiased estimator $\hat{H}_0$ satisfies:

$$\mathrm{Var}(\hat{H}_0) \geq \frac{1}{I(H)}. \tag{13}$$

To compute $I(H)$, we differentiate the likelihood function $\ell(H)$ and use the asymptotic relation, i.e.

$$\frac{\partial \ell(H)}{\partial H} = \sum_{i=1}^{n} \frac{\partial}{\partial H} \ln f(X_i; H) \approx -\sum_{i=1}^{n} \frac{(X_i - \mu)}{V(H)} \frac{\partial V(H)}{\partial H}. \tag{14}$$

Replacing $V(H)$ as in (10) and its derivative $\frac{\partial V(H)}{\partial H} = 2a^{2(H-H_0)} \ln a$, the score function simplifies to:

$$\frac{\partial \ell(H)}{\partial H} \approx -\frac{2 \ln a}{V(H)} \sum_{i=1}^{n} (X_i - \mu). \tag{15}$$

The Fisher information is obtained as:

$$I(H) = \mathbb{E}\left[\left(\frac{\partial \ell(H)}{\partial H}\right)^2\right]. \tag{16}$$

that is, using the expectation properties of Gaussian variables

$$I(H) = \frac{4(\ln a)^2}{[V(H)]^2} \sum_{i=1}^{n} \mathbb{E}[(X_i - \mu)^2] = \frac{4(\ln a)^2}{[V(H)]^2} V(H) = \frac{4n(\ln a)^2}{V(H)}. \tag{17}$$

From the Cramér-Rao bound, we finally have

$$\mathrm{Var}(\hat{H}_0) \geq \frac{1}{I(H)} = \frac{V(H)}{4n(\ln a)^2} = \frac{a^{2(H_0-H)}}{4n(\ln a)^2}. \tag{18}$$

Rewriting in terms of empirical variance estimators,

$$\mathrm{Var}(\hat{H}_0) = \frac{2\pi e}{(\ln a)^2} \left(\frac{1}{\sqrt{n}} + \frac{1}{\sqrt{m}}\right)^2. \tag{19}$$

which matches the previously derived result.

Thus, extracting two rescaled and randomized samples $Z_{1,\cdot}$ and $Z_{2,\cdot}$, of length n and m respectively and referred to time scales $\underline{a}$ and $\overline{a}$, the empirical diameter is

$$\hat{\delta}_{Z_t}(\Psi_H) = \sup_{x\in\mathbb{R}} |\Phi_{1,n}(x) - \Phi_{2,m}(x)| = \sup_{x\in\mathbb{R}} \left| \frac{1}{n}\sum_{i=1}^{n} \mathbb{1}_{\underline{a}^{-H} Z_{1,i}\leq x} - \frac{1}{m}\sum_{i=1}^{m} \mathbb{1}_{\overline{a}^{-H} Z_{2,i}\leq x} \right| \tag{20}$$

and the estimated self-similarity parameter is $\hat{H}_0 = \arg\min_{H\in(0,1]} \hat{\delta}_{Z_t}(\Psi_H)$, provided that $\hat{\delta}_{X_t}(\Psi_H)$ can be considered negligible by comparison with the critical value of the KS test [2].

3 Implementation and Data Analysis

Estimator (4) was employed to dynamically estimate the value of the parameter H using a rolling window of 500 data of:

- the logarithm of the Cboe VIX Index, spanning from January 2, 1990, to January 7, 2025, for a total of 8,860 observations;
- the logarithm of 5-min realized volatility of the S&P 500 (SPX), from 03/01/2000 to 27/06/2018, totaling 4,713 observations.

The choice of the moving window length was driven by the need to have a sufficiently large number of observations to reconstruct the distribution to which the random permutations would be applied.

We analyzed both the VIX and the 5-min realized volatility to assess to what extent our estimator is able to capture the different behaviour between the roughness of the two sequences. It is documented in literature (see, e.g., [9,13]) that the VIX index has a smoother behavior compared to realized volatility. This is more than a simple stylized fact and can be attributed to both theoretical and structural reasons rooted in stochastic modeling. An intuitive explanation of such a difference can be given as follows (in the following σ_t denotes the instantaneous spot volatility, $\hat{\sigma}_t$ the realized (historical) volatility over a past interval, and VIX_t the implied volatility index, approximately representing the market's expectation of 30-day future volatility). Both realized volatility and VIX are transformations of the spot volatility process.

The realized volatility

$$\hat{\sigma}_t^2 \approx \int_{t-1}^{t} \sigma_s^2 \, ds. \tag{21}$$

is a backward-looking average of past volatility computed from intraday returns, capturing actual price fluctuations. The Itô isometry links high-frequency returns to quadratic variation, but market microstructure noise and jumps introduce a $O_p(n^{-1/2})$ term, lowering H. The roughness depends on the sampling frequency: being contaminated by bid-ask bounce and price discreteness which induce negative autocorrelation in returns, higher frequencies (e.g., 5-min intervals) capture more microstructure and short-term noise, leading to very rough estimates ($H \approx 0.05$–0.15), while lower frequencies (e.g., daily) average out noise and smooth over short-term fluctuations, yielding higher Hurst estimates ($H \approx 0.15$–0.25).

The VIX

$$\mathrm{VIX}_t^2 \approx \mathbb{E}_t^{\mathbb{Q}}\left[\frac{1}{T}\int_t^{t+T} \sigma_s^2\, ds\right] \tag{22}$$

is a forward-looking, risk-neutral expectation over a future interval of time and is derived from S&P 500 options prices, reflecting the market's expectation of 30-day volatility. It aggregates information across strikes and maturities, smoothing out high-frequency noise. Even if the instantaneous volatility σ_t is rough ($H \approx 0.1$), the integral is a bounded linear operator that increases regularity over $[t, t+\tau]$ (if σ_t^2 has Hölder exponent H, the integral has Hölder exponent $H+1$, reducing roughness). In addition, risk-neutral pricing introduces persistence via volatility risk premiums, further smoothing VIX_t.

In rough volatility models, such as the Rough Fractional Stochastic Volatility (RFSV) model, instantaneous log-volatility is modeled as:

$$\log \sigma_t = \nu \int_{-\infty}^{t} (t-s)^{H-1/2}\, dW_s,$$

where W_s is a standard Brownian motion and ν is a scaling factor which determines the amplitude of fluctuations in $\log \sigma_t$ (higher ν indicates more variability in $\log \sigma_t$ which turns into a greater volatility of volatility) . Obviously, $H \in (0, 0.5)$ implies roughness because the sample paths are Hölder continuous of order less than H. Let its roughness be $H_{\text{spot}} = H$.

The realized volatility (21) involves a Volterra-type integral operator and it is a known result from real analysis that if $f \in C^{\alpha}$, then

$$g(t) := \int_t^{t+\Delta} f(s)\, ds \in C^{\alpha+1}.$$

Therefore, averaging increases Hölder regularity and hence $H_{\text{realized}} > H_{\text{spot}}$.

VIX index further involves conditional expectation (22). Therefore, it involves two smoothing operations: the time integration, which (as already shown) increases regularity; and the conditional expectation under a filtration, which smooth out irregularities (e.g., via Doob's inequality). Therefore,

$$H_{\text{VIX}} > H_{\text{realized}} > H_{\text{spot}}.$$

Estimator (20) was applied with parameters $\underline{a} = 1$, $\overline{a} = 50$, which entails a 95%-confidence interval of $\hat{H}_0 \pm 0.0577$. In order to reduce the noise affecting the estimates, data were filtered using the following methods: (a) smoothing with 5obs-moving average; (b) Savitzky-Golay; (c) Low-Pass Butterworth; (d) Wavelet Denoising; (e) Kalman, with MLE to calibrate process and measurement

Table 1. Comparison of smoothing and filtering methods.

Filter	Smoothing Quality	Adaptability to Dynamics	Noise Rejection	Computational Complexity	Parameter Estimation/Tuning
(a)	Basic/Low-frequency emphasis	Poor/fixed window	Moderate (good for white noise)	Very low	Window size chosen manually
(b)	Good (preserves shape and peaks)	Poor for non-stationary dynamics	Moderate/high (can amplify noise)	Low to moderate	Window size and polynomial order
(c)	High (better than MA)	Fixed frequency response	High/strong attenuation above cutoff	Moderate (filtering algorithm)	Choice of cutoff frequency and filter order
(d)	Very high/multi-scale analysis	Good (adaptive in time-frequency)	Excellent/localized in time and frequency	High	Thresholding method and wavelet basis must be selected
(e)	High (recursive estimation)	Excellent/tracks time-varying states	Dynamically adjusts to uncertainty	Moderate to high (recursive, matrix operations)	MLE used to calibrate process and measurement noise covariances

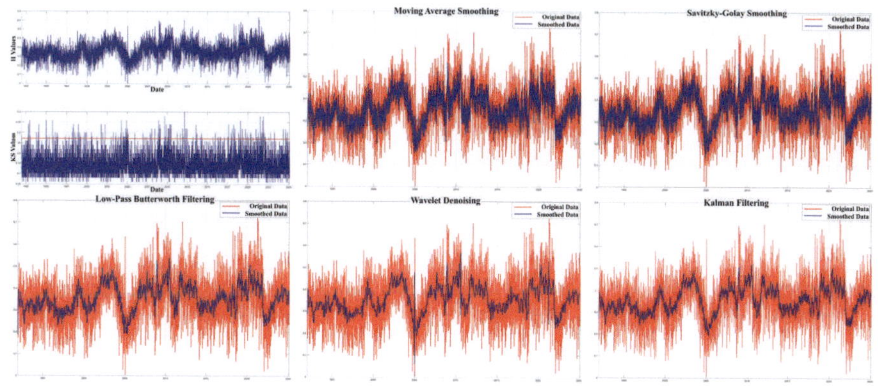

Fig. 1. Top left: $\hat{H}_0(t)$ of the VIX and KS values (the red line indicates the KS critical value at 95% confidence). In the other panels the red paths indicate the estimates and the blue path the filtered sequences. Top mid: 5obs-moving average smoothing. Top right: Low-Pass Butterworth. Bottom left: Savitzky-Golay. Bottom mid: Wavelet denoising. Bottom right: Kalman filter. (Color figure online)

noises. Table 1 summarizes the characteristics of each filter. Figure 1 shows the results of the dynamical analysis.

This comparison was motivated by prior findings (see, e.g., [9,13]), showing that implied volatility typically exhibits a higher value of the Hurst parameter than historical volatility. It was thus of interest to assess whether this difference also emerges when using our estimator, which is based on the entire distribution rather than on the scaling of individual moments. The estimates are visualized in Fig. 2, while Table 2 summarizes the estimation results.

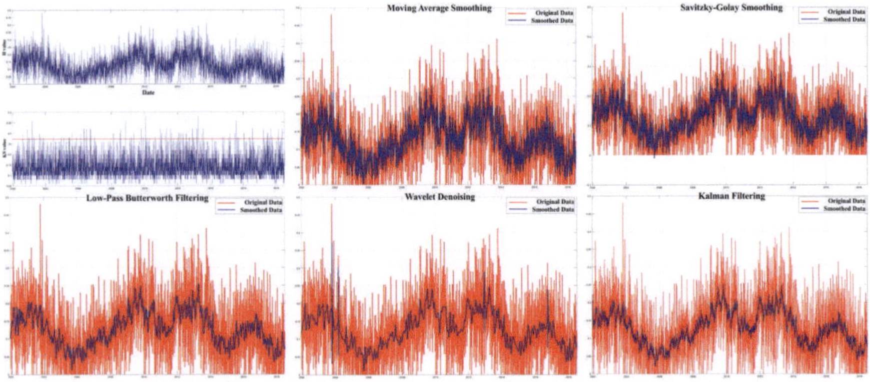

Fig. 2. Top left: $\hat{H}_0(t)$ of the 5min-log realized volatility on SPX and KS values (the red line indicates the KS critical value at 95% confidence). In the other panels the red paths indicate the estimates and the blue path the filtered sequences. Top mid: 5obs-moving average smoothing. Top right: Low-Pass Butterworth. Bottom left: Savitzky-Golay. Bottom mid: Wavelet denoising. Bottom right: Kalman filter.

Table 2. Sample statistics of the analyzed sequences and GOF parameters for the fitted models (in parenthesis the standard errors for the estimated parameters and the critical values and the p-values for the GOF tests.

Feature	VIX		5min-RV	
Sample Size	8360		4213	
Mean	0.3547		0.1363	
Standard Deviation	0.0924		0.0715	
Skewness	0.0202		0.4399	
Kurtosis	3.1401		2.9735	
Support	[0.004,0.741]		[0.001,0.482]	
Candidate Distribution		**Normal**		**Weibull**
Parameter Estimates		mean: 0.3547		Scale: 0.1517
		(0.001011)		(0.001314)
		sigma: 0.092431		Shape: 1.9155
		(0.000715)		(0.02433)
Log-Likelihood		8045.74		4955.26
AIC		−16087.48		−9906.52
BIC		−16073.42		−9893.82
Anderson-Darling (A^2)		0.3517		
		(75.18, p-value: 0.4722)		
KS test		0.0064		
		(0.0148, p-value: 0.8775)		

3.1 Discussion

Firstly, as can be seen from the values of both the Anderson-Darling and the Kolmogorov Smirnov test in Table 2, for the VIX $\hat{H}_0$ follows a normal distribution with a mean of 0.355 and a standard deviation of 0.092. It should be noted that the value 0.092 seems to be consistent with the theoretical standard deviation calculated using Eq. (9), which yields a value of 0.097 given the parameters used in the analysis. Since the estimation is based on a rolling window with overlapping samples, the resulting sequence of parameter estimates is inherently autocorrelated due to the substantial overlap in the data used for each estimate. This serial dependence induces a smoothing effect, which can lead to an empirical variance of the estimates that is artificially lower than the theoretical variance expected under the assumption of a constant parameter.

Although the lower empirical variance relative to its theoretical counterpart could be interpreted as evidence in favor of parameter constancy for $\hat{H}_0$, a closer visual inspection of the estimated series suggests otherwise. The sequence exhibits marked cyclical behavior and localized trends, which are inconsistent with the assumption of stability (Fig. 1). These features are particularly pronounced during specific subperiods, such as the prolonged upward trends from 08/1998 to 05/1999 or from 06/2000 to 04/2002 or even from 03/2005 to 01/2007, and the sustained downward movements observed over 05/1999-06/2000, 07/2003–01/2005, and more recently, 11/2021–02/2022. Such dynamics may reflect underlying structural changes or shifts in the data-generating process, warranting further investigation through formal tests of parameter stability.

As documented also in previous studies, the values of $\hat{H}_0(t)$ for the VIX are significantly higher than those obtained by applying the same method to the 5-min realized volatility. In the period we were able to examine, the analysis reveals an average value of 0.1353 with a standard deviation of 0.0715 (Table 2). The values are in line with estimates in, e.g., [6, 10, 11] and many other studies. Even if a cyclical pattern emerges also in this case (Fig. 2), the distribution of the estimated $\hat{H}_0(t)$ is not Gaussian in this case; rather, it appears closer to a Weibull. The difference in the value of $\hat{H}_0$ between the VIX and the realized volatility is consistent with the findings in [9, 13], who examine the implied volatility-based approximations of the spot volatility. In particular, using at-the-money options on the S&P500 index with short maturity, [13] confirms roughness, but with a parameter of order 0.3, larger than that usually obtained from historical data (values less than 0.2), and explains the difference with the smoothing effect due to the remaining time to maturity of the considered options. The difference is reasonable and stems primarily from the nature of the VIX as a forward-looking measure derived from option prices, which inherently smooths short-term fluctuations through market expectations and averaging across strikes. In contrast, historical volatility is computed from high-frequency return data, capturing microstructure noise, jumps, and short-term market dynamics that contribute to a rougher appearance.

4 Conclusion

We examined the temporal evolution and distributional properties of the local Hurst exponent $\hat{H}_0(t)$, as a means of characterizing the roughness and persistence of financial volatility over time. By applying a moving-window estimation procedure based on a new method to estimate the self-similarity parameter of a fractional Brownian motion, we have provided empirical insights into the dynamic behavior of the Hurst exponent of the VIX index. The analysis suggests that meaningful deviations from a constant-parameter model characterize the dynamics. In particular, the estimated sequences exhibit pronounced local trends and cyclicality, indicating that the roughness of volatility may itself be a time-varying phenomenon. Moreover, our results confirm that the VIX exhibits significantly higher average Hurst values relative to realized volatility. This contrast may reflect fundamental differences between retrospective (realized) and prospective (implied) measures of market uncertainty, suggesting that the informational content and structure of volatility depend on the source and nature of the underlying data.

While these findings are preliminary, they point to the need for further research on the dynamic and potentially endogenous nature of volatility roughness. The results also raise broader methodological questions about the adequacy of fixed-parameter rough volatility models in capturing real-world dynamics, and whether extensions to accommodate local scaling behavior may yield better empirical performance.

References

1. Abry, P., Gonçalves, P., Veitch, D.: Wavelet-based estimation of the self-similarity parameter of long-range dependent signals. Stat. Methodol. **6**(1), 33–52 (2009)
2. Angelini, D., Bianchi, S.: Kolmogorov–Smirnov estimation of self-similarity in long-range dependent fractional processes. Physica D: Nonlinear Phenomena **476**, 134,697 (2025). https://doi.org/10.1016/j.physd.2025.134697. https://www.sciencedirect.com/science/article/pii/S0167278925001745
3. Bayer, C., Horvath, B., Muguruza, A.: Deep calibration of rough stochastic volatility models. Quant. Finan. **21**(1), 25–41 (2021)
4. Bennedsen, M., Pakkanen, M.S., Lunde, A.: Learning the roughness of stochastic volatility from option prices. J. Financ. Economet. **19**(3), 403–431 (2021)
5. Bianchi, S.: A new distribution-based test of self-similarity. Fractals **12**(03), 331–346 (2004). https://doi.org/10.1142/S0218348X04002586
6. Brandi, G., Di Matteo, T.: Multiscaling and rough volatility: an empirical investigation. Int. Rev. Financ. Anal. **84**, 102,324 (2022)
7. Coeurjolly, J.F.: Simulation and identification of the fractional Brownian motion: a bibliographical and comparative study. J. Stat. Softw. **5**, 1–53 (2000)
8. Fama, E.F.: Efficient capital markets: a review of theory and empirical work. J. Financ. **25**(2), 383–417 (1970)
9. Floc'h, F.L.: Roughness of the implied volatility. arXiv:2207.04930 (2022). https://arxiv.org/abs/2207.04930

10. Fukasawa, M., Takabatake, T., Westphal, R.: Consistent estimation for fractional stochastic volatility model under high-frequency asymptotics. Math. Financ. **32**(4), 1086–1132 (2022)
11. Gatheral, J., Jaisson, T., Rosenbaum, M.: Volatility is rough. Quant. Finan. **18**(6), 933–949 (2018)
12. Lacaze, B., Roviras, D.: Effect of random permutations applied to random sequences and related applications. Signal Process. **82**(6), 821–831 (2002)
13. Livieri, G., Mouti, S., Pallavicini, A., M.R.: Rough volatility: evidence from option prices. IISE Trans. **50**(9), 767–776 (2018)
14. Rosenbaum, M., Morel, V.: Estimation of roughness parameters: application to rough volatility modeling. Finance Stochast. **23**(2), 501–533 (2019)

A Comparison of Data-Driven Synthetic Performance Indicators for Default Prediction

Roberto Casarin, Fausto Corradin, and Antonio Peruzzi(✉)

Ca' Foscari University of Venice, Venice, Italy
{r.casarin,fausto.corradin,antonio.peruzzi}@unive.it

Abstract. This paper compares two approaches to constructing synthetic performance indicators for corporate default prediction: the expert-informed Synthetic Performance Indicator (ISP) and a Neural Network Synthetic Performance Indicator (NNISP). Both are built upon a common set of economic and financial indices from a large panel of firms operating in the Triveneto macro-region of Italy, including not only large companies but also small and medium enterprises (SMEs). The ISP relies on expert elicitation to assign weights to eight selected indicators, while the NNISP is obtained by calibrating the set of composition weights by training a Neural Network to predict a firm's liquidity. We assess the predictive power of each score using multiple classification models and benchmark them against the full set of indicators. We show that the ISP offers stable and interpretable performance across model types, while the NNISP exhibits more flexibility in non-linear settings. These findings highlight a trade-off between expert-based interpretability and machine-driven adaptability, pointing to the potential of combined methodologies for future applications in credit risk modeling.

Keywords: Default · Probability of Default · Logit · Neural Networks

1 Introduction

Understanding and predicting financial risk remains a central challenge for financial institutions, policymakers, and investors. See, for example, Miller and Reuer

A. Peruzzi—This study was funded by the European Union - NextGenerationEU, in the framework of the GRINS - Growing Resilient, INclusive and Sustainable project (GRINS PE00000018 - CUP H73C22000930001), National Recovery and Resilience Plan (NRRP) - PE9 - Mission 4, C2, Intervention 1.3. The views and opinions expressed are solely those of the authors and do not necessarily reflect those of the European Union, nor can the European Union be held responsible for them. The authors gratefully acknowledge the support of the *Ordine dei Commercialisti e degli Esperti Contabili* (ODCEC) of Treviso, Italy, in the development of the ISP index and their valuable contribution to the dissemination of the results.

M. La Rocca et al. (Eds.): WMAF 2025, *New Perspectives in Mathematical and Statistical Methods for Actuarial Sciences and Finance*, pp. 74–85, 2025.
https://doi.org/10.1007/978-3-032-05551-4_7

(1996) for an introduction to risk evaluation from a behavioral decision theory, finance, and management theory perspective. In this paper, we focus on credit risk, since accurate estimates of default probabilities support effective risk management, inform lending decisions, and play a critical role in regulatory frameworks. The analysis includes not only large companies but also small and medium enterprises (SMEs) contributing to the literature on credit risk evaluation for SMEs (Behr and Guttler, 2007; Bartik et al., 2020). See also Ciampi et al. (2021) and Cheraghali and Molnár (2024) for up-to-date literature reviews.

Since the foundational work of Altman (1968), a rich body of research has explored how firm-level financial indicators can be used to predict insolvency, initially through statistical tools such as linear discriminant analysis (Lo, 1986) and later through Logit, Probit (Kliešťik et al., 2015), survival models (De Leonardis and Rocci, 2008), and deep learning approaches, such as Neural Networks (Altman et al.,1994 Ciampi and Gordini, 2013; Wu et al., 2022). Over time, the field has embraced more sophisticated techniques to address issues of dimensionality and overfitting. Regularisation methods, like Lasso, Ridge, and Elastic Net (Westgaard and Van der Wijst, 2001; Tibshirani, 1996; Zou and Hastie, 2005), allow for improved stability, especially when dealing with large sets of correlated variables. Nonetheless, a core difficulty persists: condensing financial information into a single, robust scoring system that generalizes across firms and sectors.

Traditional solutions have often relied on variable selection strategies such as factor analysis, which seeks to identify latent financial dimensions like liquidity, profitability, or leverage. While useful, such approaches are typically sample-dependent and prone to instability. As a result, researchers and practitioners have increasingly looked to composite scoring indices that synthesize relevant indicators into a summary measure of firm performance. This paper offers a comparative assessment between two composite scoring models for firms' performance: the *Synthetic Performance Indicator* (ISP) and the *Neural Network Synthetic Performance Indicator* (NNISP). The former is an expert-driven index combining eight financial indicators over a five-year horizon, selected and weighted via input from chartered accountants. The latter is a data-driven index trained directly on the same set of economic and financial indicators.

The novelty of this approach lies in using artificial Neural Networks to calibrate the composition weights of the indicators using liquidity as a proxy for the firm's target performance. We consider liquidity, as there is a huge debate between practitioners and academics about its relevance as a proxy of a firm's survival (Li, 2024). Some argue that liquidity indices may contain useful contextual information that may enhance default prediction (Giordani et al., 2014). While the ISP reflects structured expert knowledge, the NNISP is entirely data-driven, allowing us to examine the benefits and limitations of both approaches in default prediction tasks. Using a common set of modeling tools – including Logit regression, Lasso, and a Feedforward Neural Network – we compare the out-of-sample performance of the ISP, the full indicator set, and the neural score.

The structure of the paper is as follows. Section 2 introduces the dataset and the set of forty firm-level indicators used in the analysis. Section 3 presents the construction of both the ISP and the NNISP. Section 4 evaluates and compares their performance in predicting firm defaults. Section 5 summarizes the key findings and discusses streams of future research.

2 Data

This section presents the dataset used to construct and compare two synthetic indicators for default prediction: the ISP, based on expert knowledge, and the NNISP, a data-driven index obtained via Neural Networks. Both indicators are derived from firm-level balance sheet data. The data refer to firms in the Italian Triveneto macro-region and cover a five-year window.

2.1 Source and Coverage

The dataset is extracted from the AIDA – Bureau Van Dijk database (Bureau van Dijk, 2023), as of October 17, 2024. It includes over 180,000 firms operating in the Triveneto macro-region, representing approximately 80% (by economic value) of all enterprises in the macro-region. The geographical coverage spans the following NUTS-3 administrative units Veneto (Belluno, Padova, Rovigo, Treviso, Venice, Verona, Vicenza), Trentino-Alto Adige (Bolzano, Trento) and Friuli-Venezia Giulia (Gorizia, Pordenone, Trieste, Udine).

Our analysis focuses solely on firms classified as *Active*, i.e., companies that are currently in operation and not undergoing insolvency or bankruptcy proceedings. To ensure temporal consistency and data quality, we apply the following inclusion criteria: i) registered office in one of the listed provinces; ii) annual revenues greater than one million euros; iii) complete and non-missing balance sheet data for each of the five fiscal years, 2019–2023. After filtering, the final sample consists of 14,768 firms. Figure 1 illustrates the spatial distribution of the sample across provinces, with shading proportional to the number of firms in each area.

2.2 Economic and Financial Indicators

For each firm, we collect a standardized set of economic and financial indicators derived from the balance sheet and income statement. These indicators span core economic and financial performance dimensions, including profitability, solvency, liquidity, growth, and efficiency.

Some of these indicators serve as inputs for both the expert-based ISP and the neural-network-based scoring approach. The current ratio (current assets to current liabilities) is also used as a proxy target variable for training the second score. In the next section, we detail the construction of each synthetic indicator and outline the modeling strategies employed to evaluate their predictive performance.

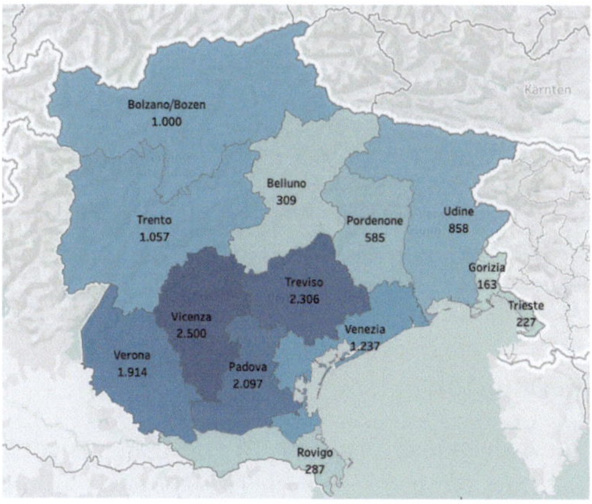

Fig. 1. Map of the Triveneto macro-region and its NUTS-3 provinces. Darker shades indicate larger number of firms included in the final sample.

3 Methodology

This section presents the methodology used to construct and evaluate two alternative synthetic indicators for default prediction, ISP and NNISP. Both indicators are compared against the full set of financial variables used in their construction.

3.1 ISP Construction

The ISP is a composite index designed to capture a firm's long-term performance and financial soundness. It aggregates eight financial indicators – deemed the most relevant by the pool of Triveneto professional accountants – into a single score. Its construction follows a three-step procedure:

1. **Normalization of Indicators:** For each year and each variable, values are rescaled on a 0–1,000 scale, assigning the maximum score to the best-performing firm and 0 to the worst.
2. **Assignment of Weights:** The relative importance of each indicator is derived from a structured survey of over 900 professional accountants (ODCEC members) across the 13 provinces in the Triveneto region. The resulting weights reflect expert consensus and follow the methodology established by the European Commission's Joint Research Centre for composite indicators (JRC, 2008).
3. **Dynamic Weighting:** To reward long-lasting performance, indicators are calculated annually over the past five years and then aggregated using time-decaying weights: 2023: 0.33, 2022: 0.27, 2021: 0.20, 2020: 0.13, and 2019: 0.07.

The ISP is structured into two sub-indices. The *Economic Performance Indicator* (ISP-E) captures profitability and growth through ROE, ROI, EBITDA-to-Revenue, and revenue growth and the *Financial Stability Indicator* (ISP-F), focuses on capital structure and debt sustainability via leverage, debt-to-equity ratio, asset growth, and the ratio of net financial position to EBITDA. Each sub-index is computed as a weighted average of its constituent indicators. The overall ISP score is a convex combination of ISP-E and ISP-F, with respective weights of 49% and 51%.

3.2 NNISP Construction

To develop a fully data-driven alternative to the ISP, we construct a composite score using a Feedforward Neural Network (FNN) (McCulloch and Pitts, 1943; Rosenblatt, 1958; Haykin, 1994) trained to predict a key measure of short-term liquidity: the *current ratio*, defined as

$$\textit{Current Ratio}_i = \frac{\textit{Current Asset}_i}{\textit{Current Liability}_i}, i = 1, \ldots, N \tag{1}$$

with N denoting the number of firms. As discussed in Sect. 1, the current ratio, commonly used as a proxy for liquidity, captures essential aspects of a firm's short-term financial stability and may provide valuable insights into its overall financial health (Giordani et al., 2014; Ross et al., 2018).

The Neural Network learns to assign relevance to each input indicator – among the 8 indicators identified by the professional accountants over five years – through optimization, and these learned weights are used to derive a synthetic performance index analogous in structure to the ISP. A Feedforward Neural Network is a function approximator composed of multiple layers of interconnected nodes (neurons), where each neuron computes a nonlinear transformation of a weighted sum of its inputs. Mathematically, a Neural Network with L layers computes a mapping $\hat{y} = f(\mathbf{x}) : \mathbb{R}^p \to \mathbb{R}$ of the form:

$$\begin{aligned}
\mathbf{h}^{(1)} &= \sigma\left(\mathbf{W}^{(1)}\mathbf{x} + \mathbf{b}^{(1)}\right), \\
\mathbf{h}^{(2)} &= \sigma\left(\mathbf{W}^{(2)}\mathbf{h}^{(1)} + \mathbf{b}^{(2)}\right), \\
&\vdots \\
\hat{y} &= \mathbf{W}^{(L)}\mathbf{h}^{(L-1)} + \mathbf{b}^{(L)},
\end{aligned} \tag{2}$$

where:

- $\mathbf{x} \in \mathbb{R}^p$ is the input vector of $p = 40$ (the set of 8 economic and financial indicators observed for 5 years),
- $\mathbf{W}^{(l)}$ and $\mathbf{b}^{(l)}$ are the weight matrices and bias vectors at layer l,
- $\sigma(\cdot)$ is an element-wise nonlinear activation function (e.g., ReLU or sigmoid),
- $\hat{y}$ is the predicted output, here representing the current ratio.

The network parameters are estimated by minimizing the mean squared error (MSE) between the predicted and observed values of the current ratio on a training dataset. See Calin 2020) for a review of deep learning architectures.

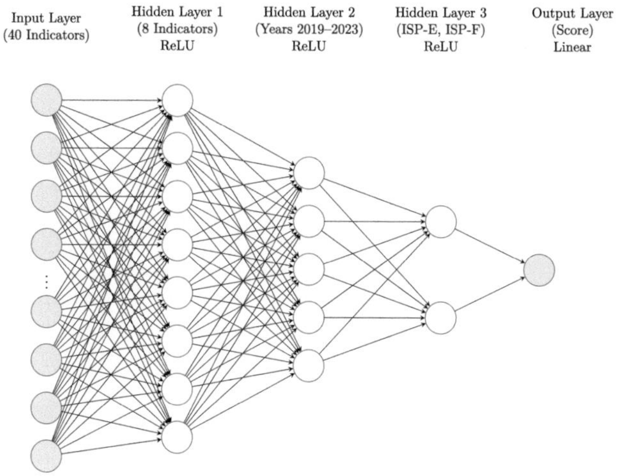

Fig. 2. Example of Neural Network architecture for the construction of the Neural Network Synthetic Performance Indicator. The input layer uses as variables the 8 indicators identified by the pool of professional accountants and observed over 5 years.

To ensure comparability with the ISP, one can opt for a Neural Network architecture that mirrors its structure.

The network can be composed of three hidden layers, each reflecting a structural dimension of the ISP: The first hidden layer includes 8 neurons, corresponding to the 8 indicators used in the ISP (4 related to economic performance, 4 to financial stability); The second hidden layer includes 5 neurons, each representing one of the five years of data (2019–2023); The third hidden layer contains 2 neurons that summarize the economic and financial components, mimicking the ISP-E and ISP-F structure.

The output layer consists of a single neuron with a linear activation function, returning the predicted current ratio. Figure 2 provides a graphical representation of this example of Neural Network architecture. Of course, other architectures may be suitable. For example, one could design a single-hidden-layer NN with 40 nodes and provide an optimization procedure to make the weights comparable with those of ISP.

Once the network is trained, we extract the learned weights connecting the input layer to the final prediction. These weights are then normalized and aggregated, tracing the contributions through the layers, into a single composite score for each firm. The resulting synthetic indicator reflects how much each original financial input contributes to explaining short-term liquidity, according to the trained model.

In practice, the score for firm i is computed by forward propagating its input vector $\mathbf{x}_i$ through the trained network, and interpreting the result as a learned scoring function. Formally:

$$\text{NNISP}_i = f_{\boldsymbol{\theta}}(\mathbf{x}_i), \tag{3}$$

where $f_{\boldsymbol{\theta}}$ denotes the trained network and $\boldsymbol{\theta} = \{\mathbf{W}^{(l)}, \mathbf{b}^{(l)}\}_{l=1}^{L}$ are the estimated parameters.

This synthetic performance score, entirely derived from data, provides a natural benchmark against which to compare the expert-informed ISP in terms of default prediction accuracy.

3.3 Performance Assessment

To evaluate the effectiveness of the two synthetic indicators—the expert-based ISP and the data-driven NNISP – we assess their predictive ability in forecasting firm defaults. Specifically, we compare the classification performance of ISP and NNISP against that of the full set of 40 financial indicators originally used to construct both indices.

Let us denote with $\mathcal{I}_0 = \{i = 1, \ldots, N,\ s.t.$ firm i was in a condition of severe legal or financial distress in 2023$\}$. The outcome variable in our analysis is a binary indicator Y_i, defined as:

$$Y_i = \begin{cases} 1 & \text{if firm } i \in \mathcal{I}_0, \\ 0 & \text{otherwise.} \end{cases}$$

Our operative definition of financial distress includes bankruptcy, liquidation, and insolvency proceedings. We consider four predictive modeling approaches:

1. **Logit Regression**: a standard logistic regression model using either ISP, NNISP, or the full set of 40 indicators as inputs (Westgaard and Van der Wijst, 2001);
2. **Lasso**: a penalized logistic model that performs variable selection by shrinking coefficients of less informative indicators (Tibshirani, 1996);
3. **Elastic Net**: a hybrid regularization technique that combines the penalties of Lasso and Ridge regression, offering improved stability in the presence of correlated predictors (Zou and Hastie, 2005);
4. **Feedforward Neural Network (FNN)**: a multi-layer neural network trained directly on the predictors to classify default risk (McCulloch and Pitts, 1943; Rosenblatt, 1958; Haykin, 1994).

Each model is estimated with the same training and test sets to ensure comparability across specifications. For the ISP and NNISP, we use them as single regressors. For the 40-indicator baseline, we use the raw indicators as input features. Model performance is assessed based on standard classification metrics, including Area Under the Curve (AUC), accuracy, sensitivity, and specificity.

This evaluation allows us to assess whether the dimensionality reduction achieved by the synthetic indicators – either expert-driven or learned –leads to a loss of predictive power relative to the full feature set, and which approach offers the best trade-off between parsimony and predictive accuracy.

4 Results

Table 1 reports the in-sample classification performance of four models (Logit, Elastic Net, Two-Stage Adaptive Lasso, and Neural Network) applied to three different sets of predictors: the full set of 40 financial indicators, the Synthetic Performance Indicator (ISP), and the Neural Network Synthetic Performance Indicator (NNISP). Overall, the Neural Network trained on the full set of 40 indicators is the best-performing model, as it attains the highest AUC (0.7725), along with the highest sensitivity (0.7174) and specificity (0.7154). This confirms the advantage of flexible, non-linear models when high-dimensional financial information is available.

Among the composite indicators, the ISP generally outperforms the NNISP in regularized linear models (Elastic Net, Lasso), particularly in terms of robustness and balance between sensitivity and specificity. However, the NNISP exhibits improved performance compared to its linear counterparts when used as input to the Neural Network (AUC = 0.6927), indicating that its structure is better exploited in non-linear settings.

Notably, even though both ISP and NNISP are dimensionality-reduced indicators, the ISP appears to offer more consistent predictive power across modeling approaches. This suggests that expert-driven aggregation may confer greater generalization ability in small-data or structured contexts.

Concerning the out-of-sample exercises, the upper panel of Table 2 presents the confusion matrices for out-of-sample predictions obtained using both the Logit model and Neural Network classifier across the three types of inputs: the full set of 40 financial indicators, the expert-based ISP, and the data-driven NNISP.

The Logit model trained on the full set of 40 indicators is the best-performing in terms of true positive rate, correctly identifying 146 out of 230 defaulted firms. The Neural Network using the same inputs performs slightly worse in this respect (136 defaults), although it improves specificity, reducing the number of false positives from 4436 to 4018.

When comparing the composite scores, the ISP tends to outperform the NNISP, particularly in linear settings. For example, under the Logit model, the ISP correctly identifies 142 defaults (vs. 140 for NNISP) and yields fewer false positives (5203 vs. 5412). Interestingly, the Neural Network's performance improves markedly when used with NNISP, reaching 124 correct default predictions compared to 121 with ISP, suggesting that the NNISP benefits more from non-linear modeling.

However, both synthetic scores come with a relevant trade-off: they reduce model complexity and improve interpretability, while losing in terms of classification accuracy when compared to the full set of indicators. Nonetheless, the ISP maintains a stable performance across model types, highlighting the importance of expert-informed feature aggregation.

The bottom panel of Table 2 reports the metrics of the classification task and confirms that the full set of 40 indicators is the best performing, particularly when used with a Neural Network, which achieves the highest Correct

Table 1. In-sample classification performance across models and indicator sets.

(a) Full Set of 40 Indicators				
Metric	**Logit**	**Elastic Net**	**2-Stage Lasso**	**Neural Net**
Sensitivity	0.6957	0.6652	0.6652	0.7174
Specificity	0.6925	0.6954	0.6916	0.7154
False Positive Rate	0.9654	0.9666	0.9670	0.9617
False Negative Rate	0.0069	0.0076	0.0076	0.0062
RCC (Correct Classification Rate)	0.6926	0.6950	0.6912	0.7155
AUC (ROC)	0.7406	0.7371	0.7392	0.7725
(b) Synthetic Performance Indicator (ISP)				
Sensitivity	0.6435	0.6435	0.6435	0.6522
Specificity	0.6182	0.6182	0.6182	0.6623
False Positive Rate	0.9740	0.9740	0.9740	0.9704
False Negative Rate	0.0090	0.0090	0.0090	0.0082
RCC	0.6186	0.6186	0.6186	0.6622
AUC (ROC)	0.6630	0.6630	0.6630	0.7181
(c) Neural Network Performance Indicator (NNISP)				
Sensitivity	0.6087	0.4957	0.6304	0.6348
Specificity	0.6304	0.7542	0.6159	0.6368
False Positive Rate	0.9746	0.9691	0.9747	0.9731
False Negative Rate	0.0097	0.0105	0.0094	0.0090
RCC	0.6300	0.7502	0.6161	0.6368
AUC (ROC)	0.6604	0.6604	0.6604	0.6927

Classification Rate (RCC = 0.7216) and the best trade-off between sensitivity and specificity. Among the two composite indicators, the ISP outperforms the NNISP with the Logit model, while the NNISP outperforms the ISP in terms of Correct Classification Rate.

Table 3 reports the out-of-sample classification results by size: SMEs and large firms. Across all input types and models, predictive accuracy is higher for SMEs, largely due to the larger sample size and greater variability. The Logit model using the 40 indicators achieves the best balance for SMEs, while the Neural Network performs slightly better in reducing false negatives. It is worth noting that results for large firms are considerably less stable, given the small number of defaults observed for large firms.

To sum up, our results confirm that while models trained on all 40 indicators offer the highest predictive accuracy, both composite indicators serve as reliable, lower-dimensional alternatives. Moreover, the NNISP shows its potential when combined with flexible non-linear models, while ISP may be most useful for small and medium enterprises where context knowledge plays a key role.

Table 2. Out-of-sample confusion matrices and performance metrics for Logit and Neural Network classifiers using 40 indicators, ISP, and NNISP.

Confusion Matrices							
		40 Indicators		ISP		NNISP	
Model	**Actual**	**Pred = 1**	**Pred = 0**	**Pred = 1**	**Pred = 0**	**Pred = 1**	**Pred = 0**
Logit	$Y = 1$	146	84	142	88	140	90
	$Y = 0$	4436	10102	5203	9335	5412	9126
Neural Net	$Y = 1$	136	94	121	109	124	106
	$Y = 0$	4018	10520	4808	9730	5081	9457
Performance Metrics							
Metric		**Logit**	**Neural Net**	**Logit**	**Neural Net**	**Logit**	**Neural Net**
Sensitivity		0.6348	0.5913	0.6130	0.5435	0.6087	0.5391
Specificity		0.6949	0.7236	0.6294	0.6495	0.6277	0.6505
False Positive Rate		0.9681	0.9673	0.9745	0.9761	0.9748	0.9762
False Negative Rate		0.0083	0.0089	0.0096	0.0110	0.0098	0.0111
Correct Classification Rate (RCC)		0.6939	0.7216	0.6292	0.6479	0.6274	0.6488

Table 3. Out-of-sample confusion matrices and performance metrics by firm size and model type.

Confusion Matrices—Small and Medium Enterprises (SMEs)							
		40 Indicators		ISP		NNISP	
Model	Actual	Pred = 1	Pred = 0	Pred = 1	Pred = 0	Pred = 1	Pred = 0
Logit	$Y = 1$	141	84	141	84	138	87
	$Y = 0$	4327	9563	5213	8677	5554	8336
Neural Net	$Y = 1$	134	91	133	92	124	101
	$Y = 0$	4076	9814	5146	8744	5079	8811
Confusion Matrices—Large Companies							
		40 Indicators		**ISP**		**NNISP**	
Model	**Actual**	**Pred = 1**	**Pred = 0**	**Pred = 1**	**Pred = 0**	**Pred = 1**	**Pred = 0**
Logit	$Y = 1$	0	7	5	2	3	4
	$Y = 0$	17	629	96	550	109	537
Neural Net	$Y = 1$	0	7	0	7	1	6
	$Y = 0$	11	635	15	631	20	626
Performance Metrics—Small and Medium Enterprises (SMEs)							
Metric		**Logit**	**Neural Net**	**Logit**	**Neural Net**	**Logit**	**Neural Net**
	Sensitivity	0.6267	0.5956	0.6267	0.5911	0.6133	0.5511
	Specificity	0.6885	0.7066	0.6247	0.6295	0.6001	0.6343
	False Positive Rate	0.3115	0.2934	0.3753	0.3705	0.3999	0.3657
	False Negative Rate	0.3733	0.4044	0.3733	0.4089	0.3867	0.4489
	Correct Classification Rate (RCC)	0.6875	0.7048	0.6247	0.6289	0.6004	0.6330
Performance Metrics—Large Companies							
Metric		**Logit**	**Neural Net**	**Logit**	**Neural Net**	**Logit**	**Neural Net**
	Sensitivity	0.0000	0.0000	0.7143	0.0000	0.4286	0.1429
	Specificity	0.9737	0.9830	0.8514	0.9768	0.8313	0.9686
	False Positive Rate	0.0263	0.0170	0.1486	0.0232	0.1687	0.0314
	False Negative Rate	1.0000	1.0000	0.2857	1.0000	0.5714	0.8571
	Correct Classification Rate (RCC)	0.9632	0.9724	0.8499	0.9663	0.8270	0.9602

5 Conclusions

This paper has compared two alternative approaches to constructing composite indicators for default prediction: the expert-driven *Synthetic Performance Indicator* (ISP) and a data-driven *Neural Network Synthetic Performance Indicator* (NNISP). Both indicators were derived from a common set of forty firm-level financial ratios covering key aspects of profitability, liquidity, growth, and leverage.

Our results show that both ISP and NNISP represent a parsimonious but competitive alternative to the use of a large set of indicators. The ISP performs consistently well across model classes, especially in linear and regularized settings, reflecting the stability and interpretability afforded by expert knowledge. In contrast, the NNISP shows greater variation in performance but benefits more when used with flexible non-linear classifiers such as Neural Networks. This suggests that while the NNISP may be less robust overall, it can offer advantages in more complex learning environments. Nonetheless, there is an evident trade-off between parsimony and accuracy as the models trained on the full set of indices perform best overall.

Despite these promising results, some limitations remain. First, the ISP's reliance on fixed weights derived from expert elicitation may reduce its adaptability to changing economic conditions. Conversely, while the NNISP is fully data-driven, its construction hinges on the use of a single target variable (the current ratio), which may not fully capture broader aspects of financial performance.

Future research could look at other approaches that combine expert-informed weighting with data-driven adjustments. In addition, the scoring assessment could be extended beyond the default classification. Finally, the integration of non-financial data, such as governance indicators or ESG metrics, may further enhance the predictive capacity of the synthetic scoring systems.

References

Altman, E.I.: Financial ratios, discriminant analysis and the prediction of corporate bankruptcy. J. Financ. **23**(4), 589–609 (1968)

Altman, E.I., Marco, G., Varetto, F.: Corporate distress diagnosis: comparisons using linear discriminant analysis and neural networks (the Italian Experience). J. Banking Finance **18**(3), 505–529 (1994)

Bartik, A.W., Bertrand, M., Cullen, Z., Glaeser, E.L., Luca, M., Stanton, C.: The impact of COVID-19 on small business outcomes and expectations. Proc. Natl. Acad. Sci. **117**(30), 17656–17666 (2020)

Behr, P., Güttler, A.: Credit risk assessment and relationship lending: an empirical analysis of German small and medium-sized enterprises. J. Small Bus. Manage. **45**(2), 194–213 (2007)

van Dijk, B.: AIDA Database (2023). Accessed 17 Oct 2025

Calin, O.: Deep Learning Architectures. Springer (2020)

Cheraghali, H., Molnár, P.: SME default prediction: a systematic methodology-focused review. J. Small Bus. Manage. **62**(6), 2847–2905 (2024)

Ciampi, F., Giannozzi, A., Marzi, G., Altman, E.I.: Rethinking SME Default Prediction: a systematic literature review and future perspectives. Scientometrics **126**, 2141–2188 (2021)

Ciampi, F., Gordini, N.: Small enterprise default prediction modeling through artificial neural networks: an empirical analysis of Italian small enterprises. J. Small Bus. Manage. **51**(1), 23–45 (2013)

De Leonardis, D., Rocci, R.: Assessing the default risk by means of a discrete-time survival analysis approach. Appl. Stoch. Model. Bus. Ind. **24**(4), 291–306 (2008)

Giordani, P., Jacobson, T., Von Schedvin, E., Villani, M.: Taking the twists into account: predicting firm bankruptcy risk with splines of financial ratios. J. Financ. Quant. Anal. **49**(4), 1071–1099 (2014)

Haykin, S.: Neural Networks: A Comprehensive Foundation. Prentice Hall PTR (1994)

JRC. Handbook on Constructing Composite Indicators: Methodology and User Guide. OECD Publishing (2008)

Klieštik, T., Kočišová, K., Mišanková, M.: Logit and Probit model used for prediction of financial health of a company. Procedia Econ. Finance **23**, 850–855 (2015)

Li, K.: Liquidity Ratios and Corporate Failures. Account. Finance **64**(1), 1111–1134 (2024)

Lo, A.W.: Logit versus discriminant analysis: a specification test and application to corporate bankruptcies. J. Econometrics **31**(2), 151–178 (1986)

McCulloch, W.S., Pitts, W.: A logical calculus of the ideas immanent in nervous activity. Bull. Math. Biophys. **5**, 115–133 (1943)

Miller, K.D., Reuer, J.J.: Measuring organizational downside risk. Strateg. Manag. J. **17**(9), 671–691 (1996)

Rosenblatt, F.: The perceptron: a probabilistic model for information storage and organization in the brain. Psychol. Rev. **65**(6), 386 (1958)

Ross, S.A., Westerfield, R.W., Jordan, B.: Corporate Finance, 2010. Edition Hardcover (2018)

Tibshirani, R.: Regression shrinkage and selection via the lasso. J. R. Stat. Soc. Ser. B Stat Methodol. **58**(1), 267–288 (1996)

Westgaard, S., Van der Wijst, N.: Default probabilities in a corporate bank portfolio: a logistic model approach. Eur. J. Oper. Res. **135**(2), 338–349 (2001)

Wu, D., Ma, X., Olson, D.L.: Financial distress prediction using integrated Z-score and multilayer perceptron neural networks. Decis. Support Syst. **159**, 113814 (2022)

Zou, H., Hastie, T.: Regularization and variable selection via the elastic net. J. R. Stat. Soc. Ser. B Stat Methodol. **67**(2), 301–320 (2005)

Rethinking the Indexation of Retirement Age: Cohort vs. Period Life Expectancy

Mariarosaria Coppola[1(✉)], Maria Russolillo[2], and Rosaria Simone[1]

[1] Department of Political Sciences, Federico II University, Naples, Italy
{m.coppola,rosaria.simone}@unina.it

[2] Department of Economics and Statistics, University of Salerno, Fisciano, Italy
mrussolillo@unisa.it

Abstract. Over the past two centuries, life expectancy has risen steadily in industrialized countries, challenging the financial sustainability of pension systems. In systems with indexed pension age, a key issue is whether to use period life expectancy (PLE) or cohort life expectancy (CLE). CLE relies on real cohort data and captures future mortality improvements; requiring a full lifespan of data, it can be estimated only for extinct cohorts. To address this, CLE is often estimated using life tables combining past and expected future mortality. In contrast, PLE is based on mortality rates observed in a single period. It is simpler but less accurate than CLE, particularly when assessing the retirement systems sustainability. The aim of this paper is twofold: measuring the gap between CLE and PLE at age 65 for Italian population using Lee-Carter and Renshaw-Haberman models for mortality projections, and assessing how this gap affects the pension age shift.

Keywords: cohort life expectancy · period life expectancy · LC model · RH model · pension age shift

1 Introduction

Life expectancy in industrialized countries has been climbing at an almost linear pace for more than 180 years, gaining roughly three months per calendar year [8]. This longevity revolution is widening the gap between the exit from the labor market and death, creating pressure on the solvency and intergenerational fairness of public and private pension systems.

The accurate measurement of this phenomenon depends crucially on the type of life table used to summarize survival perspectives. Most actuarial practice still relies on period life expectancy (PLE), which, by construction, assumes that today's mortality regime will prevail forever; it therefore ignores future mortality improvements that actual cohorts will experience. PLE reflects the average number of years a hypothetical individual would live if they experienced throughout their life the age-specific mortality rates observed in a given calendar year. It is a synthetic measure that does not account for future changes in mortality, and therefore any temporary shocks (e.g., due to pandemics) can significantly distort PLE, as it projects those transient conditions across an entire lifespan.

M. La Rocca et al. (Eds.): WMAF 2025, *New Perspectives in Mathematical and Statistical Methods for Actuarial Sciences and Finance*, pp. 86–95, 2025.
https://doi.org/10.1007/978-3-032-05551-4_8

In contrast, cohort life expectancy (CLE) follows a real birth cohort throughout its lifetime, incorporating projected improvements or deteriorations in mortality as the cohort ages. It thus captures expected future changes in survival and is more aligned with how mortality is likely to evolve over time. This makes CLE a more accurate reflection of lifetime longevity, especially in populations experiencing sustained improvements in life expectancy.

In 2006, [4] formalized the link between PLE and CLE, demonstrating that the former is essentially a lagged indicator of the latter. The authors showed that, in populations with a sustained decline in mortality, current PLE matches the life expectancy of cohorts born 40–50 years earlier; the vertical distance between the two curves, the so-called *gap*, quantifies the extra years that cohorts gain by living through future improvements.

Other studies confirm that this gap is far from negligible when translated into pension liabilities. Using data from Portugal and Spain, [1] find that replacing CLE with PLE at age 65 reduces the expected duration of the pension period from one to three years and implicitly penalizes current pensioners by up to 30 per cent of their accumulated pension wealth. In a recent paper [3], the authors propose an indexation mechanism for retirement age that keeps the Expected Pension Period Duration ($EPPD$) constant across cohorts. The mechanism computes the minimum forward shift (lag) in the *State Pension age* (SPA) such that cohort specific life expectancy does not exceed a benchmark $EPPD$ derived from an initial reference cohort. Applied to Italian data, their simulations illustrate how model assembling can smooth the volatility of lags and how gender specific indexing mitigates distributional inequalities.

Building on this debate, the present study extends the period–versus–cohort comparison to Italian birth cohorts from 1952 to 2022. Mortality is fitted and projected with both the Lee–Carter (LC) and Renshaw–Haberman (RH) models, and the analysis pursues two main objectives:

- Quantify the gap between cohort and period life expectancy at age 65;
- Assess its impact on the SPA shift required to keep a constant benchmark $EPPD$.

The remainder of the article is organized as follows. Section 2 derives the formulas for life expectancy and for the constant $EPPD$ indexation rules. Section 3 reports the cohort–period gaps and the implied retirement-age shifts for Italian cohorts 1952–2022. Section 4 concludes.

2 Cohort and Period Life Expectancy: Definitions and Implications for Indexed Retirement Age

2.1 Cohort and Period Life Table

In demography, life tables are essential tools for analyzing mortality patterns and computing life expectancy. Life tables can be constructed according to two different approaches: period and cohort. The former furnishes the so-called period life table (or static life table) representing the mortality experienced by a different cohort during a single period of time (generally not longer than three years).

In other terms, period life table supposes a hypothetical cohort that experiences current mortality rates throughout its entire life. The latter provides a cohort life table (or dynamic life table) that follows a real cohort (individuals born in the same year) throughout their lifetime, using age-specific mortality rates that the cohort actually experiences as it ages.

The main difference between period and cohort mortality tables lies in the fact that period tables reflect cross-sectional mortality, while cohort tables reflect longitudinal mortality.

Table 1. Mortality rates by ages and calendar years

ages	Calendar years				
	...	$t-1$	t	$t+1$	...
0	.....	$q_0(t-1)$	$q_0(t)$	$q_0(t+1)$	....
1	.....	$q_1(t-1)$	$q_1(t)$	$q_1(t+1)$	.....
......	.....	.....	......	.....	.....
x	.....	$q_x(t-1)$	$q_x(t)$	$q_x(t+1)$	.....
$x+1$	.....	$q_{x+1}(t-1)$	$q_{x+1}(t)$	$q_{x+1}(t+1)$	.....
......	.....	.....	......	.....	.....
$\omega-1$	.....	$q_{\omega-1}(t-1)$	$q_{\omega-1}(t)$	$q_{\omega-1}(t+1)$	.....

This can be better explained by Table 1, where $q_x(t)$ represents the probability of death within 1 year for an individual aged x in year t.

The matrix in Table 1 can be read in different ways [6, 7]. The elements of the generic column form a cross-sectional mortality table, as they refer to individuals of various ages observed in the same calendar year t. The generic diagonal, on the other hand, represents the mortality table for individuals born in the same year t that is, a cohort table.

These two different approaches lead to different estimates of life expectancy. As explained in [1], knowing and applying the best estimate of the remaining life expectancy of individuals is vital for pension policy in the public and private sectors, as an incorrect estimate can lead to significant financial misalignments, unfair policy outcomes, or inefficient resource allocation.

2.2 Retirement Age Indexed to Life Expectancy: Cohort Vs Period Approach

Life expectancy at age x is the average number of years a person aged x is expected to live, based on a specific set of mortality conditions.

In the period life tables framework, life expectancy for an individual aged x in the calendar year t, assuming that current mortality rates remain constant, can be calculated as follows [1]:

$$e_x^{Period}(t) = \sum_{n=1}^{\omega-x} {}_n p_x^P(t) = \sum_{n=1}^{\omega-x} \exp\left(-\sum_{j=0}^{n-1} m_{x+j}(t)\right) \tag{1}$$

where ${}_n p_x^t$ represents the n-year survival probability for an individual aged x in year t, $m_x(t)$ denotes the central death rate at age x in the calendar year t, and ω is the limiting age (the maximum possible age a person can live). In that case PLE reflects the mortality experienced by a hypothetical cohort spending their entire life during one period of time.

In a similar way, the CLE can be expressed as follows:

$$e_x^{Cohort}(t) = \sum_{n=1}^{\omega-x} {}_n p_x^C(t) = \sum_{n=1}^{\omega-x} \exp\left(-\sum_{j=0}^{n-1} m_{x+j}(t+j)\right) \tag{2}$$

Equation 2 highlights how age and time evolve simultaneously in CLE computation, specifically at each step j, the individual grows older while time moves forward accordingly.

As shown by [4], cohort life expectancy typically exceeds period life expectancy in populations where mortality improves over time. The decline in mortality rates that has affected developed countries throughout the 20th and 21th centuries is expected to continue but at a slower pace in older ages [2,8]. As a consequence, considering the gap

$$e_x^{\text{Gap}}(t) = e_x^{Cohort}(t) - e_x^{Period}(t) \tag{3}$$

we expect it will decrease at older ages.

As highlighted in [3], this issue is particularly relevant in pension systems where the retirement age is indexed to life expectancy. When life expectancy is underestimated, it leads to the incorrect assumption that retirees will live fewer years than they actually will. Consequently, the system will pay pensions for a longer period than planned, generating financial imbalances and undermining the sustainability of the system, especially in pay-as-you-go schemes.

This dynamic directly affects the $EPPD$, defined in [3] as the number of years over which an individual is expected to receive pension benefits. If life expectancy is underestimated, the $EPPD$ will be longer than predicted and the system will have to disburse pension benefits over more years than expected. In [3], the authors calculate the shift to apply to SPA to keep $EPPD$ at the budgeted level. When period life expectancy is used to guide such adjustments, it tends to understate future survival because it ignores projected cohort-specific mortality improvements. As a result, the shift in retirement age based on period data is often less pronounced than it should be. Retirees, therefore, tend to retire too early relative to their actual longevity, leading to longer $EPPD$ than anticipated.

Accordingly, the primary goal of this study is to assess the impact that adopting a cohort-based approach—rather than a period-based—has on the adjustment of the retirement age in a pension system featuring an indexation mechanism, such as the one proposed in [3].

Let us suppose that the SPA - that is the earliest age at which individuals can begin receiving their State Pension – is fixed at x_0^* for all cohorts. Then in general terms $e_{x_0^*}$ is the life expectancy for an individual aged x_0^* in the year t who receives a monthly payment B as long as he/she survives, and it follows that: $EPPD = e_{x_0^*}$. This approach can be more precisely expressed by distinguishing between period and cohort framework.

In the cohort framework, let $e_{x_0^*,C}^{(\mathrm{M,Cohort})}$ denote the life expectancy of an individual belonging to cohort C, aged x_0^* on the first January of year t under the mortality model M. Assuming the cohort C^* as benchmark, the $EPPD$ of an individual belonging to C^* aged x_0^* in t_0^* follows:

$$EPPD^{*(M,Cohort)} = e_{x_0^*}^{(\mathrm{M,Cohort})}(t_0^*) \tag{4}$$

being: $C^* = t_0^* - x_0^*$ the year of birth.

Similarly, following a period approach, the EPPD referring to the year t_0^* is given by:

$$EPPD^{*(M,Period)} = e_{x_0^*}^{(\mathrm{M,\ Period})}(t_0^*) \tag{5}$$

According to the longevity trend, the real life expectancy at age x_0^* will change for future generations. We determine the age x_0^*+j at which life expectancy does not exceed the $EPPD^{(M,C^*)}$ for the first time.

Similarly to [3], we define the shift to apply to the SPA, x_0^*, for the cohort C aged x_0^* at time t under the mortality model M as follows:

$$Shift_{\mathrm{SPA}}^{(\mathrm{M,Cohort})} = \inf_{j\geq 0}\left\{j : e_{x_0^*+j}^{(M,Cohort)}(t) \leq EPPD^{*(M,Cohort)}\right\} \tag{6}$$

Under the same assumptions in the period framework, we can define:

$$Shift_{\mathrm{SPA}}^{(\mathrm{M,Period})} = \inf_{j\geq 0}\left\{j : e_{x_0^*+j}^{(\mathrm{M,Period})}(t) \leq EPPD^{*(M,Period)}\right\} \tag{7}$$

Now we can calculate the gap between the SPA shift based on the period and cohort life table as follows:

$$GAP_{\mathrm{Shift}} = Shift_{\mathrm{SPA}}^{\mathrm{Cohort}} - Shift_{\mathrm{SPA}}^{\mathrm{Period}} \tag{8}$$

3 Results

Using exposure and death counts from the latest release of the Human Mortality Database, which covers the Italian population from 1872 to 2022, we fit Italian mortality data for 1952–2022 with two well-known stochastic mortality models. Building on the findings in [3], we use the LC model as our benchmark and pair it with the more flexible RH specification, whose superior fit and forecast accuracy for Italy are well documented [3]. Mortality is then projected 30 years ahead ($H = 30$) under both models; within the RH framework the cohort effect is assumed to evolve according to an AR(1) process for illustration purposes[1]

[1] Various ARIMA specifications have been tested and lead to analogous results.

3.1 Comparison Between PLE and CLE

Our analysis focuses on the time intervals between 1952 and 2022, so that the last birth cohort covered by our investigation is 1987 (individuals born in 1987 will turn 65 in 2052). Figure 1 shows the well-known dominance of cohort over period life expectancy at age 65 and its gradual convergence, especially for women, over

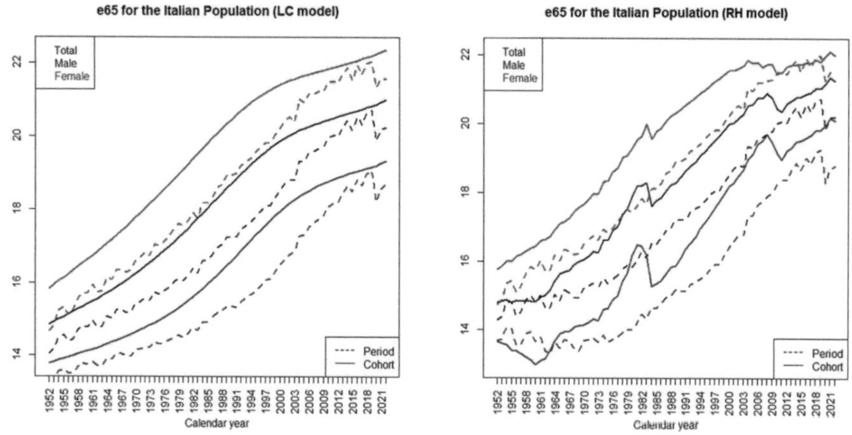

Fig. 1. Comparisons between cohort and period life expectancy at 65 years, under LC and RH model for fitting and prediction.

Table 2. Expected residual life at 65, under LC model

Calendar year	Total		Male		Female	
	Period	Cohort	Period	Cohort	Period	Cohort
1952	14.04	14.85	13.27	13.78	14.67	15.83
1956	14.37	15.13	13.50	13.94	15.11	16.23
1960	14.77	15.41	13.74	14.11	15.69	16.63
1964	15.08	15.72	13.92	14.31	16.15	17.06
1968	15.17	16.05	13.99	14.51	16.27	17.51
1972	15.56	16.44	14.23	14.76	16.81	18.02
1976	15.77	16.87	14.37	15.05	17.12	18.55
1980	16.05	17.35	14.50	15.40	17.54	19.09
1984	16.60	17.89	14.91	15.83	18.17	19.67
1988	17.03	18.44	15.22	16.32	18.66	20.21
1992	17.45	18.98	15.48	16.87	19.20	20.70
1996	17.96	19.47	15.92	17.44	19.72	21.10
2000	18.54	19.87	16.50	17.95	20.23	21.40
2004	19.33	20.17	17.26	18.36	21.01	21.61
2008	19.67	20.37	17.74	18.65	21.22	21.76
2012	20.01	20.54	18.21	18.87	21.46	21.91
2016	20.54	20.70	18.81	19.05	21.93	22.07
2020	19.88	20.87	18.21	19.20	21.33	22.25

Table 3. Expected residual life at 65, under RH model

Calendar year	Total		Male		Female	
	Period	Cohort	Period	Cohort	Period	Cohort
1952	14.28	14.81	13.68	13.66	14.75	15.77
1956	14.41	14.86	13.54	13.38	15.11	16.14
1960	14.82	14.82	13.67	12.98	15.75	16.44
1964	15.04	15.42	13.64	13.66	16.18	16.92
1968	14.95	15.92	13.47	14.09	16.19	17.45
1972	15.39	16.30	13.76	14.32	16.75	17.97
1976	15.48	16.91	13.72	14.88	16.99	18.61
1980	15.84	17.88	14.00	16.15	17.44	19.30
1984	16.53	17.61	14.66	15.28	18.12	19.58
1988	17.00	18.24	15.02	15.84	18.66	20.25
1992	17.46	18.78	15.34	16.55	19.22	20.67
1996	17.96	19.38	15.79	17.36	19.75	21.12
2000	18.52	20.00	16.42	18.20	20.26	21.56
2004	19.37	20.55	17.32	19.00	21.05	21.87
2008	19.72	20.91	17.90	19.70	21.26	21.75
2012	20.05	20.60	18.38	19.23	21.46	21.68
2016	20.60	20.88	19.03	19.61	21.92	21.80
2020	19.86	21.18	18.29	19.97	21.26	21.99

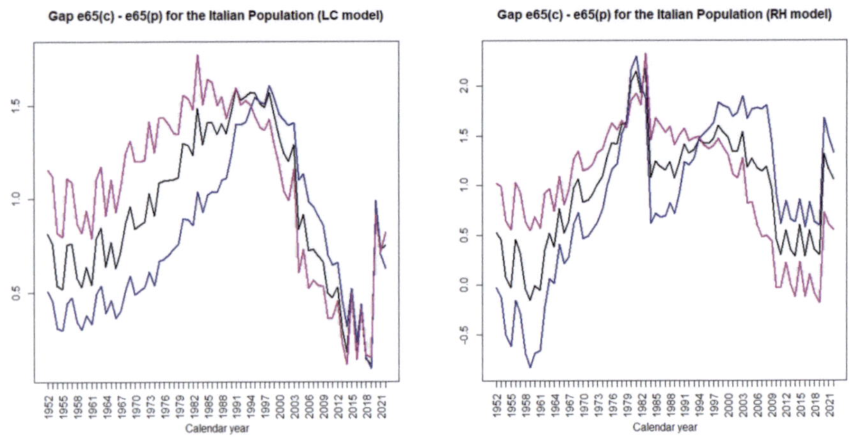

Fig. 2. GAP $e_{65}^{(c)}(t) - e65^{(p)}(t)$ for total, male and female Italian population, assuming LC and RH model.

the last three decades. Tables 2 and 3 provide a detailed quantification of the extent of this convergence. For the sake of simplicity, values are reported at four-year intervals from 1952 to 2022, with the final year presented in the tables being 2020.

Under the LC model (Table 2), the cohort-period differential for the total population widens from roughly 0.8 years in 1952 to a peak of 1.6 years around

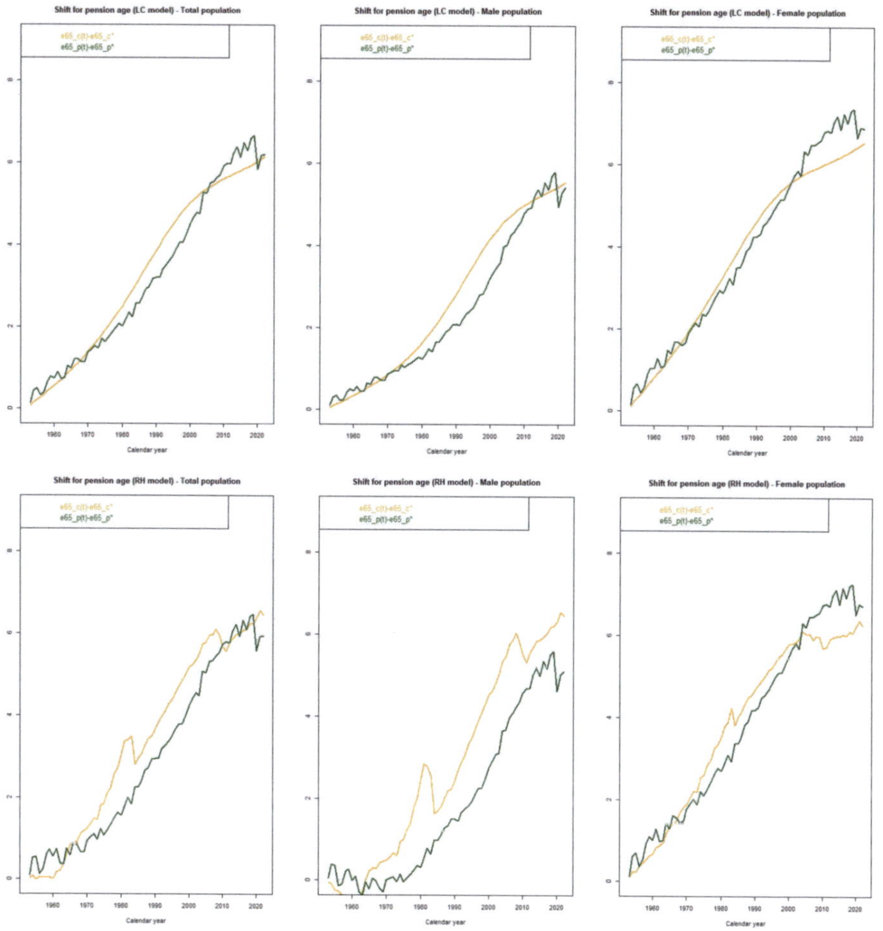

Fig. 3. Shift in pension age to be applied to $EPPD^{\star}$ assuming LC and RH model

1980, then collapses to 0.2 years by 2016. The RH model (Table 3) enlarges the mid-sample gap to about 2 years in 1980, but also shows that it narrows to only 0.3 years in 2016. The narrowing is especially rapid for women: their cohort–period gap drops below half a year about ten years before men reach the same target.

Figure 2 converts this visual evidence into the gap $e_{65}^{(c)}(t) - e_{65}^{(p)}(t)$ for both the LC and RH models: in either case, the gap for the female population is shown to increase slowly than for the male population. This circumstance is due to the fact that women experience longevity to a greater extent than men, and thus a weaker improvement in life expectancy can be foreseen, given the already higher life expectancy. For this reason, we posit that period and cohort life expectancy tend to become closer more rapidly for women than for men.

The policy relevance emerges in Fig. 3, where we compute the shift in pension age $e^{\cdot}_{65}(t) - EPPD^{\star}$ required to keep the $EPPD^{\star}$ at its 1952 benchmark. Period tables systematically understate this shift; for the most recent cohorts, the shortfall still amounts to about two years for men and one-and-a-half for women, with RH prescribing slightly larger corrections than LC. Taken together, these results confirm that the narrowing gap and the flattening of both period and cohort $e_{65}(t)$ curves imply progressively smaller postponements of retirement age for future cohorts, particularly for women, as the system approaches a longevity plateau.

4 Conclusions

Drawing on our analysis, we show that choosing between *period* and *cohort* life tables is a decisive policy instrument to safeguard the long-term sustainability and fairness of pension systems. For Italy, the cohort–period gap in residual life expectancy at age 65 widened to roughly 1.5–2 years for the 1980 birth cohorts but has since narrowed to less than half a year, especially for women, indicating both a deceleration of longevity improvements and the onset of a possible "longevity plateau". The observed narrowing of this gap indicates that longevity risk may be approaching a saturation point, as the annual gains in both period and cohort indicators have noticeably slowed in recent years. Yet even this smaller gap continues to translate into sizeable differences in the SPA: indexation rules based on period tables still underestimate the forward shift required to keep the $EPPD$ constant by about two years for men and one-and-a-half years for women in the most recent cohorts.

Two policy implications emerge. First, a cohort-based indexation mechanism remains indispensable until period and cohort indicators have fully converged. Second, as longevity gains decelerate, the magnitude—and hence the political cost—of each successive SPA adjustment is likely to fall, offering policymakers a window to embed more transparent and gender-sensitive rules without imposing abrupt changes on future retirees.

Preliminary findings will be further developed through a more in-depth analysis of the modeling framework, drawing on assembling techniques as advocated by [3]. This will enable the derivation of an average shift in pension age corresponding to period life expectancy, and consequently in its gap with cohort life expectancy, yielding a more robust approach with respect to model choice and predictive variability. Finally, an extensive retrospective study of the relationship between period and cohort-based life expectancies, conducted separately for men and women, may offer useful insights into the extent to which period life expectancy can be regarded as a proxy for cohort life expectancy, and whether this relationship has undergone structural changes over time.

References

1. Ayuso, M., Bravo, J.M., Holzmann, R.: Getting life expectancy estimates right for pension policy: period versus cohort approach. J. Pension Econ. Finance **20**(2), 212–231 (2021)
2. Vallin, J., Meslé, F.: The segmented trend line of highest life expectancies. Demogr. Res. **21**(5), 145–158 (2009)
3. Coppola, M., Russolillo, M., Simone, R.: An indexation mechanism for retirement age: analysis of the gender gap. Risks. **7**(1), 21 (2019)
4. Goldstein, J.R., Wachter, K.W.: Relationships between period and cohort life expectancy: gaps and lags. Popul. Stud. (Camb). **60**(3), 257–69 (2006)
5. Guillot, M.: Period versus cohort life expectancy. In: Rogers, R., Crimmins, E. (eds) International Handbook of Adult Mortality. International Handbooks of Population, vol. 2. Springer, Dordrecht (2011)
6. Pitcco, E.: Matematica e tecnica attuariale delle assicurazioni sulla durata di vita. Luglio Ed. (2022)
7. Pitacco, E., Denuit, M., Haberman, S., Olivieri, A.: Modelling longevity dynamics for pensions and annuity business. Oxford University Press (2009)
8. Oeppen, J., Vaupel, J.W.: Demography. Broken limits to life expectancy. Science **296**(5570), 1029–31 (2002)

Calibrating Temperature Models with Neural Networks for Weather Derivatives

Stefania Corsaro[1], Vincenzo Di Sauro[2(✉)], Zelda Marino[1], and Salvatore Scognamiglio[1]

[1] Department of Management and Quantitative Studies, Parthenope University of Naples, 80133 Naples, NA, Italy
{stefania.corsaro,zelda.marino,salvatore.scognamiglio}@uniparthenope.it
[2] Department of Economics and Legal Studies, Parthenope University of Naples, 80133 Naples, NA, Italy
vincenzo.disauro001@studenti.uniparthenope.it

Abstract. Weather significantly affects business activities, making hedging against related risks crucial; weather derivatives can help mitigate financial impacts. Most of the derivatives currently traded are linked to a temperature; having a good model for its evolution is the backbone of effective weather derivative pricing. This paper presents a novel neural network approach for jointly calibrating the mean and variance of temperature for weather derivatives pricing. We also address the challenge of explainability in neural networks by designing the architecture to replace the approach proposed in [1]. Additionally, we explore potential extensions of the model to improve forecasting accuracy further. Through numerical experiments with weather station data from Fiumicino Maccarese, we illustrate the model's application in pricing Heating Degree Day futures.

Keywords: Deep Neural Networks · Explainable Deep Learning · climate risk · weather derivatives

1 Introduction

The weather derivatives market has grown significantly in recent years, driven by increasing climate variability and businesses required to manage weather-related risks. According to the CME Group, one of the largest platforms for weather risk instruments, the market was valued at around USD25 billion in 2023, including exchange-traded and over-the-counter contracts. These financial instruments allow one to mitigate climate risks that strongly affect some business activities like agriculture, energy and retail sectors. Weather derivative contracts are typically written on climate variables such as temperature or rainfall, measured at a specific reference weather station. For this reason, in response to this rising demand, the CME Group has introduced contracts for new cities,

M. La Rocca et al. (Eds.): WMAF 2025, *New Perspectives in Mathematical and Statistical Methods for Actuarial Sciences and Finance*, pp. 96–107, 2025.
https://doi.org/10.1007/978-3-032-05551-4_9

enabling the creation of contracts tailored to local weather conditions. Among these contracts, those based on temperature indices, such as Heating Degree Days (HDD) and Cooling Degree Days (CDD) remain the most widely adopted [2], which quantify heating and cooling demands based on deviations from a designated baseline temperature. Consequently, Accurate modelling of temperature dynamics is vital for fair pricing and risk assessment. Various models have been proposed to capture the complex temperature dynamics and their impact on financial contracts. Among the earliest, Alaton et al. [3] model temperature as a mean-reverting process driven by a standard Brownian motion, assuming a constant monthly variance. Benth and Šaltyt.e-Benth [1] extend this framework by introducing a functional form for the volatility component with a seasonal structure. Building on these foundations, other authors have explored alternatives to better capture specific temperature dynamics features. For instance, Brody et al. [4] use a different stochastic driver, the fractional Brownian motion, to account for long-memory effects, while Zapranis and Alexandridis [5] incorporate time-dependent mean reversion speeds. We have chosen to replicate the model proposed by Benth and Šaltytė-Benth in [1], due to its balance between mathematical tractability and practical applicability, effectively captures mean-reverting and seasonal volatility in temperature, which is crucial for pricing weather derivatives. In this paper, we introduce a neural network-based approach for simultaneously modelling and forecasting the mean and variance of temperature. This design is inspired by the approach proposed by Nix and Weigend [6], where a neural network is trained to output both the expected value and the variance of a target variable under the assumption of a Gaussian distribution. Our motivation for this choice arises from growing empirical evidence demonstrating neural networks' ability to capture complex, non linear dependencies in time series data compared to traditional approaches. Improved volatility estimation contributes to uncertainty quantification, which is crucial for pricing weather derivatives and risk management. This paper is organised as follows: Sect. 2 introduces temperature weather derivatives, in Sect. 3, the Neural Network architecture we use in our evaluation is introduced. Section 4 presents our neural network model for estimating simultaneously the daily mean and variance of temperature and explores one potential extension. Lastly, Sect. 5 discusses the numerical results and the model's performance, and Sect. 6 wraps up the study with concluding remarks.

2 Weather Derivative Contracts

Weather derivatives are financial instruments used as a risk management strategy to mitigate potential hazards associated with adverse or unexpected weather conditions. To effectively incorporate weather data into these derivatives, quantification is essential, which is achieved by establishing indices representing rainfall, precipitation, or snowfall. However, more than 90% are linked to temperature-related indices such as HDD and CDD. The Heating Degree Days (HDD) index is utilised during winter to measure how many degrees the daily temperature falls below a predetermined base temperature. Conversely, the CDD index measures the degrees by which the daily temperature exceeds the base temperature and is used during summer. HDDs and CDDs can be expressed as follows for day t:

$$HDD_t = \max\{T_{ref} - T(t), 0\}$$
$$CDD_t = \max\{T(t) - T_{ref}, 0\}$$

where $T(t)$ represents the daily average temperature, and T_{ref} is the temperature threshold set in Europe, equal to 18 °C. The HDD and CDD indices are determined by summing daily values over a specified period, such as a month, a summer, or any other pre-defined duration. Here we will focus on temperature futures; the price at time $t \leq \tau_1$ written on the HDD index over a specified period $[\tau_1, \tau_2]$, assuming a constant continuously compounding interest rate r, is defined as the $\mathcal{F}_t$-adapted stochastic process $F_{\text{HDD}}(t, \tau_1, \tau_2)$ satisfying:

$$F_{\text{HDD}}(t, \tau_1, \tau_2) = e^{-r(\tau_2 - t)} \mathbb{E}_{\mathcal{Q}} \left[\int_{\tau_1}^{\tau_2} \max(c - T(\tau), 0)\, d\tau - \Big| \mathcal{F}_t \right]$$

where $\mathcal{Q}$ is a risk-neutral probability.

3 Neural Networks

Since our approach is based on the use of deep neural networks, in this section, we will provide a brief description. In the following, we consider the Fully-Connected Network layer and the Embedding layer, since they are the blocks composing the network architecture used in this paper. A Fully-Connected Network (FCN) is a type of neural network where each neuron in one layer is connected to every neuron in the subsequent layer, allowing for comprehensive interaction between all layers. We denote as $\boldsymbol{x} = (x_1, x_2, \dots, x_{q_0})' \in \mathbb{R}^{q_0}$ the vector of the input features. A Fully-Connected Network (FCN) layer with $q_1 \in \mathbb{N}$ units is a vectorial function that maps $\boldsymbol{x}$ to in a q_1-dimensional real-valued space:

$$\boldsymbol{z}_1 : \mathbb{R}^{q_0} \to \mathbb{R}^{q_1}, \qquad \boldsymbol{x} \mapsto \boldsymbol{z}_1(\boldsymbol{x}) = \left(z_1^{(1)}(\boldsymbol{x}), z_1^{(2)}(\boldsymbol{x}), \dots, z_1^{(q_1)}(\boldsymbol{x}) \right).$$

The output of the layer is a set of new features $\boldsymbol{z}(\boldsymbol{x})$ which are non-linear function of $\boldsymbol{x}$:

$$\boldsymbol{z}_1(\boldsymbol{x}) = \psi_1 \Big(\boldsymbol{b}_1 + W_1 \boldsymbol{x} \Big),$$

where $\psi_1 : \mathbb{R} \mapsto \mathbb{R}$ is the activation function, $W_1 \in \mathbb{R}^{q_1 \times q_0}$ the full matrix of weights and $\boldsymbol{b}_1 \in \mathbb{R}^{q_1}$ is the vector of biases. In the case of $\mathcal{Y} \subseteq \mathbb{R}$, the output of a shallow NN reads:

$$y = \psi_o \left(b_o + \sum_{k=1}^{q_1} w_o^{(k)} z_1^{(k)}(\boldsymbol{x}) \right),$$

where $z_1^{(k)}(\boldsymbol{x})$ represent the k-th element of the vector $\boldsymbol{z}_1(\boldsymbol{x})$, $b_o \in \mathbb{R}$ and $w_o^{(k)} \in \mathbb{R}, k = 1, \dots, q_1$, are output layer' coefficients, and $\psi_o : \mathbb{R} \to \mathbb{R}$ is the activation function of the output layer. If the network is deep, the vector $z_1(\boldsymbol{x})$ is used as input in the next layer for computing new features and so for the following layers. An Embedding Network (EN) layer creates a low-dimensional representation of categorical variables. The hyperparameter $q_{\mathcal{L}}$ determines the embedding size, transforming categories into a $q_{\mathcal{L}}$-dimensional real-valued space. The coordinates for each level in this space are learned parameters that require training. Let $d_1, d_2, \ldots, d_{n_{\mathcal{D}}}$ denote the categories for the qualitative variable. These representations are then passed to subsequent layers, which can be of various types, that utilise the information from these vectors to compute the quantity of interest.

4 Joint Calibration of the Mean and Volatility Temperature Models

Accurate modelling of daily average temperatures is crucial for pricing weather derivatives and managing climate-related risks. A significant contribution to this field is the model proposed by Benth et al. [1], which explicitly incorporates seasonality into both the mean and variance components of the temperature process. Since we aim to replicate the structure of Benth's model using a neural network, we review this model at the beginning of the section. The primary focus, however, is on our proposed neural network (NNT_MV) model, which is designed to jointly estimate the expected value and variance of daily temperature. Our analysis is based on data collected from a single weather station, capturing temperature dynamics at a specific geographic location. The section is structured as follows: first, we outline the benchmark model in Sect. 4.1; then, in Sect. 4.2, we introduce the proposed NNT_MV model; finally, Sect. 4.3 presents an extension of our approach.

4.1 Temperature Model

This subsection introduces the approach proposed by Benth et al. [1], which provides a robust framework for modelling daily temperature dynamics at a specific geographical point. Their model uses a sinusoidal representation of seasonal patterns in temperature variance, addressing the limitations of more straightforward frameworks that assume constant volatility. In that framework, the authors propose a stochastic differential equation to model the temperature process $T(t)$:

$$dT(t) = d\theta(t) - \alpha\,(T(t) - \theta(t))\,dt + \sigma(t)dW(t), \quad t \geq 0, \tag{1}$$

where $\theta(t)$ represents the deterministic seasonal trend, $\alpha \in \mathbb{R}$ is the mean reversion speed, $\sigma(t)$ denotes the time-varying volatility, and $W(t)$ is a standard Brownian motion. The term $\theta(t)$ typically encapsulates long-term seasonal effects through a trigonometric representation, while $\sigma^2(t)$ is modelled as a composite of sinusoidal components, thereby capturing the intra-annual variability in temperature uncertainty. In that framework, seasonality is incorporated into both the daily average temperature $\theta(t)$ and the daily variance of the temperature $\sigma^2(t)$ by truncated Fourier series, which allow for a good fit for both the seasonality and the variance component while keeping the number of parameters relatively low.

$$\theta(t) = A_\theta + B_\theta t + C_\theta \sin(\omega t + \varphi_\theta), \tag{2}$$

$$\sigma^2(t) = A_\sigma + \sum_{i=1}^{I} C_{i,\sigma} \sin(i\omega t + \varphi_{i,\sigma}) \tag{3}$$

In the model described by (2), the parameter A_θ means the average temperature during the observation period, representing the long-term mean temperature. B_θ captures the marginal daily effect in the temperature trend. C_θ denotes the amplitude of seasonality, reflecting the difference between the warmest and coldest days throughout the year and emphasising the variation in temperature across seasons. The term φ_θ represents the phase of seasonality, indicating the point at which the monotonicity of the seasonal temperature curve changes, thus aligning the seasonal pattern with the calendar year. The same is true in equation (3), A_σ represents the mean volatility, capturing the baseline level of volatility over time, $C_{i,\sigma}$ reflect the amplitudes of the sinusoidal components that depict seasonal variations in volatility and the terms $\varphi_{i,\sigma}$ are the seasonal volatility phases, illustrating how the seasonal volatility pattern develops over time. The angular frequency ω is set to $\frac{2\pi}{365}$, which corresponds to a yearly cycle for daily data. Models (2),(3) can be also rearranged as follow:

$$\theta(t, \underline{a}) = a_0 + a_1 t + a_2 \sin(\omega t) + a_3 \cos(\omega t) \tag{4}$$

$$\sigma^2(t, \underline{b}) = b_0 + \sum_{i=1}^{I} (b_i \sin(\omega t i) + b_{i+I} \cos(\omega t i)) \tag{5}$$

where $\underline{a} = (a_0, a_1, a_2, a_3)$ and $\underline{b} = (b_0, b_i, \ldots, b_{i+I})$ denote the vectors of parameters to be estimated for the mean and variance models, respectively. In this formulation, the models appear linear in the parameters a_0, a_1, a_2, a_3, b_0, b_i and b_{i+I}, for $i \in \{1, \ldots, I\}$. The model (4) can be fitted using observed average temperature data by solving the linear optimisation problem described in (6). The same is true for (5), which enables us to fit the model using a least-squares optimisation approach. The procedure begins by computing empirical values of the daily variance, obtained by aggregating observations across all years for each specific calendar day, resulting in 365 variance estimates. We fit the model with these empirical variance estimates and the chosen truncation level by solving the optimisation problem shown in (7). The parameters can be estimated by solving the following least squares problems:

$$\min_{\underline{a} \in \mathbb{R}^4} \quad \sum_t \left(\tilde{\theta}(t) - a_0 - a_1 t - a_2 \sin(\omega t) - a_3 \cos(\omega t)\right)^2, \tag{6}$$

$$\min_{\underline{b} \in \mathbb{R}^{2I+1}} \quad \sum_t \left(\tilde{\sigma}^2(t) - b_0 - \sum_{i=1}^{I} [b_i \sin(\omega i t) + b_{i+I} \cos(\omega i t)]\right)^2. \tag{7}$$

From the estimated coefficients, the original model parameters can be recovered as follows:

Table 1. Parameter transformations from estimated harmonic coefficients.

$\hat{A}_\theta = \hat{a}_0;$	$\hat{A}_\sigma = \hat{b}_0;$
$\hat{B}_\theta = \hat{a}_1;$	
$\hat{C}_\theta = \sqrt{\hat{a}_2^2 + \hat{a}_3^2};$	$\hat{C}_{i\sigma} = \sqrt{\hat{b}_i^2 + \hat{b}_{i+I}^2};$
$\hat{\varphi}_\theta = \arctan\left(\frac{\hat{a}_3}{\hat{a}_2}\right) - \pi;$	$\hat{\varphi}_{i\sigma} = \arctan\left(\frac{\hat{b}_{i+I}}{\hat{b}_i}\right) - \pi.$

Finally, the authors estimate the value of α by regressing today's deseasonalised temperature against the deseasonalised temperature of the previous day, leading to the following equation:

$$T(t) - \theta(t) = k(T(t-1) - \theta(t-1)), \quad \text{where} \quad \alpha = 1 - k.$$

In this setting, the expected value $\theta(t)$ and variance $\sigma^2(t)$ of daily average temperature are modelled independently using distinct optimisation procedures. Although this approach is interpretable and effectively captures key seasonal patterns, it has different limitations. Firstly, separately estimating the expected value and variance can lead to inconsistencies, ignoring potential interactions between the mean and variance of the temperature, which are often intrinsically linked in real-world data, resulting in less accurate predictions and uncertainty estimates. Secondly, since variance is not directly observable, it must be derived from the data, which requires strong assumptions. If these are incorrectly specified, they can result in unreliable uncertainty estimates. To address this limitation, we propose an alternative approach.

4.2 Joint Calibration via Neural Network

We propose a neural network based approach, defined as the *Neural Network Temperature Mean and Variance* model (NNT_MV), to estimate both the expected value $\theta(t)$ and the variance $\sigma^2(t)$ of temperature at each time step based on input data while assuming a Gaussian distribution for observations. This method enhances understanding of future temperature dynamics and enables the construction of confidence intervals. We developed a neural network architecture replicating the parametric structure outlined in [1] for both mean and variance. This design allows us to maintain explainability, since neural networks are solely employed to learn these parameters, we preserve the interpretability of the original model while taking advantage of the flexibility and generalisation capabilities of deep learning. The network consists of two subnetworks: one dedicated to estimating the mean $\theta(t)$, and the other focused on variance $\sigma^2(t)$.

Each subnetwork is designed to replicate the functional form outlined in 4 and 5 at the output layers of our networks; these components are concatenated into a final output layer that simultaneously provides both the expected value and variance of temperature at each time step. Therefore, we propose using neural networks to accurately represent and learn the parameters describing the seasonal components. We formulate this as a supervised learning problem, where these parameters are approximated using a deep neural network composed of an appropriate number of FCN layers. The estimates of the parameters collected in the vectors $\underline{a}$ and $\underline{b}$ are obtained by solving the following optimization problem:

$$\left(\hat{\underline{a}}, \hat{\underline{b}}\right) = \underset{\underline{a}\in\mathbb{R}^4, \underline{b}\in\mathbb{R}^{2I+1}}{\arg\min} \sum_t \left[\frac{1}{2}\log\left(2\pi\sigma^2(t,\underline{b})\right) + \frac{\left(\tilde{T}(t) - \theta(t,\underline{a})\right)^2}{2\sigma^2(t,\underline{b})}\right]. \quad (8)$$

This approach provides a significant advantage by allowing the simultaneous estimation of both the mean and variance via a single optimisation problem. This joint learning paradigm improves computational efficiency and simplifies the overall estimation process. Once the optimal parameters have been obtained, one derives the optimal estimates of the parameters using the original relations described in Table 1 This method efficiently estimates the mean and variance of temperature dynamics within a neural network framework. We improve predictive accuracy and uncertainty quantification by using the power of neural networks while preserving the interpretability of parametric models.

4.3 Extending the NNT_MV Model

The NNT_MV model presented in this paper provides significant flexibility in temperature modelling and can be extended in numerous ways. Acknowledging that other variables may also influence the model's parameters is crucial. For example, the parameters controlling the influence of the trigonometric component should vary over time rather than remain constant. A possible approach to implement this is to make the parameters $a_{(\theta_2)}$ and $a_{(\theta_3)}$ contingent upon the month. This modification enhances the model's adaptability, resulting in a more precise representation of temperature dynamics. We refer to this extension as the Monthly-Dependent (MD_NNT_MV) model, enhancing adaptability and improving temperature dynamics representation. The month embedding is defined as:

$$\boldsymbol{e}_{a_2} : \mathcal{M} \to \mathbb{R}^{q^{a_2}}, \quad m \mapsto \boldsymbol{e}_{a_2}(m),$$

where $\mathcal{M} = \{1, \ldots, 12\}$ denotes the set of months, and $\boldsymbol{e}_{a_2}(m)$ is the embedding vector of dimension q^{a_2} associated to month m. This embedding is used as the input to the neural subnetwork that estimates the parameter a_2. In the Sect. 5, we will present the results of the numerical experiments conducted to compare the reference model with our proposed model and its extension, utilizing data obtained from a weather station. In the following section, we conducted numerical experiments comparing the three models presented in this paper.

5 Numerical Analysis

This section presents the results obtained by applying the benchmark model introduced by [1] and our neural network approach NNT_MV and its extension MD_NNT_MV to temperature data from the Fiumicino Maccarese weather station. All experiments were conducted within the R programming environment, utilising the `keras` package to implement and train neural networks.

5.1 Temperature Modelling

The numerical experiments conducted in this study leverage the Fiumicino Maccarese weather station dataset provided by the Lazio Region Agrometerology Service ARSIAL. Observations have been systematically collected since 2004 and are continually updated. To simulate average temperature, we used this dataset, which compiles daily data on air temperature measured at an altitude of 2 m. The data was divided into a training data set containing the daily average temperatures from the years in $T_{\text{train}} = \{t \in \mathcal{T} \mid 2015 \leq t \leq 2023\}$ and a testing data set defined as $T_{\text{test}} = \{t \in \mathcal{T} \mid t = 2024\}$. This partitioning approach guarantees that the model is trained on historical data and subsequently evaluated on the most recent observations, allowing us to assess that our calibration and subsequent evaluations accurately reflect the latest climatic conditions and trends. To maintain consistency in the number of observations for each year, data corresponding to February 29 were excluded from the analysis. To capture the seasonal pattern in volatility, we have to set the truncation parameter I, determining the number of sinusoidal components included in the model. In Benth et al. [1], the order was chosen arbitrarily; here, we follow the same approach and we set $I = 4$. Given our focus on evaluating the accuracy of the predictions through point and interval forecasts, we involve a range of metrics. For point forecast accuracy, we use the global MSE, while for interval forecasts, we aim for a coverage probability of $\alpha_{\text{conf}} = 0.95$. We consider the Prediction Interval Coverage Probability (PICP), which measures the proportion of true values that fall within predicted confidence intervals and the Mean Prediction Interval Width (MPIW) to account for the width of the confidence intervals. A well-calibrated model should have a PICP close to α_{conf} and minimise the MPIW to reduce the uncertainty in forecasting. Table 2 shows the in-sample and out-of-sample performance of the BENTH, NNT_MV and MD_NNT_MV models across MSE, PICP, and MPIW. Each metric provides a distinct perspective on the predictive capabilities of the models and their handling of uncertainty. The performance comparison indicates significant differences among the three models, BENTH, NNT_MV, and MD_NNT_MV, in terms of both point and interval forecasts. The NNT_MV model demonstrates better generalisation capabilities than the BENTH model. Though NNT_MV exhibits a marginal increase in in-sample mean squared error, its out-of-sample MSE is significantly lower, showing improved performance on unseen data. Furthermore, adding a monthly component in the MD_NNT_MV allows it to achieve the lowest MSE in and out-of-sample. The neural network-based models better achieve the interval forecast results. Undercover is prevalent in the BENTH model and significantly differs from the reference value. On the other hand, the difference is significantly reduced in the NNT_MV model and with further improvement in the MD_NNT_MV model. About interval width, the NNT_MV model produces slightly wider intervals than the BENTH model, which suggests a more careful and better-calibrated treatment of uncertainty. This increase in width is not necessarily a limitation; instead, it reflects the model's ability to recognise a higher level of uncertainty in its predictions. As a result, coverage improves, as shown by the less deviation from the target confidence level. By widening its prediction intervals, the NNT_MV model avoids underestimating the uncertainty characterising the BENTH model. On the other hand, the MD_NNT_MV model further enhances coverage while reducing the average width of its prediction intervals, highlighting its ability to efficiently capture uncertainty, yielding well-calibrated narrower intervals than those produced by the other models. Figure 1 presents a comparative visualisation of the observed daily mean temperatures for 2024, represented by the black line, alongside the forecasts generated by the BENTH and NNT_MV models. The predicted mean temperatures are illustrated with solid lines—blue for the BENTH model, orange for the NNT_MV model and green for the MD_NNT_MV model. The shaded areas surrounding these lines also indicate the corresponding 95% prediction intervals, effectively capturing the models' uncertainty estimation.

Table 2. performance of the different models in terms of *MSE*, $|PICP - \alpha_{\mathrm{conf}}|$ and *MPIW*, in sample and out of sample; The $|PICP - \alpha_{\mathrm{conf}}|$ values represent the absolute difference between the observed PICP and the reference value $\alpha_{\mathrm{conf}} = 0.95$.

Model	MSE		$\lvert PICP - \alpha_{\mathrm{conf}}\rvert$		MPIW	
	In-Sample	Out-of-Sample	In-Sample	Out-of-Sample	In-Sample	Out-of-Sample
BENTH	5.3999	6.2106	0.0103	0.0486	8.7073	8.7073
NNT_MV	5.4900	5.8002	0.0019	0.0158	8.9531	8.9531
MD_NNT_MV	4.9794	5.1757	0.0014	0.0075	8.5600	8.5600

For each model, the predicted temperature is given together with a confidence interval; thus, the plots are colour-coded to compare predictions and their uncertainty band easily. Both neural models significantly outperform the BENTH model in modelling seasonal dynamics and short-term fluctuations, capturing high variability in observed temperature values. Moreover, the NNT_MV already performs better than the BENTH; still, the extension MD_NNT_MV outperforms, intensely smoothing the curve and improving alignment with the observed values. Empirical evidence shows that, concerning uncertainty quantification, the confidence intervals for the neural network models are, on average, more sensitive and thinner due to the greater flexibility of the proposed approach.

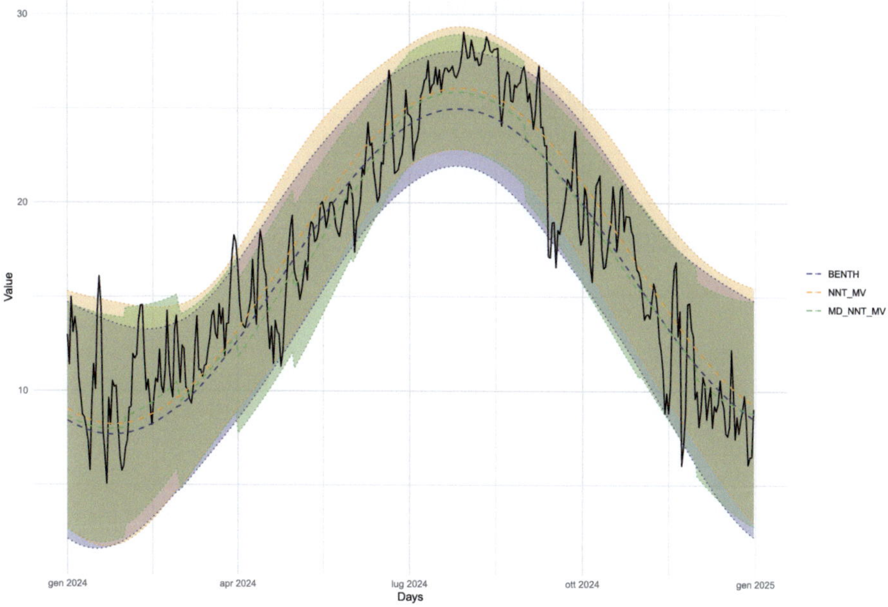

Fig. 1. Forecast of daily mean temperature for 2024 with 95% prediction intervals using the BENTH, NNT_MV and MD_NNT_MV models.

5.2 Weather Derivatives Pricing

Let us implement the futures price written on the HDD across the three models: the BENTH model and two neural network models (NNT_MV and MD_NNT_MV). Following the approach proposed in [1], we specify the market price of risk $\gamma = 0$, and we compute the constant temperature volatility, defined as the standard deviation of the residuals from the respective temperature models, and finally, we estimated the mean reverting parameters as detailed in Sect. 4.1. We focused on temperature data from the Fiumicino Maccarese weather station with a valuation date of January 1, 2023, for contracts covering the winter period from December 15, 2023, to March 15, 2024. Table 3 reports the HDD-futures prices computed using the pricing formula presented in 2, we also include the corresponding price obtained under the constant volatility assumption for each model.

For each model, the futures price under time-varying volatility is lower than it's computed under constant volatility. This suggests that explicitly modelling the temporal structure of volatility, such as seasonal fluctuations, reduces the expected value of accumulated HDDs. Thus, future prices avoid overestimating the contribution of extreme temperature deviations. When comparing the time-varying results across models, the BENTH model produces the highest futures price, while the neural network-based models yield significantly lower values. The MD_NNT_MV model achieves an intermediate price between BENTH and

Table 3. HDD-futures prices estimated with time-varying volatility and under constant volatility.

Model	HDD-futures price	
	Time-varying volatility	Constant volatility
BENTH	858.5348	858.9041
NNT_MV	806.4863	807.2834
MD_NNT_MV	813.6026	814.4010

NNT_MV, aligns with the performance metrics results presented in the previous section. This model achieved the lowest Mean Squared Error and outperformed all other models in interval forecast accuracy, with the smallest divergence from the 95% confidence level and the narrowest MPIW. The MD_NNT_MV HDD-futures price appears justified, resulting from a more precise uncertainty quantification, and it can be viewed as a more adaptable tool for pricing weather derivatives.

6 Conclusions

This study presents a novel neural network-based temperature model for pricing weather derivatives, which jointly models the conditional mean and variance of daily average temperature. Our approach enhances Benth's framework by directly modelling volatility on temperature data using a single optimization scheme. The network architecture replicates the structure proposed in [1] at the output layer, ensuring transparency and interpretability, which is crucial for real-world applications where explainability is essential. We conduct numerical experiments using data from the Fiumicino Maccarese weather station to demonstrate our model's effectiveness. It outperforms the traditional approach in both point and interval forecasting. Additionally, our model is applied to pricing HDD-futures, showing improved precision and robustness in derivative valuation through joint modelling of conditional moments. Future developments will focus on incorporating altitude as an input feature, and exploring architectures like recurrent neural networks or LSTM networks to better capture temporal dependencies in temperature data.

Acknowledgments. This Work is Funded by European Union - NextGenerationEU funded by the European Union - NextGenerationEU under the National Recovery and Resilience Plan (PNRR) M42 program MUR PRIN 2022 "Modeling and valuation of financial instruments for climate and energy risk mitigation" - proposal code [2022FPLY97] CUP[I53D23002700006].

References

1. Benth, F., Benth, J.: The volatility of temperature and pricing of weather derivatives. Quant. Finance **7**(5), 553–561 (2007). https://doi.org/10.1080/14697680601155334
2. Hung-Hsi, H., Yung-Ming, S., Pei-Syun, L.: HDD and CDD option pricing with market price of weather risk for Taiwan. J. Futures Markets: Futures, Options, Other Derivative Products **28**(8), 790–814 (2008). https://doi.org/10.1002/fut.20337
3. Alaton, P., Djehiche, B. and Stillberger, D.: On modelling and pricing weather derivatives. Appl. Math. Finance **9**(1), 1–20 (2002). https://doi.org/10.1080/13504860210132897
4. Brody, D.C., Syroka, J., Zervos, M.: Dynamical pricing of weather derivatives. Quant. Finance **2**(3), 189 (2002). https://doi.org/10.1088/1469-7688/2/3/302
5. Zapranis, A. and Alexandridis, A.: Modelling the temperature time-dependent speed of mean reversion in the context of weather derivatives pricing. Appl. Math. Finance **15**(4), 355–386 (2008). https://doi.org/10.1080/13504860802006065
6. Nix, D.A., Weigend, A.S.: Estimating the mean and variance of the target probability distribution. In: Proceedings of 1994 IEEE International Conference on Neural Networks, vol. 1, pp. 55–60. IEEE (1994). https://doi.org/10.1109/ICNN.1994.374138
7. Guo, C., Berkhahn, F.: Entity embeddings of categorical variables (2016), arXiv preprint arXiv:1604.06737

Modeling Health and Disability Trajectories in Later Life: A Multi-state Approach Using HRS Data

Domenico De Giovanni[1], Massimiliano Menzietti[2], Marco Pirra[1], and Fabio Viviano[1](✉)

[1] Department of Economics, Statistics and Finance, University of Calabria, Ponte Bucci Cubo 0C, 87036 Rende (Cosenza), Italy
{ddegiovanni,marco.pirra,fabio.viviano}@unical.it

[2] Department of Economics and Statistics, University of Salerno, Via Giovanni Paolo II, 84084 Fisciano (Salerno), Italy
mmenzietti@unisa.it

Abstract. This paper develops a continuous-time multi-state Markov model to analyze health and disability transitions in later life using data from the Health and Retirement Study (HRS). We define five distinct states combining health and functional status, allowing for 12 possible transitions including recovery and death. Transition intensities are estimated via a proportional hazards framework, incorporating individual-specific and time-dependent covariates. Model selection is guided by likelihood-based criteria, and results highlight significant heterogeneity by age, gender, and baseline condition. We derive one-year transition probabilities and compute dynamic life expectancies by initial state. Our findings reveal persistent disparities in ageing trajectories and underscore the importance of distinguishing between health and disability dimensions. The model offers practical insights for projecting future health burdens and designing long-term care strategies.

Keywords: Multi-state Markov models · Functional disability · Health conditions · Health and Retirement Study

1 Introduction

Understanding the dynamics of health transitions and their implications for longevity is essential for the design of sustainable health, pension, and long-term care systems. In ageing populations, individuals often experience complex trajectories involving both morbidity and functional decline, which cannot be captured by conventional survival models. To address this, we develop and estimate a continuous-time multi-state Markov model that jointly captures transitions across health and disability states and ultimately to death.

Using rich panel data from the Health and Retirement Study (HRS), we specify a five-state framework that distinguishes between good and ill health as well

M. La Rocca et al. (Eds.): WMAF 2025, *New Perspectives in Mathematical and Statistical Methods for Actuarial Sciences and Finance*, pp. 108–119, 2025.
https://doi.org/10.1007/978-3-032-05551-4_10

as functional disability. The model incorporates covariate-dependent transition intensities using a proportional hazards structure, allowing for heterogeneity by age, gender, and calendar time. We examine state-specific life expectancies, transition patterns, and the evolving burden of poor health and disability from age 65 onward. Our findings offer important insights into individual ageing processes and their implications for health expectancy and care planning. The remainder of the paper is organized as follows. Section 2 introduces the multi-state model structure and the associated transition intensity framework. Section 3 describes the dataset and outlines the methodological approach used for estimation. Section 4 presents the numerical results, including model calibration, transition probabilities, and life expectancy estimates. Section 5 concludes with a discussion of key findings and policy implications.

2 The Multi-state Model

We adopt a continuous-time, finite-state Markov model to represent individual transitions across joint health and disability states in later life. The structure, similarly to that of Sherris and Wei [6], comprises five mutually exclusive states:

- State 1: Good health and non-disabled ($\mathrm{H\bar{D}}$)
- State 2: Ill health and non-disabled ($\mathrm{I\bar{D}}$)
- State 3: Good health and disabled (HD)
- State 4: Ill health and disabled (ID)
- State 5: Death (absorbing state)

This state structure allows for a nuanced depiction of health trajectories by separating health status (good vs. ill) from disability status (disabled vs. non-disabled). The model permits 12 transitions, capturing deterioration, partial or full recovery, and death. Transitions from the ill-health state to the good health state ($\mathrm{ID} \rightarrow \mathrm{H\bar{D}}$ and $\mathrm{ID} \rightarrow \mathrm{HD}$) are not allowed, as having experienced a serious illness is by nature considered an irreversible condition.

2.1 Health and Disability Groupings

For interpretability and analysis, we define several overlapping state groupings:

- **Alive states:** All transient (non-death) states: $\{\mathrm{H\bar{D}}, \mathrm{I\bar{D}}, \mathrm{HD}, \mathrm{ID}\}$
- **Non-disabled:** $\{\mathrm{H\bar{D}}, \mathrm{I\bar{D}}\}$
- **Disabled:** $\{\mathrm{HD}, \mathrm{ID}\}$
- **Good health:** $\{\mathrm{H\bar{D}}, \mathrm{HD}\}$
- **Ill health:** $\{\mathrm{I\bar{D}}, \mathrm{ID}\}$

These classifications enable targeted analysis of specific health-dependency combinations relevant for ageing and long-term care research (Figs. 2 and 3).

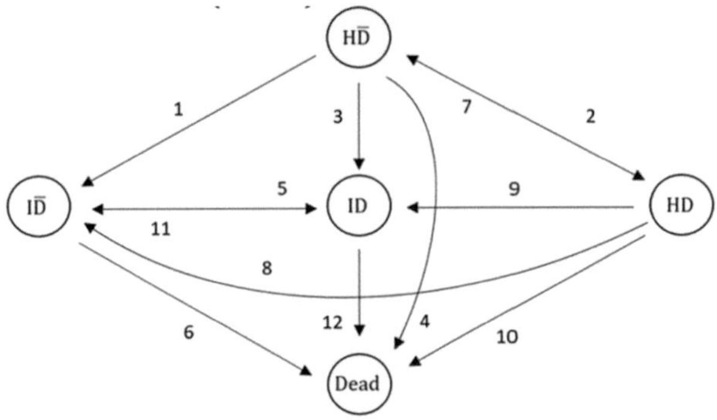

Fig. 1. Full transition diagram among the five states. Arrows indicate allowed transitions; numbering corresponds to unique transition types used in estimation.

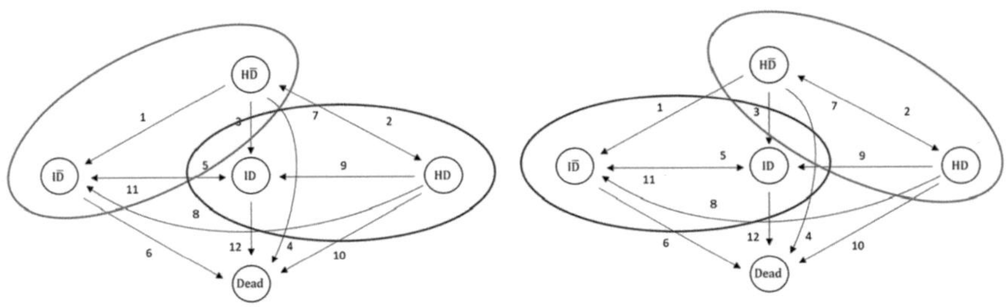

Fig. 2. State groupings: Left panel shows non-disabled (left circle) vs. disabled (right circle) states; right panel shows good health (right circle) vs. ill health (left circle) states. These classifications are used for subgroup-specific life expectancy and transition analysis.

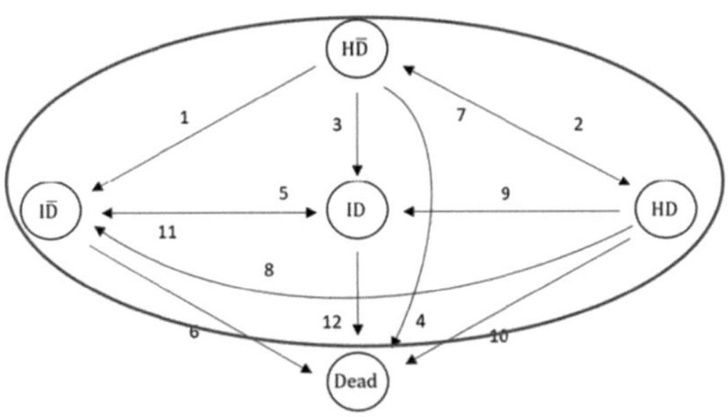

Fig. 3. Overlay of all four alive states used to define transitions and subgroup trajectories over time. This representation serves as a unified framework for aggregating life expectancy components.

2.2 Modeling Assumptions and Transition Intensities

Let $S(t)$ denote the state occupied by an individual at time t. The transitions between states are governed by transition intensities $q_{rs}(t, X(t))$, which represent the instantaneous risk of moving from state r to s, potentially depending on some covariates $X(t)$:

$$q_{rs}(t, X(t)) = \lim_{\delta_t \to 0} \frac{P(S(t+\delta_t) = s \mid S(t) = r)}{\delta_t} \tag{1}$$

These intensities populate the generator matrix $Q(t, X(t))$, where each row sums to zero:

$$q_{rr} = -\sum_{s \neq r} q_{rs} \tag{2}$$

We assume a Markov process, meaning that the transition intensities are independent of the history of the process:

$$q_{rs}(t, X(t), \mathcal{F}_t) = q_{rs}(t, X(t)) \tag{3}$$

In the time-homogeneous case, sojourn times are exponentially distributed, and the probability of transitioning from state r to s is proportional to $-q_{rs}/q_{rr}$. For the time-inhomogeneous case, intensities may vary with calendar time, age, or other covariates.

In this paper, we assume that transition intensities $q_{rs}(t, X(t))$ are piecewise constant. We also allow the model to incorporate covariate information on transition rates. To this end, we adopt a proportional hazards specification for covariate-dependent intensities. For each transition $r \to s$, the intensity is modeled as:

$$q_{rs}(t, X(t)) = q_{rs}^{(0)} \exp(\beta_{rs}^T X(t)). \tag{4}$$

This allows the transition process to flexibly respond to individual-specific and time-varying characteristics.

We implement estimation using the `msm` package in `R`, which supports likelihood-based estimation of multi-state models, matrix exponential computation of transition probabilities, and standard hypothesis tests for covariate effects [3,4].

3 Data and Methodology

The Health and Retirement Study (HRS), conducted by the University of Michigan, is a nationally representative longitudinal panel study of older Americans. Funded by the National Institute on Aging and the Social Security Administration, the survey has been collecting biennial data on individuals aged 50 and above since 1992. The HRS includes a broad array of information covering demographics, socioeconomic status, income, assets, employment, health conditions, healthcare expenditures, and family structure.

A particular strength of the HRS is its rich set of variables related to health conditions and functional status. Health indicators include binary responses to diagnoses of high blood pressure, diabetes, cancer (excluding skin cancer), chronic lung diseases, cardiovascular conditions, psychiatric disorders, and arthritis. Functional limitations are assessed through self-reported difficulties in Activities of Daily Living (ADLs), such as walking across a room, dressing, bathing, eating, getting in or out of bed, and using the toilet. For our analysis, we recode the ADL indicators into a binary measure (0 = no limitations; 1 = at least one limitation), aggregated across the six tasks. Table 1 lists the selected variables from the RAND HRS Longitudinal File used in the analysis.

Table 1. Subset of selected variables from the RAND HRS Longitudinal File

Variable	Description
HHIDPN	Household and person identifier
RAGENDER	Gender
RABMONTH, RABYEAR	Birth month and year
RADMONTH, RADYEAR	Death month and year
WAVE	Interview wave number
RxIWSTAT, RxIWENDM, RxIWENDY	Interview status and end date
RxAGEM, RxAGEY	Age at interview (months and years)
RxWALKR – RxTOILT	ADL difficulty indicators
RwHIBPE – RwARTHRE	Chronic health condition indicators

Following established literature [2,5], we classify individuals into discrete health states: those reporting at most one ADL limitation are categorized as being in the *Non-disabled* state; those with two or more limitations are classified as *Disabled*. In addition, we consider individuals to be in *Ill health* if they ever had at least one of the following illnesses: diabetes, cancer, lung disease, heart problems and stroke. Otherwise, they are considered in *Good health*. Mortality data are used to identify transitions into the absorbing *Death* state.

3.1 Sample and Transition Data

For the present study, we use data from wave 3 (1998) to wave 14 (2020) of the HRS. Earlier waves are excluded due to inconsistencies in the formulation of ADL-related questions. The most recent data available at the time of analysis is from the 2020 release of the RAND HRS Longitudinal File (Version 2, May 2024). The final cleaned dataset comprises 24703 individuals, of whom 57% are female. Table 2 provides descriptive statistics for each observed transition between health and disability states in the five-state model introduced in Sect. 2. Each row corresponds to a unique transition as defined in the transition diagram (Fig. 1).

Table 2. Descriptive statistics on the transitions

Transition	N observed	% Female	Age Female	StDev	Age Male	StDev
1 → 2	6006	55.7%	66.30	10.64	66.26	9.42
1 → 3	1724	68.9%	70.82	13.55	67.76	12.11
1 → 4	532	63.7%	73.78	13.18	69.48	12.16
1 → 5	1652	52.2%	77.34	12.14	73.39	11.42
2 → 4	4931	61.0%	73.07	11.86	72.86	11.29
2 → 5	5778	47.3%	77.07	10.98	76.32	10.03
3 → 1	1091	68.8%	66.26	12.01	63.27	10.06
3 → 2	199	68.3%	67.87	11.51	64.73	10.16
3 → 4	349	72.2%	72.35	14.02	72.05	12.70
3 → 5	624	69.1%	85.08	11.54	78.72	11.93
4 → 2	2785	64.3%	68.84	11.59	68.39	10.82
4 → 5	3637	60.6%	81.76	10.73	79.28	10.55

The table reports the number of transitions observed (*N observed*), the proportion of female individuals (% Female), and the average age and standard deviation for males and females at the time of transition. These statistics offer insight into the demographic and age-related characteristics associated with different transition pathways. Transitions from state 1 to 3 (onset of disability in good health) and from 3 to 1 (recovery) exhibit notably higher proportions of female individuals, suggesting that disability episodes and their resolution may be more prevalent or better captured among women. Transitions to the absorbing state 5 (Death), particularly from states 3 and 4, are associated with higher average ages, consistent with expected end-of-life patterns. Moreover, transitions from more severe states (e.g., 3→4, 4→5) tend to involve older populations and show substantial gender differences in age at transition, further emphasizing heterogeneity in ageing and decline trajectories.

3.2 Estimation of Transition Intensities

We estimate continuous-time transition intensities across five health states using the `msm` package in R [3]. This package facilitates the fitting of general multi-state models to panel data and computes the implied transition probabilities via matrix exponentiation of the estimated intensity matrix. The model accommodates covariate effects on transition intensities and allows for specification of different covariate sets by transition. Model selection, transition probability estimation, and subsequent derivation of life expectancies are discussed in the following sections.

4 Numerical Analysis

The transition intensity model applied in the analysis follows a proportional hazards specification, allowing for covariate-dependent transition rates between states. Specifically, the transition intensity from state r to state s for individual i at time t, given covariates $X^i(t)$, is defined as:

$$q_{rs}(t, X^i(t)) = q_{rs}^{(0)} \exp\left\{\beta_{rs}^{\text{gender}} G_i + \beta_{rs}^{\text{age}} x_i(t) + \beta_{rs}^{\text{time}} t\right\},$$

where G_i denotes the gender indicator, $x_i(t)$ represents age at time t, and $\beta_{rs}^{\cdot}$ are transition-specific coefficients. This formulation accommodates both time-invariant and time-varying covariates, enabling flexible modeling of heterogeneous health trajectories.

4.1 Model Comparison and Selection

To assess the relative performance of different model specifications, we estimated a sequence of nested models with progressively richer covariate structures. Table 3 reports the log-likelihood values, degrees of freedom, and Akaike Information Criterion (AIC) scores for each specification.

Table 3. Model comparison: log-likelihoods, degrees of freedom (df), and AIC.

	logliks	df	AIC
Model 0: $q_{rs}(t, X^i(t)) = q_{rs}^{(0)}$	−97688.18	12.00	195400.40
Model 1: $q_{rs}(t, X^i(t)) = q_{rs}^{(0)} \exp\left\{\beta_{rs}^{\text{gender}} G_i\right\}$	−97442.96	24.00	194933.90
Model 2: $q_{rs}(t, X^i(t)) = q_{rs}^{(0)} \exp\left\{\beta_{rs}^{\text{gender}} G_i + \beta_{rs}^{\text{age}} x_i(t)\right\}$	−92667.21	36.00	185406.40
Model 3: $q_{rs}(t, X^i(t)) = q_{rs}^{(0)} \exp\left\{\beta_{rs}^{\text{gender}} G_i + \beta_{rs}^{\text{age}} x_i(t) + \beta_{rs}^{\text{time}} t\right\}$	−92567.80	48.00	185231.60
Model 4: Covariates selection by transition	−92571.79	36.00	185215.60

Among the specifications, **Model 4**, which allows transition-specific covariate effects, achieves the lowest AIC. This model is therefore retained for the main empirical analysis.

4.2 Covariate Selection by Transition

The adopted specification allows different sets of covariates to influence the hazard of each transition. Table 4 details the covariates included for each transition. Most transitions are influenced by gender and age, while calendar time is incorporated selectively where temporal variation in transition intensities is empirically supported.

This structure offers a balanced trade-off between parsimony and flexibility, reflecting the heterogeneous nature of health transitions observed in later life.

Table 4. Covariates selected for each transition

Transition	Covariates
$1 \to 2$	Gender + Age
$1 \to 3$	Gender + Age
$1 \to 4$	Age + time
$1 \to 5$	Gender + Age + time
$2 \to 4$	Gender + Age + time
$2 \to 5$	Gender + Age + time
$3 \to 1$	Gender + Age
$3 \to 5$	Gender + Age + time
$4 \to 2$	Gender + Age
$4 \to 5$	Gender + Age

4.3 One-Year Transition Probabilities at Age 65

Table 5 and Table 6 report the estimated 1-year transition probabilities across health and disability states for males and females, respectively, at the baseline age of 65 in the year 2024. These probabilities reflect the likelihood of transitioning from one state to another within a single year and are derived from the fitted multi-state model incorporating covariate effects.

Table 5. 1-year transition probabilities for males (age 65, 2024)

Male	$H\bar{D}$	$I\bar{D}$	HD	ID	Dead
$H\bar{D}$	0.9328	0.0444	0.0130	0.0013	0.0085
$I\bar{D}$	0.0000	0.9510	0.0000	0.0299	0.0191
HD	0.2202	0.0235	0.6302	0.0554	0.0706
ID	0.0000	0.1877	0.0000	0.7111	0.1013
Dead	0.0000	0.0000	0.0000	0.0000	1.0000

For both sexes, the highest persistence is observed in the initial state, with healthy non-disabled individuals ($H\bar{D}$) having over a 93% chance of remaining in that state after one year. Transition probabilities to death are relatively low across all non-absorbing states, consistent with baseline survival expectations at age 65. Differences between males and females emerge particularly in the transitions from the disabled states (HD, ID), where females tend to exhibit slightly higher probabilities of remaining in their current state and lower mortality risks, consistent with their generally higher longevity. These probabilities provide the foundation for projecting future state distributions and estimating life expectancy metrics.

Table 6. 1-year transition probabilities for females (age 65, 2024)

Female	H$\bar{\text{D}}$	I$\bar{\text{D}}$	HD	ID	Dead
H$\bar{\text{D}}$	0.9419	0.0352	0.0161	0.0015	0.0053
I$\bar{\text{D}}$	0.0000	0.9434	0.0000	0.0416	0.0151
HD	0.2066	0.0220	0.6680	0.0587	0.0448
ID	0.0000	0.1765	0.0000	0.7525	0.0711
Dead	0.0000	0.0000	0.0000	0.0000	1.0000

4.4 Dynamic Patterns of Transition and Survival

Figures 4, 5 and 6 illustrate the projected dynamics of health status and survival for 65-year-old males, conditional on their initial health and disability state. These visualizations capture both the expected longevity and the evolving structure of life lived in different states over the projection horizon 2024–2069. The results for the female population and other ages can be easily derived from the developed model and exhibit similar patterns.

Figure 4 traces the evolution of selected 1-year transition probabilities on a logarithmic scale for a 65-year-old male. Transition risks generally remain low, particularly for transitions into death from healthy states. However, mortality and deterioration risks are substantially elevated among those initially in disability, and for some states the probabilities exhibit gradual temporal shifts. Notably, the death probability for individuals in state ID slightly declines over the projection horizon, possibly reflecting anticipated medical advances or cohort effects.

To compute the expected time spent in a subset S of the four transient states before absorption, i.e. $S \subseteq \{1, 2, 3, 4\}$, we use a forward approach based on the product of stochastic transition matrices. In our setting, transition probabilities are derived from time-dependent transition intensities $q_{rs}(t, X^i(t))$, estimated via the Cox proportional hazards model. These intensities define a continuous-time Markov process with generator matrix:

$$\mathbf{Q}(t, X^i(t)) = [q_{rs}(t, X^i(t))]$$

To obtain discrete-time transition matrices over yearly intervals, we apply the matrix exponential:

$$\mathbf{P}(t) \approx \exp\left(\mathbf{Q}(t, X^i(t)) \cdot \varDelta t\right), \quad \varDelta t = 1$$

Thus, the transition matrices $\mathbf{P}(t)$ used in the forward computations are obtained by exponentiating the estimated generator matrices.

Once the discrete-time transition matrices $\mathbf{P}(t)$ are obtained, the state occupancy probabilities at each future time t can be computed recursively from an initial distribution $\boldsymbol{\pi}(t_0)$ using matrix multiplication:

$$\boldsymbol{\pi}(t) = \boldsymbol{\pi}(t_0) \cdot \mathbf{P}(t_0) \cdot \mathbf{P}(t_0 + 1) \cdots \mathbf{P}(t - 1)$$

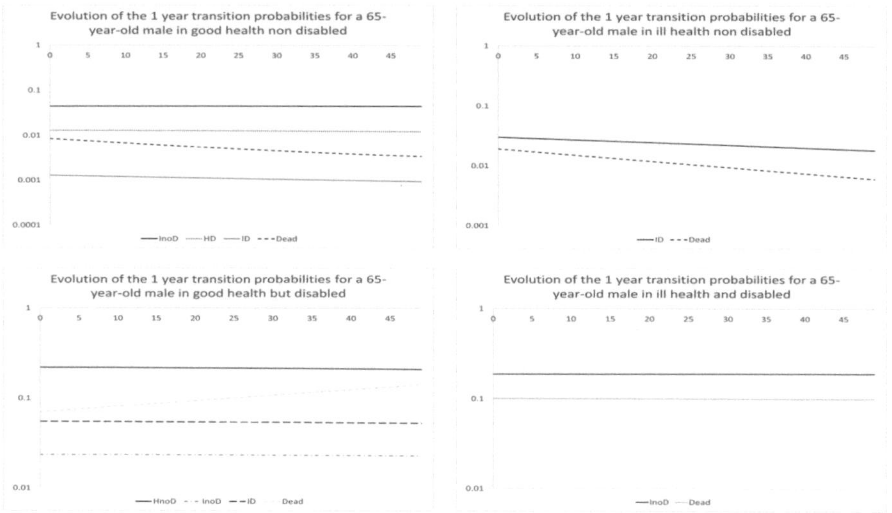

Fig. 4. Evolution of the 1-year transition probabilities (log scale) for a 65-year-old male across different health and disability states.

The expected time spent in a subset of transient states $S \subseteq \{1, 2, 3, 4\}$ before absorption (into state 5) is then given by:

$$T_S(i, t_0) = \sum_{t=t_0}^{T-1} \sum_{j \in S} \pi_j(t),$$

where T is the time of last projection.

Figure 5 shows the evolution of total and state-specific life expectancies. As expected, individuals in good health and not disabled ($H\bar{D}$) at baseline enjoy the longest total life expectancy, with the majority of remaining life spent in favorable health states. For other initial states, although total life expectancy increases slightly over time, the gains are concentrated in less favorable states (e.g., disabled or in ill health), particularly for those starting in HD or ID. The separation of LE(H) and LE($\bar{D}$) trajectories underscores the importance of jointly considering both health and functional status when assessing morbidity compression.

Finally, Fig. 6 provides a complementary view, decomposing the total expected remaining life into proportions spent in each state. The stacked area charts reveal that the share of life spent in the healthy and non-disabled state ($H\bar{D}$) is highest for those who begin in that state, but it declines steadily with age. Conversely, individuals starting in ill health or with disability (ID) spend a growing proportion of their remaining life in adverse states. This stratification highlights persistent disparities in ageing trajectories by initial condition, despite universal survival improvements over time.

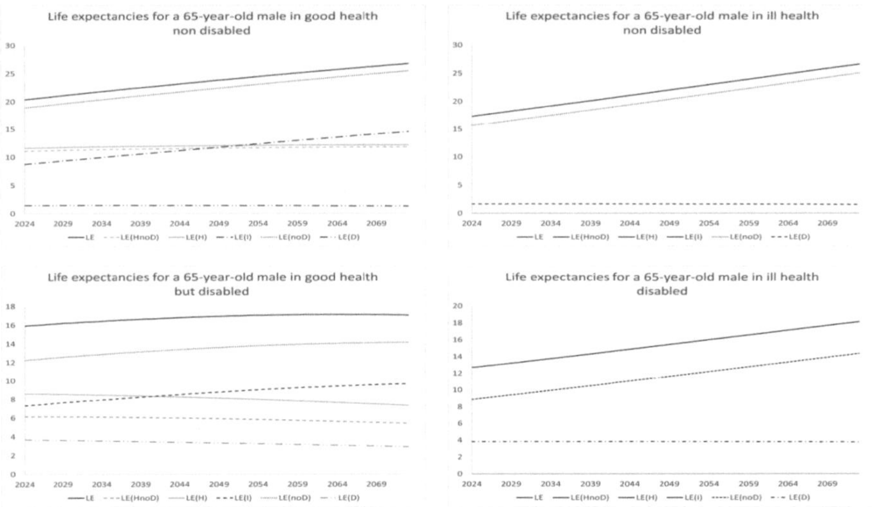

Fig. 5. Life expectancies from age 65 for males under different health and disability states (2024–2069).

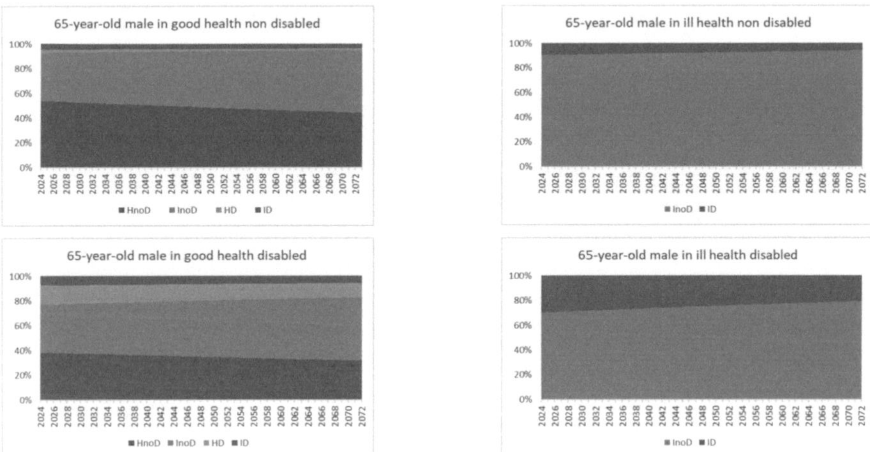

Fig. 6. The distribution of life expectancy at age 65 for males in different alive states.

Together, these figures emphasize the differentiated ageing pathways shaped by initial health and disability status and demonstrate how transition models can be leveraged to project both survival and quality of life measures across time.

5 Concluding Remarks

This study presents a comprehensive multi-state model of health and disability dynamics in later life, estimated using high-quality longitudinal data from the HRS. By distinguishing between multiple alive states defined by combinations of health status and disability, we provide a detailed picture of the transitions that older individuals are likely to experience and how these patterns vary with individual characteristics.

The results underscore the importance of accounting for both health and functional limitations when assessing life expectancy and the expected distribution of time spent in various health states. In particular, initial health and disability status play a critical role in shaping future trajectories, highlighting persistent disparities in healthy ageing.

Our framework can be used to inform projections of care needs, long-term insurance product design, and the evaluation of interventions aimed at extending healthy life expectancy. Future work may incorporate further refinements such as stochastic duration effects, more granular disease classifications, or the integration of healthcare utilization measures.

References

1. Cox, D.R.: The Theory of Stochastic Processes. Routledge, London (2017)
2. Fong, J.H., Shao, A.W., Sherris, M.: Multistate actuarial models of functional disability. North Am. Actuar. J. **19**(1), 41–59 (2015)
3. Jackson, C.: Multi-state models for panel data: the msm package for R. J. Stat. Softw. **38**, 1–28 (2011)
4. Marshall, G., Jones, R.H.: Multi-state models and diabetic retinopathy. Stat. Med. **14**(18), 1975–1983 (1995)
5. Shao, A.W., Sherris, M., Fong, J.H.: Product pricing and solvency capital requirements for long-term care insurance. Scand. Actuar. J. **2017**(2), 175–208 (2017)
6. Sherris, M., Wei, P.: A multi-state model of functional disability and health status in the presence of systematic trend and uncertainty. North Am. Actuar. J. **25**(1), 17–39 (2021)

Backtesting Expected Shortfall for Bitcoin: A Joint Combined LSTM-Based Approach

Giovanni De Luca, Anna Pia Di Iorio(✉), and Andrea Montanino

University of Naples Parthenope, Naples, Italy
annapia.diiorio001@studenti.uniparthenope.it

Abstract. This work aims to identify the most accurate model in passing the joint-combined backtesting procedure for Value-at-Risk and Expected Shortfall forecasts for Bitcoin. First, GARCH and Markov Switching GARCH are estimated and used to forecast the corresponding VaR and ES. Next, the Long Short-Term Memory model is applied to refine these risk measures. Finally, four models (GARCH, Markov-Switching GARCH, Joint-Combined, Long-Short Term Memory Joint-Combined) are compared based on average loss and backtesting performances. Results suggest that the LSTM-Joint-Combined model apparently represents the best model delivering the lowest average predictive loss across the evaluated settings. Furthermore, it considerably enhances the efficacy of the JC approach.

Keywords: Backtesting · Cryptocurrencies · Expected Shortfall · Joint-Combined Regression · LSTM · Markov-Switching GARCH

1 Introduction

Since Nakamoto's Manifesto (Nakamoto, 2008) the cryptocurrency market has boomed, gaining the attention of the entire financial industry, and reaching a global market cap of $3.45 trillion as of May 20, 2025 (CoinGecko, 2025). Bitcoin is undoubtedly the cryptocurrency with the highest market capitalization.

Several studies argue that Bitcoin does not exhibit the characteristics of a real currency (Cheah and Fry, 2015; Blau, 2017; Li et al. 2019). Instead, it is considered more a technology-based product due to its decentralized nature (White et al., 2020) or an asset used both for portfolio diversification (Ahsan et al., 2024; Tarchella et al. 2024), and speculative strategies (Glaser et al.2014; Baur et al. 2018; Krause 2025).

Bitcoin's price has repeatedly exhibited signs of irrational euphoria (Pavlidis 2023; Corbet et al. 2018), and high volatility (Katsiampa2017; Balcilar et al. 2017; Bariviera 2017; Antar 2025; Zhou et al.2025). Therefore, it is necessary to accurately model and forecast its volatility, and to develop suitable methodologies for its measurement too.

M. La Rocca et al. (Eds.): WMAF 2025, *New Perspectives in Mathematical and Statistical Methods for Actuarial Sciences and Finance*, pp. 120–131, 2025.
https://doi.org/10.1007/978-3-032-05551-4_11

Moreover, equally noteworthy is the measuring of extreme losses (Artzner et al. 1999) as well as their validation through backtesting procedures. In fact, without backtesting, there is no empirical evidence that the model accurately captures the risk. Widely used metrics for risk measurement are the Value at Risk (VaR) and the Expected Shortfall (ES). The VaR allows to estimate the maximum potential loss of a given portfolio over a specific time period, at a given confidence level (Morgan et al. 1997), but it violates the subadditivity principle. So the ES or Conditional VaR (CVaR) allows to estimates the average loss in the worst-case scenarios (those beyond the VaR) making it the primary risk measure within the market risk framework defined by the Basel Accords Basel Committee on Banking Supervision (2019).

Many studies have focused on the development of risk measurement methods and their validation techniques. Kupiec et al. (1995) developed the Proportion of Failure (POT) test (also called unconditional coverage test) to assess whether the frequency of VaR breaches is consistent with the declared confidence level. Christoffersen (1998) extended the Kupiec test by also checking whether VaR breaches are temporally independent, the so called conditional coverage test. However, ES is not characterized by the elicitability. According to Osband (1985), a functional is said to be elicitable if there exists a scoring function that is strictly consistent for it. Specifically, ES is not the solution to a traditional loss minimization problem (Gneiting 2011), and no scoring functions are available that are consistent with it.

More recently, Fissler and Ziegel (2016) established that the pair VaR - ES is jointly elicitable under mild regularity conditions, extending the concept of elicitability to higher-order (multivariate) functionals. They have shown that a functional is elicitable when considering the pair consisting of the quantile and the ES at the same probability level. The general form of the joint scoring function for VaR and ES is defined as follows:

$$\begin{aligned} S(y, v, e) = {} & (I\{y \leq v\} - \alpha)G_1(v) - I\{y \leq v\}G_1(y) + \\ & + G_2(e)\left(e - v + \frac{(v-y)I\{y \leq v\}}{\alpha}\right) - G_2(e) + a(y) \end{aligned}$$

where y denotes a random variable, v is an estimate of the VaR, and e is an estimate of the ES. Precisely, when the function G_1 is twice continuously differentiable, G_2 is three times continuously differentiable, with $G_2' = G_2$, the derivative G_2' is strictly positive, and both G_1 and G_2 are strictly increasing and integrable functions, then a strictly consistent scoring function S can be constructed. This class of functions is known as the *Fissler-Ziegel family*, and they were proposed to ensure joint consistency in the evaluation of VaR and ES forecasts. Given a probability level $\alpha \in (0, 1)$, as shown below:

$$\begin{cases} Y = X_q' \theta_0^q + \mu^q \\ Y = X_e' \theta_0^e + \mu^e \end{cases}$$

where $Q_\alpha(\mu^q|X) = 0$ and $ES_\alpha(\mu^e|X) = 0$, the model is parameterized by $\theta_0 = (\theta_0^{q\prime}, \theta_0^{e\prime})' \in \Theta \subset \mathbb{R}^k$, where the parameter space Θ is compact, convex, and has a non-empty interior, i.e., $\text{int}(\Theta) \neq \emptyset$.

So, the conditional VaR and conditional ES of a random variable, given the covariates X, are expressed as follows:

$$\begin{cases} Q_\alpha(Y|X) & = X_q'\theta_0^q \\ ES_\alpha(Y|X) & = X_e'\theta_0^e \end{cases}$$

Bayer and Dimitriadis (2022), based on Mincer and Zarnowitz (1969) method, proposed regression-based tests that uniquely allow for evaluating the ES forecasts independently, overcoming challenges due to ES non-elicitability.

In particular, they proposed a joint regression framework to model the conditional ES at level τ as a linear function, $ES(Y_t|\mathcal{F}_{t-1}) = \gamma_1 + \gamma_2\hat{e}_t$, and the conditional quantile as $Q(Y_t|\mathcal{F}_{t-1}) = \beta_1 + \beta_2\hat{p}_t$, jointly estimating the regression parameters (β, γ) using the Fissler and Ziegel joint scoring function for ES and VaR.

Associated with cryptocurrencies analyses, Gkillas and Katsiampa (2018), using Extreme Value Theory (EVT), estimated the VaR and ES as risk measures for five of the most prominent cryptocurrencies. Taylor (2019) employed quantile regression to estimate the ES from Value at Risk forecasts, using maximum likelihood based on an asymmetric Laplace (AL) density, while Caporale and Zekokh (2019) determined the most accurate models for estimating the volatility of four major cryptocurrencies. They included Markov-Switching GARCH (MS-GARCH) variants, to predict VaR and Expected Shortfall ES with the aim to identify the best-performing models using rigorous backtesting and the Model Confidence Set (MCS) procedure (Hansen et al. 2011).

To the best of our knowledge, the work of Lu et al. (2021) represents the only initiative in the academic literature aimed at predicting cryptocurrency risk measures using a combined LSTMJoint Combined regression approach. Therefore, this paper, building on that work, wants to identify the most effective model, among a selected set, in passing VaR and Expected Shortfall ES backtesting, while accounting for potential regime shifts in the Bitcoin time series. Indeed, standard GARCH models may yield non-robust estimates in the presence of regime changes in the historical series (Bauwens et al. 2014; 2010). In this context, a MS-GARCH model, as proposed Ardia et al. (2018), appears to be a more suitable approach.

This work also aims to contribute to the empirical literature on forecast evaluation methods for Expected Shortfall. The remainder of the paper is structured as follows: Sect. 2 presents the data. Section 3 describes the methodology. Section 4 introduces results. Finally, Sect. 5 concludes the paper, explaining also future improvements.

2 Data

Logarithmic returns of daily Bitcoin closed prices, obtained from the CoinMarketCap, are used. The time series spans from January 1, 2019, to March 31, 2025, resulting in a total of 2282 observations, yielding 2281 returns:

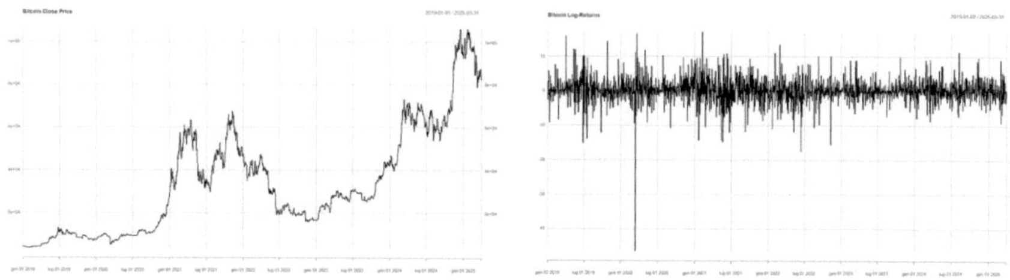

(a) Bitcoin Prices (b) Logarithmic Returns

Fig. 1. Time Series of Bitcoin Close Prices (Fig. 1a) and Bitcoin Logarithmic Returns (Fig. 1b)

Figure 1 shows clear evidence of volatility clustering in Bitcoin prices, signaling presence of heteroskedasticity in the time series.

Descriptive statistics of the Bitcoin logarithmic returns are reported in Table 1 and it is simple to see that the distribution is left-skewed and exhibits strong leptokurtosis.

Table 1. Descriptive statistics: the minimum, first quartile (Q1), median, mean, third quartile (Q3), maximum, standard deviation, skewness, kurtosis, and the p-value of the Jarque-Bera (JB) test of Bitcoin returns from January 1, 2019, to March 31, 2025.

D.S.	Bitcoin
Minimum	–46.47
Q1	–1.31
Median	0.07
Mean	0.13
Q3	1.60
Maximum	17.18
S.D.	3.40
Skewness	–1.08
Kurtosis	18.10
JB-test	0.00

3 Methodology

The methodology is structured in three main steps. First, the volatility models are estimated and used to forecast the corresponding VaR and ES. Secondly, the Long-Short Term Memory (LSTM) model is applied to refine these risk measures. Finally, four models (GARCH, MS-GARCH, J-C, LSTM-J-C) are compared based on average loss and backtesting performances.

3.1 Volatility Models

In order to estimate volatility and compute the VaR and ES, two models are taken into account: the standard GARCH (Bollerslev 1986) model and the Markov-Switching GARCH (Haas et al. 2004). MS-GARCH model allows for changes in the volatility dynamics through a regime-switching mechanism governed by a hidden Markov process. This framework enables the model to capture abrupt changes in volatility patterns, providing a potentially more robust estimation of risk measures such as VaR and ES, especially during periods of market turbulence (Bauwens et al. 2014; 2010). Following Ardia et al. (2018; 2019), let y_t denote the time series of Bitcoin logarithmic returns at time t, the general specification of the Markov-Switching model is formulated as follows:

$$y_t \mid (s_t = k, I_{t-1}) \sim \mathcal{D}(0, h_{k,t}, \xi_k)$$

where $\mathcal{D}(0, h_{k,t}, \xi_k)$ denotes a continuous distribution with zero mean, time-varying variance $h_{k,t}$, and additional shape parameters contained in the vector ξ_k. According to Ardia et al. (2018), the latent variable s_t, defined over a discrete state space $\{1, \ldots, K\}$, evolves according to a first-order, unobserved, homogeneous and ergodic Markov chain with transition probability matrix $P \equiv \{p_{i,j}\}_{i,j=1}^{K}$, where

$$p_{i,j} \equiv P[s_t = j \mid s_{t-1} = i].$$

I_{t-1} denotes the information available at time $t-1$, such that $I_{t-1} \equiv \{y_{t-i}, i > 0\}$. It is assumed that the conditional variance of y_t follows a GARCH-type model (Haas et al. 2004). In particular, conditional on the regime $s_t = k$, the variance $h_{k,t}$ is specified as a function of past returns and of a regime-dependent parameter vector θ_k:

$$h_{k,t} \equiv h(y_{t-1}, h_{k,t-1}, \theta_k)$$

where $h(\cdot)$ is an I_{t-1}-measurable function that defines a filter for the conditional variance and ensures its positiveness. It is further assumed that $h_{k,1} \equiv \bar{h}_k \quad (k = 1, \ldots, K)$, where $\bar{h}_k$ denotes a fixed initial level of variance for regime k, set equal to the unconditional variance in regime k. In this study, we consider a two-regime MS-GARCH model, where the first regime is described by a GARCH(1,1) (Bollerslev 1986) specification with a normal distribution:

$$h_{1,t} \equiv \alpha_{0,1} + \alpha_{1,1} y_{t-1}^2 + \beta_1 h_{1,t-1}$$

while the second regime is described by an EGARCH(1,1) (Nelson 1991) specification with t di student distribution:

$$\ln(h_{2,t}) \equiv \alpha_{0,2} + \alpha_{1,2}\left(|\eta_{2,t-1}| - \mathbb{E}\left[|\eta_{2,t-1}|\right]\right) + \alpha_{2,2}\eta_{2,t-1} + \beta_2 \ln(h_{2,t-1})$$

To ensure positiveness in the first regime, it is required that $\alpha_{0,1} > 0$, $\alpha_{1,1} > 0$, and $\beta_1 \geq 0$. Covariance-stationarity in this regime is obtained by requiring that $\alpha_{1,1} + \beta_1 < 1$. In the second regime, positiviness is automatically ensured by the model specification. Covariance-stationarity in each regime is obtained by requiring that $\beta_2 < 1$. For both volatility models, a fixed 1000 day rolling window was used to generate, through one-step-ahead forecasts, the next 1281 VaR and ES values. These will serve as features for the LSTM model. The LSTM model functions as a nonlinear memory filter since it receives the VaR and ES series computed by GARCH and MS-GARCH, and emits LSTM-adjusted VaR and ES forecasts. These refined estimates feed the joint-regression-combined framework (that joint regression is computed also without LSTM application for comparison purposes), enabling joint VaR-ES estimation that bypasses ES's non-elicitability and delivers the highest backtest accuracy among all benchmark models.

3.2 Long-Short Term Memory Model

In the context of Recurrent Neural Network, LSTM model was proposed by Hochreiter and Schmidhuber (1997). The purpose of the LSTM model was to circumvent the issues of long term dependencies and the exploding gradient's problem. This model belongs to the category of particular articulated network architecture, specialised in representing and modelling sequential data. This particular type of network introduces a gates' mechanism that regulates the amount of information transferred between temporal states and enables the model to dynamically select the information that is relevant for the subsequent state. Indeed, the unique structural features of neurons possess the capacity to select which information is to be retained, which is to be erased, and which to be update through the memory gate (c_t), the input gate (i_t), the forget gate (g_t) and the output gate (o_t).

The internal process of an LSTM unit can be mathematically described by the following expressions:

$$\begin{aligned}
g_t &= \sigma(U_g x_t + W_g h_{t-1} + b_g) \\
i_t &= \sigma(U_i x_t + W_i h_{t-1} + b_i) \\
\tilde{c}_t &= \tanh(U_c x_t + W_c h_{t-1} + b_c) \\
c_t &= g_t * c_{t-1} + i_t * \tilde{c}_t \\
o_t &= \sigma(U_o x_t + W_o h_{t-1} + b_o) \\
h_t &= o_t * \tanh(c_t)
\end{aligned}$$

where U e W are the weights matrices, b is the bias term, $\sigma(.)$ is the sigmoid activation function, tanh(.) is tanh activation function and the symbol * denotes the element-wise multiplication (Kim and Won 2018). Figure 2 shows the general structure of an LSTM unit's mechanism.

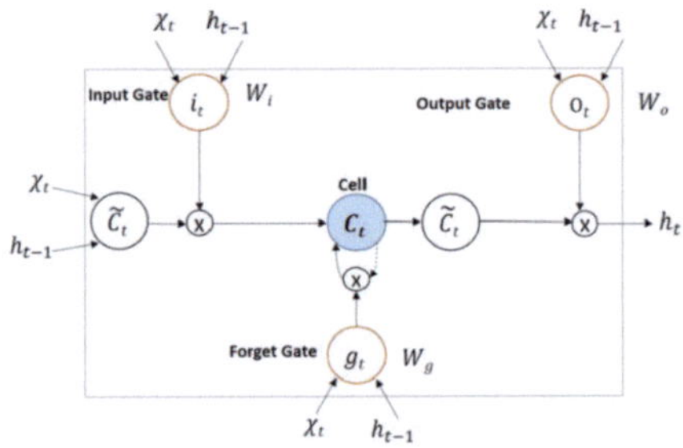

Fig. 2. General structure of an LSTM unit (García-Medina and Luu Duc Huynh 2021)

In the context of hyperparameters optimization procedures, Random Search algorithm was found to be the best fit to our setting: it covers the search space more uniformly; moreover, it's computational efficiency in relation to high-dimensional research spaces is worthy of note. We performed 4 random searches (one for each risk measure forecast). Each random search encompesses 30 trials for a total of 70 trials of randomling sampling the following hyperparameters:

1. Number of units per layer [16, 32, 64, 128]
2. Number of layers [1, 2, 3]
3. Dropout rate [0.1, 0.2, 0.3]
4. Batch size [16, 32, 64]
5. Learning rate [1e-4, 1e-3, 1e-2]
6. Epochs [30, 40, 50]
7. Activation function [Hyperbolic Tanget, Rectified Linear Unit]

3.3 Average Losses and Backtesting

Following the approach of Lu et al. (2021), and using the same rolling window size adopted for the LSTM model to ensure consistency across models, one-step-ahead forecasts of VaR and ES were generated for each model. This procedure yielded 226 out-of-sample predictions for both risk measures, thereby ensuring the comparability of all approaches. The Joint-Combined (J-C) regression model (Bayer and Dimitriadis 2022) was applied both to the original (non-adjusted) VaR and ES forecasts and to those adjusted by the LSTM. The scoring functions used were those proposed by Fissler and Ziegel (2016):

$$G_1(Z) = \{Z, 0\}$$

$$\begin{cases} G_2(Z) = -log(-Z), Z < 0 \\ G_2(Z) = -\sqrt{-Z}, Z < 0 \\ G_2(Z) = -\frac{1}{Z}, Z < 0 \\ G_2(Z) = ln(1 + \exp(Z)) \\ G_2(Z) = \exp(Z) \end{cases}$$

For each model, the joint loss function was computed, and two backtesting procedures were performed to assess the risk forecasting accuracy of the four models (GARCH, MS-GARCH, J-C, and LSTM-J-C). The two evaluation tests employed were the Conditional Calibration (CC) test Nolde and Ziegel (2017) and the Expected Shortfall Regression (ESR) test Dimitriadis and Bayer (2019); Bayer and Dimitriadis (2022).

4 Findings

Table 2 shows the average loss of the forecasts related to the implemented models, under different scoring functions, using the joint quantile and expected shortfall regression Bayer and Dimitriadis (2022).

Table 2. The average loss values of the forecasts for the GARCH, MS-GARCH, J-C, and LSTM-J-C models under different scoring function combinations

$G_2(z)$	$G_2 = -\log(z)$	$G_2 = -\sqrt{z}$	$G_2 = -1/z$	$G_2 = \ln(1 + \exp(z))$	$G_2 = \exp(z)$
$G_1 = z$					
GARCH	1.921	2.561	0.073	0.260	0.260
MS-GARCH	1.900	2.532	0.064	0.253	0.253
J-C	1.894	2.527	0.063	0.252	0.252
LSTM-J-C	1.889	2.529	0.062	0.253	0.253
$G_1 = 0$					
GARCH	1.655	2.295	0.192	–0.006	–0.006
MS-GARCH	1.641	2.273	–0.194	–0.006	–0.006
J-C	1.638	2.269	–0.194	–0.035	–0.006
LSTM-J-C	1.628	2.268	–0.198	–0.007	–0.018

Although no statistical significance testing (such as the Model Confidence Set procedure) has been conducted yet, from Table 2, when $G_1(z) = z$ the LSTM-J-C model seems to consistently achieve the lowest average losses across most specifications, exception made for the combination associated with $G_1 = z$ and $G_2 = -\sqrt{z}$, $G_2 = \ln(1 + \exp(z))$ and $G_2 = \exp(z)$, respectively, indicating superior performance in the joint estimation of VaR and ES. The GARCH model performs worse than both the MS-GARCH and J-C. The MS-GARCH and J-C models produce nearly identical loss values, suggesting similar predictive accuracy. The lowest average joint loss of 1.889 is related to the scoring function combination $G_2(z) = -\log(z)$. Conversely, the worst-performing model in this setting is attributed to GARCH, with the highest loss of 2.561 under $G_2(z) = \sqrt{z}$. When $G_1(z) = 0$, this trend is confirmed. The LSTM-J-C model continues to yield the lowest average losses (1.628), with $G_2(z) = -\log(z)$, pair, demonstrating robustness to the choice of scoring function. The worst result is again obtained by the GARCH model, which records a loss of 2.295 under $G_2(z) = -\sqrt{z}$ pair.

Table 3 reports an example of the application of the Conditional Calibration test and Expected Shortfall Regression test, for each model, related to the $G_1 = 0$ and $G_2 = -\sqrt{z}$ pair at a 95% probability level.

Table 3. Bootstrap p-value of Two-side backtest, at 95% probability level, related to Conditional Calibration (CC) test and Expected Shortfall Regression test (ESR).

Probability level	95%	
Two-side backtest	CC	**ESR**
GARCH	0.96	0.99
MS-GARCH	0.99	0.99
J-C	0.99	0.86
LSTM-J-C	0.98	0.99

It presents bootstrapped p-values (1000 iterations) of the two-sided ES backtests at the 95% probability level, using the CC and the ESR test. The results indicate that the MS-GARCH and LSTM-J-C model exhibits the strongest backtesting performance (0.99 and 0.98 p-values) related to CC test, while GARCH, MS-GARCH and LSTM-J-C model shows the strongest backtesting performance (0.99 p-value) related to ESR test. The J-C model shows, despite largely sufficient results, notably lower performance, particularly in the ESR test, where the bootstrap p-value is 0.86. These findings suggest the pivotal role of LSTM in enhancing the efficacy of the JC approach. The application of JC to the individual volatility models yields suboptimal outcomes in comparison to the JC values derived from the LSTM.

5 Conclusions and Future Research

This study aims to identify the most accurate model in passing the joint-combined backtesting procedure for VaR and ES, in line with the approaches of Dimitriadis and Bayer (2019); Bayer and Dimitriadis (2022), by comparing the predictive performance of various models and employing different scoring functions as proposed by Fissler and Ziegel (2016). The work builds upon the methodology developed by Lu et al. (2021), but extends it by considering a broader time horizon and incorporating a regime-switching volatility model (MS-GARCH) to capture potential regime shifts in Bitcoin volatility, following Ardia et al. (2018) and Caporale and Zekokh (2019). Furthermore, the study investigates whether the risk measure forecasts related to individual volatility models, such as GARCH and MS-GARCH, can be improved through an LSTM model. Results suggest that the LSTM-J-C model delivers the lowest average predictive loss across the evaluated settings. Moreover, the implementation of LSTM considerably enhances the efficacy of the JC approach, which demonstrates poorer performance when applied to to individual volatility models. However, these findings will require validation through the application of the Model Confidence Set (MCS) procedure, based on average loss obtained across different models, scoring function combinations, and confidence levels. Future extensions of the study will include also the Historical Simulation method to generate VaR and ES forecasts, enabling broader comparisons. Finally, the analysis will be expanded to cover other cryptocurrencies.

References

Ahsan, T., Zawadzki, K., Khan, M.: Cryptocurrencies as a speculative asset: How much uncertainty is included in cryptocurrency price? SAGE Open **14**(2), 21582440241256264 (2024)

Antar, M.: Quantile analysis of bitcoin returns: uncovering market dynamics. J. Risk Finance **26**(1), 122–146 (2025)

Ardia, D., Bluteau, K., Boudt, K., Catania, L.: Forecasting risk with markov-switching garch models: a large-scale performance study. Int. J. Forecast. **34**(4), 733–747 (2018)

Ardia, D., Bluteau, K., Boudt, K., Catania, L., Trottier, D.-A.: Markov-switching garch models in r: the msgarch package. J. Stat. Softw. **91**, 1–38 (2019)

Artzner, P., Delbaen, F., Eber, J.-M., Heath, D.: Coherent measures of risk. Math. Finance **9**(3), 203–228 (1999)

Balcilar, M., Bouri, E., Gupta, R., Roubaud, D.: Can volume predict bitcoin returns and volatility? a quantiles-based approach. Econ. Model. **64**, 74–81 (2017)

Bariviera, A.F.: The inefficiency of bitcoin revisited: a dynamic approach. Econ. Lett. **161**, 1–4 (2017)

Basel Committee on Banking Supervision: Explanatory note on the minimum capital requirements for market risk. Available online, Technical report, Bank for International Settlements (2019)

Baur, D.G., Hong, K., Lee, A.D.: Bitcoin: medium of exchange or speculative assets? J. Int. Finan. Markets. Inst. Money **54**, 177–189 (2018)

Bauwens, L., De Backer, B., Dufays, A.: A bayesian method of change-point estimation with recurrent regimes: application to garch models. J. Empir. Financ. **29**, 207–229 (2014)

Bauwens, L., Preminger, A., Rombouts, J.V.: Theory and inference for a markov switching garch model. Economet. J. **13**(2), 218–244 (2010)

Bayer, S., Dimitriadis, T.: Regression-based expected shortfall backtesting. J. Financ. Economet. **20**(3), 437–471 (2022)

Blau, B.M.: Price dynamics and speculative trading in bitcoin. Res. Int. Bus. Financ. **41**, 493–499 (2017)

Bollerslev, T.: Generalized autoregressive conditional heteroskedasticity. Journal of Econometrics **31**(3), 307–327 (1986)

Caporale, G., Zekokh, T.: Modelling volatility of cryptocurrencies using markov-switching garch models. Res. Inter. Bus. Finance **48**, 143–155 (2019)

Cheah, E.-T., Fry, J.: Speculative bubbles in bitcoin markets? an empirical investigation into the fundamental value of bitcoin. Econ. Lett. **130**, 32–36 (2015)

Christoffersen, P.F.: Evaluating interval forecasts. Inter. Econ. Rev., 841–862 (1998)

CoinGecko Global cryptocurrency market cap charts (2025). Accessed 20 April 2025
Corbet, S., Lucey, B., Yarovaya, L.: Datestamping the bitcoin and ethereum bubbles. Financ. Res. Lett. **26**, 81–88 (2018)
Dimitriadis, T., Bayer, S.: A joint quantile and expected shortfall regression framework (2019)
Fissler, T., Ziegel, J.F.: Higher order elicitability and osband's principle (2016)
García-Medina, A., Huynh, L.D., Luu Duc Huynh, T.: What drives bitcoin? an approach from continuous local transfer entropy and deep learning classification models. Entropy **23**(12), 1582 (2021)
Gkillas, K., Katsiampa, P.: An application of extreme value theory to cryptocurrencies. Econ. Lett. **164**, 109–111 (2018)
Glaser, F., Zimmermann, K., Haferkorn, M., Weber, M.C., Siering, M.: Bitcoin-asset or currency? revealing users' hidden intentions. Revealing Users' Hidden Intentions, 15 April. ECIS (2014)
Gneiting, T.: Making and evaluating point forecasts. J. Am. Stat. Assoc. **106**(494), 746–762 (2011)
Haas, M., Mittnik, S., Paolella, M.S.: A new approach to markov-switching garch models. J. Financ. Economet. **2**(4), 493–530 (2004)
Hansen, P.R., Lunde, A., Nason, J.M.: The model confidence set. Econometrica **79**(2), 453–497 (2011)
Hochreiter, S., Schmidhuber, J.: Long short-term memory. Neural Comput. **9**(8), 1735–1780 (1997)
Katsiampa, P.: Volatility estimation for bitcoin: a comparison of garch models. Econ. Lett. **158**, 3–6 (2017)
Kim, H.Y., Won, C.H.: Forecasting the volatility of stock price index: a hybrid model integrating lstm with multiple garch-type models. Expert Syst. Appl. **103**, 25–37 (2018)
Krause, D. (2025). Bitcoin as a reserve asset? a cautionary analysis for the federal and state governments amid bear markets and fiscal risk. A Cautionary Analysis for the Federal and State Governments Amid Bear Markets and Fiscal Risk, 7 April (2025)
Kupiec, P.H. et al.: Techniques for verifying the accuracy of risk measurement models, vol. 95. Division of Research and Statistics, Division of Monetary Affairs, Federal(1995)
Li, Z.-Z., Tao, R., Su, C.-W., Lobonţ, O.-R.: Does bitcoin bubble burst? Qual. Quant. **53**, 91–105 (2019)
Lu, X., Liu, C., Lai, K.K., Cui, H.: Risk measurement in bitcoin market by fusing lstm with the joint-regression-combined forecasting model. Kybernetes **52**(4), 1487–1502 (2021)
Mincer, J.A., Zarnowitz, V.: The evaluation of economic forecasts. In: Economic Forecasts and Expectations: Analysis of Forecasting Behavior and Performance, pp. 3–46. NBER (1969)
Morgan, J., et al.: Creditmetrics-technical document, vol.1, pp. 102–127. JP Morgan, New York (1997)
Nakamoto, S.: Bitcoin: A peer-to-peer electronic cash system (2008)
Nelson, D.B.: Conditional heteroskedasticity in asset returns: a new approach. Econometrica J. Econometric Soc., 347–370 (1991)
Nolde, N., Ziegel, J.F.: Elicitability and backtesting: perspectives for banking regulation (2017)
Osband, K.H.: Providing Incentives for Better Cost Forecasting (Prediction, Uncertainty Elicitation). University of California, Berkeley (1985)

Pavlidis, E.G.: Bubbles and crashes: a tale of quantiles. J. Time Series Analy. (2023)
Tarchella, S., Khalfaoui, R., Hammoudeh, S.: The safe haven, hedging, and diversification properties of oil, gold, and cryptocurrency for the g7 equity markets: evidence from the pre-and post-covid-19 periods. Res. Int. Bus. Financ. **67**, 102125 (2024)
Taylor, J.W.: Forecasting value at risk and expected shortfall using a semiparametric approach based on the asymmetric laplace distribution. J. Bus. Econ. Statist. **37**(1), 121–133 (2019)
White, R., Marinakis, Y., Islam, N., Walsh, S.: Is bitcoin a currency, a technology-based product, or something else? Technol. Forecast. Soc. Chang. **151**, 119877 (2020)
Zhou, Y., Xie, C., Wang, G.-J., Gong, J., Zhu, Y.: Forecasting cryptocurrency volatility: a novel framework based on the evolving multiscale graph neural network. Financial Innovation **11**(1), 87 (2025)

Understanding and Attitudes Toward Reverse Mortgage in Italy: Cognitive Dissonance and Future Concerns

Emilia Di Lorenzo and Alba Roviello(✉)

Departments of Economics and Statistics, Università degli Studi di Napoli Federico II, Via Cintia, 21, 80126 Naples, Italy
{I.emilia.dilorenzo,II.alba.roviello}@unina.it

Abstract. Reverse Mortgage (RM), known as Prestito Ipotecario Vitalizio (PIV) in Italy, is an alternative financing solution aimed at individuals aged 60 and above who typically lack access to conventional credit lines. Subscribing the RM allows elders homeowners to convert part of their home's equity into a lump sum or periodic payments preserving the right to live in the house and, differently from the bare ownership, without losing any legal rights on the property. The loan, along with accumulated interest, has to be repaid only when the borrower dies, sells the home, or permanently moves out. The Non-Negative Equity Guarantee (NNEG) protects the heirs by ensuring that, if the property's value at the time of loan repayment is insufficient to cover the outstanding debt, no additional amount is required. Despite its availability in the Italian market for several years, RM remains a relatively underutilized financial instrument. To assess current levels of awareness and identify the barriers limiting access to this property-linked financing option, we present some preliminary results obtained conducting a comprehensive qualitative and quantitative research study among potential Italian subscribers, which takes into account individuals' future concern about economic sustainability, their understanding and knowledge of supplementary pension products with priority over RM and actual interest in or use of the product. The objective is to provide meaningful insights into Italian perception of the Reverse Mortgage offering guidance for policy-makers aiming to incorporate this tool into broader social and economic policy strategies.

Keywords: Reverse Mortgage · Survey · Italian households · Financial Decisions · Cognitive Dissonance

1 Introduction

Reverse Mortgage (RM) represents a non-traditional form of financing intended for people aged 60 and over, typically excluded from standard credit access. RM provides a loan, in form of a lump sum or periodic payments, secured by a mortgage on the borrower's primary residence. The main differences with the

M. La Rocca et al. (Eds.): WMAF 2025, *New Perspectives in Mathematical and Statistical Methods for Actuarial Sciences and Finance*, pp. 132–143, 2025.
https://doi.org/10.1007/978-3-032-05551-4_12

bare ownership are that the borrower retains the totality of the legal right on the property and at borrower's death, heirs can redeem the property of the house by repaying the debt and the accumulated interest. Otherwise, the house can be sold to cover the debt and the surplus would be paid to the heirs. The Non-Negative Equity Guarantee (NNEG) ensures the heirs in the event that the value of the house at the time of debt settlement is not sufficient to redemption. Reverse Mortgage presents a pivotal opportunity to address the social vulnerability afflicting numerous elderly individuals worldwide: according to [1], the global Reverse Mortgage market size was estimated USD 1.83 billion in 2023 and it is projected to raise at a CAGR (Compound Annual Growth Rate) of 5.9% from 2024 to 2030. In the US, RM market is experiencing consistent growth, driven by demographic aging, increasing property values, and the increasing demand for sustainable retirement income solutions. After dropping to £2.6 billion in 2023 from £6.2 billion in 2022, the UK market is now starting to recover ([2]). Same trend was observed in the Australian market, where the greatest concern about future financial possibility of elders is leading to a rapid expansion of the RM market ([3]). In other countries such as Japan and South Korea, RM has attracted emerging interest as a practical solution for seniors to access home equity ([1]). Differently, RM market in Europe is still in its early stages compared to the previously mentioned markets. However, the recent demographic dynamics and the increasing life expectancy, as well as the volatility of the house market price can lead to a rise in the elders' interest in equity release products. In the Italian context, where the RM is known as Prestito Ipotecario Vitalizio (PIV), strong deterrent role is associated to personal aversion to take on debt and to emotional aspects related to the house and its inheritance scope ([4]). However, as showed in [5], elderly may obtain relevant welfare gains contracting a RM contract, especially when out-of-pocket healthcare expenditures become considerable. In this work, we present some preliminary results obtained conducting a comprehensive research study for the purpose of assessing current levels of awareness about RM among Italian population and to identify the factors hindering access to this credit facility in the Italian market. A first step in the investigation of the determinants of interest in RM in Italy was made in [4], where an empirical analysis on UniCredit's 2007 survey on household savings was performed. In details, the survey targets UniCredit's clients aged 21–75 with at least € 10,000 in deposits. At that time, it resulted to be the unique survey settled in Italy that includes one specific question on RMs. Risk aversion, debt aversion and financial literacy were also taken into consideration. The analysis revealed that, among more than 1200 respondents, approximately 60% reported no interest in RM, while the remaining 40% expressed varying levels of interest, ranging from low to very high. Notably, factors related to perceived risk showed a strong correlation with respondents' attitudes toward the product. In particular, a willingness to sell one's home as a means to supplement future income and uncertainty regarding post-retirement living standards emerged as significant predictors of higher interest. In contrast, debt aversion and greater housing equity were associated with a lack of interest. These findings suggest that RM are not widely viewed as instruments to enhance quality of life in old age, but rather as a last-resort financial solution in times of need. To gain deeper insight into the factors contributing to the low demand for this product, we conducted both qualitative and quantitative analyses among potential Italian subscribers, focusing on individuals' future concern about economic sustainability, levels of comprehension and knowledge of supplementary pension products with priority over RM and actual usage or interest in the product. Our aim is to furnish valuable insights into Italian individuals' perception of the Reverse Mortgage with the aim of providing guidance to policymakers seeking to integrate the Reverse Mortgage into their intervention strategies. Based on our findings, we furnish an economic indicator, called Net Positivity Score (NPS), which enables to measure respondents concerns for the future. The NPS constitutes a practical way to quantify levels of pessimism/optimism towards economic situation in old ages and, consequently, to assess personal propensity to consider financial support instruments such as Reverse Mortgage. Section 2 contains a brief introduction to the actuarial structure of the RMs. Section 3 outlines the methodologies of the research. Section 4 describes main findings of the research while Sect. 5 concludes.

2 Actuarial Insights in the Reverse Mortgage

The contractual structure of Reverse Mortgage is based on the equivalence principle assessed at the issue time $t = 0$, between the lender, typically a bank or insurance company, and the borrower, who is the homeowner aged x. At $t = 0$, the borrower may receive a lump sum payment, corresponding to the appraised value of the property or a percentage thereof, net of the Non-Negative Equity Guarantee (NNEG). This guarantee ensures that the outstanding debt will never exceed the market value of the property. As an alternative to the lump sum, the borrower may opt to receive fixed periodic payments throughout entire lifetime. Suppose that ω denotes the extreme age of the borrower. Lender must actively manage the risk associated with the uncertain duration of the contract, which depends on the unpredictability of the borrower's lifespan, known as *longevity risk*, the risk related to fluctuations over time in the value of the house used as a guarantee, known as *house price risk* and described by the random appreciation rate of the property $\tilde{r}_t^H$ at the generic time t, and the risk due to the volatility of interest rates, known as *financial risk* and evaluated, at the generic time t, with the rate $\tilde{r}_t^{RM}$, sum of the risk-free interest rate $\tilde{r}_t$ and the risk premium π (see also [6–9] and references therein). Denoting with H_0 the current market value of the house and with α the proportion of initial house value ($0 < \alpha \leq 1$) on which the contract is written, the loan amount on which interest accrues is given by $A_0 = \alpha H_0$. The initial lump sum LS_0, to be paid by the lender at time $t = 0$, satisfies the following relation:

$$LS_0 = R\ddot{a}_x < A_0,$$

with R as the constant annual installment disbursed by the lender to the borrower in case of periodic payments and $\ddot{a}_x$, as usual, representing the actuarial present value of a unitary life annuity paid to an x-aged individual. The amounts $(A_0 - LS_0)$ describes the cost of the NNEG, which ensures that, at a generic time t within the contractual duration, heirs have to repay the amount

$$V_t = A_t - \max(0, A_t - H_t) = \min(A_t, H_t),$$

where A_t and H_t respectively denote the loan balance and the property value at time t. This means that, as a consequence of the NNEG, the lender is short a put option on the property, while the borrower benefits from downside protection. Taking into account stochastic aspects related to above mentioned risks, we can describe the corresponding random amounts in the following way:

$$\tilde{A}_t = A_0 \prod_{j=0}^{t-1} (1 + \tilde{r_j}^{RM}) = \alpha H_0 \prod_{j=0}^{t-1} (1 + \tilde{r_j} + \pi),$$

$$\tilde{H}_t = \alpha H_0 \prod_{j=0}^{t-1} (1 + \tilde{r_j}^{HV}),$$

$$\tilde{V}_t = \tilde{A}_t - \max(0, \tilde{A}_t - \tilde{H}_t) = \min(\tilde{A}_t, \tilde{H}_t).$$

In order to determine the financial equilibrium condition at $t = 0$, denote with ${}_{t-1/1}q_x$ the probability of death during the t-th year after age x. We have that:

$$\tilde{LS}_0 = \sum_{t=1}^{\omega-x} \tilde{V}_t \left[\prod_{j=1}^{t} (1 + \tilde{r_j})^{-1} \right] {}_{t-1/1}q_x.$$

Therefore, given the mortality model P used to estimate death probabilities, the value of the lump sum received by the borrower at $t = 0$ is given by the conditional expected value

$$LS_0 = E[E[\tilde{LS}_0 | P]].$$

3 The Survey Design

Despite the availability of the RM in the Italian market, it remains a relatively underutilized financial instrument. To further investigate the reasons behind the low demand for this product, we carried out a comprehensive research study, incorporating both qualitative and quantitative methods. The analysis was conducted between May and September 2024 among potential Italian subscribers, in collaboration with Loghía Market Research Institute. The qualitative research was conducted during May 2024 through focus groups, based on an open-ended discussion guided by an experienced psychologist as moderator. In particular, three main topics were investigated: emotional, cultural, and economic significance of home ownership for Italians, debt aversion in old age and knowledge of the main supplementary pension products available in the Italian market. In details, 4 focus groups (8 respondents each) of 2 h length were conducted in two cities: Milan and Naples. Respondents were all over 60, evenly divided into two age groups: 60–70 and 71 and over. All respondents were required to be the owners of the property they lived in, and to have primary or shared decision-making authority over the family's financial affairs. The quantitative research phase was structured in light of the main insights collected during the qualitative phase of the research. A total of 904 interviews were conducted, between July 4th and September 6th, 2024, within a representative sample of potential Reverse Mortgage subscribers residing in Italy. Interviewees were recruited using a mixed-mode methodology: 50% of the interviews were conducted via CAWI (Computer Assisted Web Interviews) with respondents recruited from pre-profiled panels while 50% of the interviews were conducted via CATI/CAMI (Computer Assisted Telephone/Mobile Interviews) with respondents selected at random from telephone directories and/or by random dialing for mobile phones. The questionnaire was pivoted to analyze the perceived confidence or concern regarding the financial sustainability at old age and the familiarity with supplementary pension instruments, with a focus on the Reverse Mortgage. Interest in the product, as well as its actual use, was also subject to investigation.

4 Main Findings

4.1 Concerns About the Future

In order to provide practical information about the level of perceived lightheartedness of respondents, we constructed an economic hope indicator considering the economic resources people expect to rely on in old age. In particular, individuals were asked to answer the following question:

> *"Thinking about all the resources you expect to have available for retirement, how secure do you feel about your future on a scale of 0 (not at all secure) to 10 (completely secure)?"*

The indicator was built by subtracting the percentage of those who expressed the most optimistic scores (9-10) from the percentage of those more pessimistic (0–6), yielding the Net Positivity Score (NPS). The Net Positivity Score (NPS) is a standardized metric on a scale of –100% to +100% that allows for easy comparison across different frameworks and over time. It has to be noted that the methodology employed to build the indicator was rooted in traditional customer satisfaction surveys, where NPS stands for "Net Promoter Score" (see for example [10,11]). It quantifies an individual's willingness to recommend (and thus become ambassadors for) a specific product, service, or social initiative. Unlike mean value (which just calculates a central value among all scores), the NPS has the advantage of effectively differentiates between extremely satisfied and dissatisfied respondents. By isolating these polar ends of the spectrum, we can identify individuals most susceptible to taking action. For example, a negative NPS indicates a preponderance of pessimists over optimists, suggesting a pressing need for

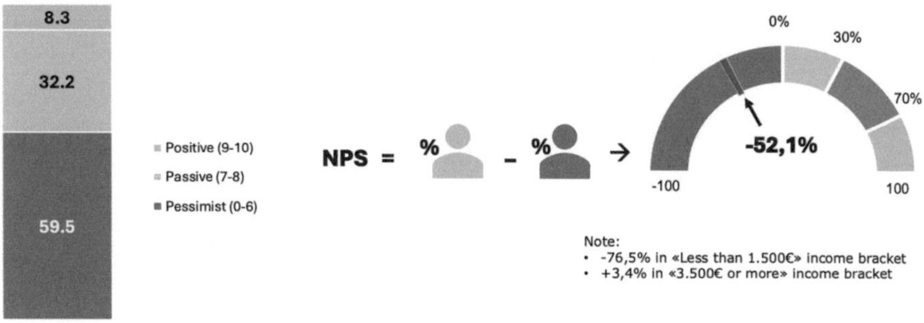

Fig. 1. Net Positivity Score - Base: Total sample (904)

improvement actions. A NPS between 0% and 30% means that positive perceptions outweigh negative ones, nonetheless proactive measures would be necessary to enhance performance. The highly desirable safe zone is among 30% and 70%. A value greater than 70% would reflect an enthusiastic attitude. In our analysis, the NPS was assessed to –51.2%, signaling significant concerns for the future, which are especially critical for those with limited current financial resources, as one can see in Fig. 1.

Interpreted as a proxy for future concerns, the NPS identified the segment of the population exhibiting the highest levels of pessimism and, consequently, a greater propensity to consider financial instruments designed to enhance their economic standing, such as Reverse Mortgage.

A pattern of cognitive dissonance was noted, which confirmed similar evidences from qualitative outcomes. On one hand, when asked about the possibility of relying on their own economic resources for the future, respondents showed a declining economic self-esteem as they envisioned more demanding financial scenarios (from maintaining independence to needing care at home or even helping their family financially). This pessimism was particularly present among those with limited current resources or greater family burdens (i.e., children, dependents). On the other hand, in spite of this prevailing sentiment of anxiety, there was a marked reluctance to embrace potential solutions that necessitate external intervention (from which, moreover, government assistance remains excluded). Therefore, while concerns about meeting substantial financial obligations were evident, there was a discernible reluctance to engage with potential solutions. This reluctance was compounded by a lack of financial foresight regarding retirement planning. This attitude inevitably involves the Reverse Mortgage, which suffers, moreover, from the lack of knowledge about it.

4.2 Knowledge and Perception of the Reverse Mortgage

Our investigation showed that the Reverse Mortgage is poorly known and understood. As one can note from Fig. 2, only 26,2% of the respondents affirms to know Reverse Mortgage, a very small proportion respect to other products. This lack of

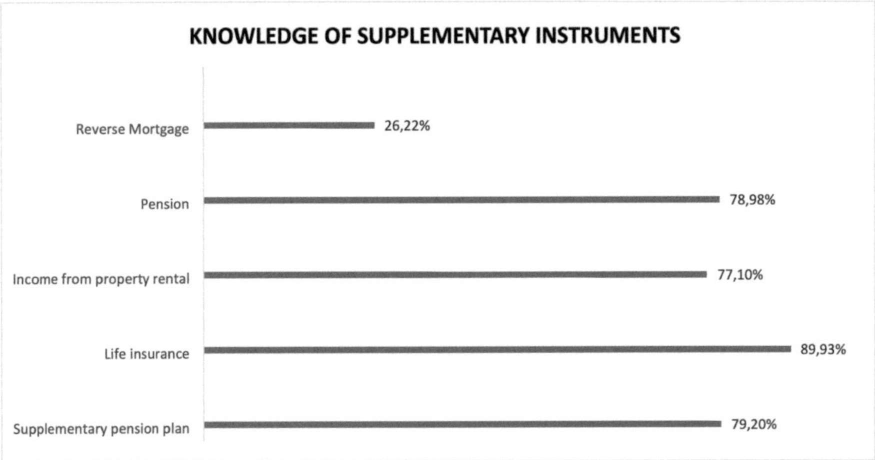

Fig. 2. Learning percentage about alternative income-generating financial instruments - Base: Total sample (904)

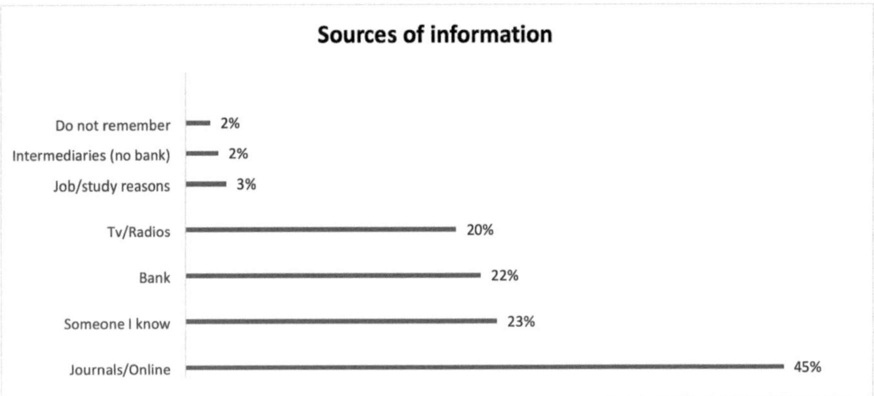

Fig. 3. Main means of communication through which respondents have become aware of the RM - Base: Total sample (904)

knowledge likely stems from an insufficient information campaign by providers; indeed, those who knew Reverse Mortgage found information independently, as shown in Fig. 3. Once explained, Reverse Mortgage was understood fairly well in its mechanisms, as shown in Figs. 4 and 5, although the explicit judgment on clarity betrays the effort required for its comprehension. However one concept seemed to be relatively more difficult to grasp than the others: the fact that children are not obliged to repay the debt and interest unless they want to. In the light of qualitative outcomes, it can be assumed that this difficulty may be more emotional than conceptual, relating to the idea of burdening heirs: RM clashes with the notion of transferring a potential problem to children, possibly

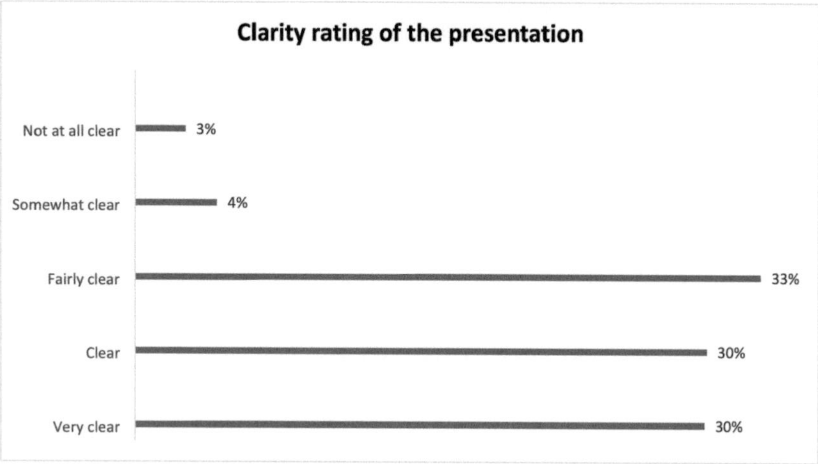

Fig. 4. Clarity and in-depth insight into the reverse mortgage mechanism - Base: Total sample (904)

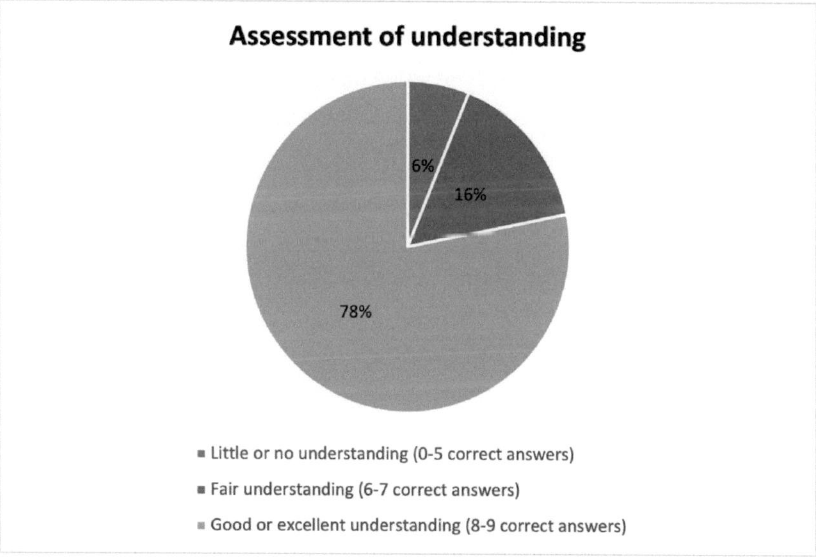

Fig. 5. Measurement of comprehension using control questions - Base: Total sample (904)

depriving them of an inheritance, and leading to a sense of failure for themselves. Only 26,3% resulted to be willing to contract the RM (see Fig. 7). Additionally, it is interesting to observe that those who have expressed greater interest in the Reverse Mortgage seem to have a better understanding of it. This suggests that a more effective information campaign could help spread awareness and usage

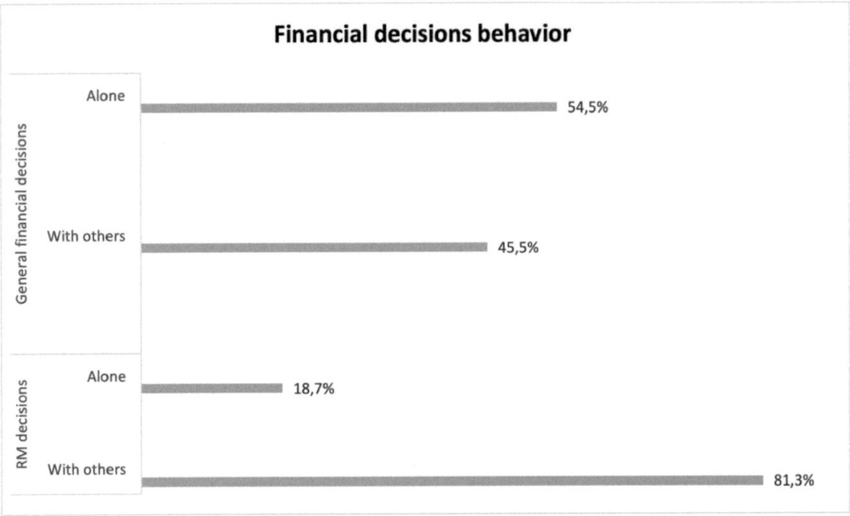

Fig. 6. Shift in attitude regarding housing-related financial decisions - Base: Total sample (904)

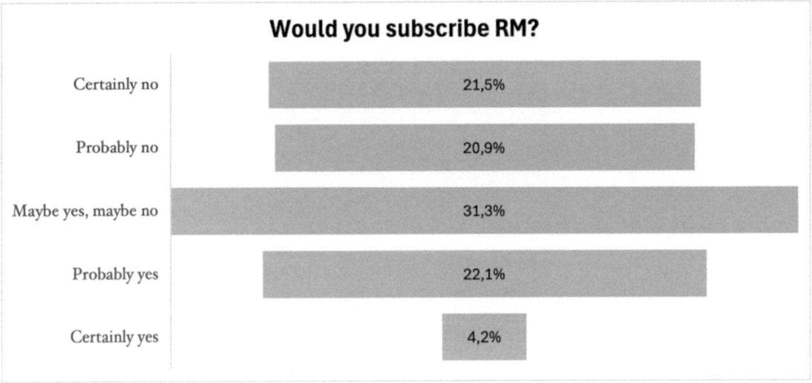

Fig. 7. Interest in contracting Reverse Mortgage - Base: Total sample (904)

of this tool, especially among open minded people and among the most vulnerable segments of the population, with lower economic hope index quantified by the Net Positivity Score: retirees with low incomes, people with high debts, and those who have to financially support their families. The involvement of heirs in the decision to take out a Reverse Mortgage emerges as a critical point. In fact, as illustrated in Fig. 6, it was observed that when the topic turned to the family home, even those who would normally make their own financial decisions autonomously felt the need to involve their family members when considering Reverse Mortgage, especially when there are children or grandchildren with

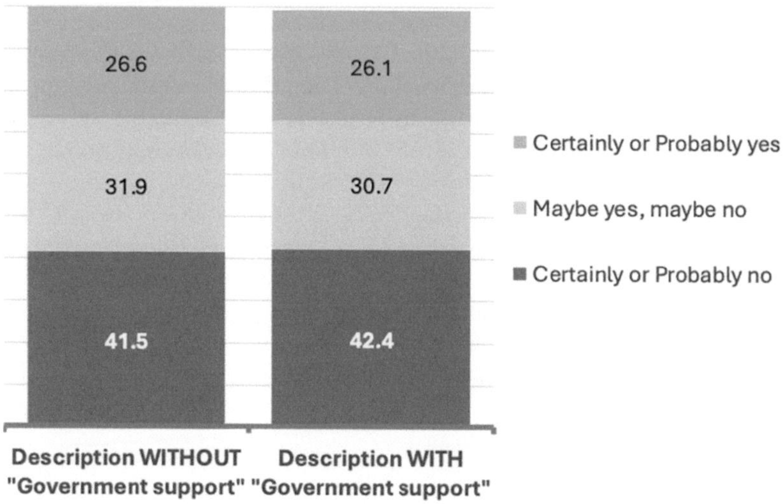

Fig. 8. Propensity towards RM related to government intervention (percentage values) - Base: Total sample (904)

whom there are strong emotional or financial ties (still close, still economically dependent, or towards whom they feel a sense of responsibility), or when there has already been expressed interest in the property. Regarding decision-making autonomy, it is worth noting that in decisions affecting the family home, and thus Reverse Mortgage, there is a tendency to overcome traditional gender-based attitudes: men, generally more assertive in financial choices, take a step back when it comes to the family home. Conversely, women, stereotypically linked to the domestic hearth, feel called upon and tend to become more assertive. Lastly, the fact that one of the descriptions mentioned potential government support did not affect the propensity towards RMs, as shown in Fig. 8.

5 Conclusion

A comprehensive analysis of focus group findings and quantitative data suggests that the Reverse Mortgage faces a number of barriers to its widespread adoption in Italy. The highly conservative attitude towards one's home and the fear of debt suggest that, even if all parties involved agree, a Reverse Mortgage is considered a last resort, to be taken into account only in cases of absolute necessity, in coherence with [4]. This perception, coupled with limited product knowledge, leads to a reluctance to engage in proper financial planning for retirement (cognitive dissonance), and the constraints imposed by involving heirs in housing-related decisions, collectively contribute to a very low level of interest in Reverse Mortgage.

Within this framework, it is imperative that campaigns designed to promote Reverse Mortgage are effectively aimed at the particular sector of population most likely to be responsive to such a solution. So, they should focus on a specific target of individuals with very low incomes and no other available resources, burdened by other debts and/or more involved in family care. In addition, a younger target audience, more open to new proposals, as demonstrated by those surveyed online, seems to show greater curiosity. These considerations along with the evidence about self-information on Reverse Mortgage suggest that there is a need to explore alternative communication strategies for Reverse Mortgages. Indeed, it is apparent that banks and providers offering Reverse Mortgage exhibit a marked reluctance to promote it. While a detailed exploration of the underlying reasons is beyond the scope of this study, it is plausible to suggest that these institutions face a number of risks associated with this product: informative campaigns through alternative channels could prove beneficial. The present analysis constitutes a first step in the comprehension of the possible motivations for limited use of RM. Additional considerations will be provided in future research. Some preliminary results may be found in [12], where a regression analysis was performed to test for perceived barriers to the adoption of the product.

References

1. Grand View Research Reverse Mortgage Market Size, Share & Trends Analysis Report 2024 - 2030 (2023). https://www.grandviewresearch.com/industry-analysis/reverse-mortgage-market-report
2. Equity Release Council, Q4 & FY 2024 lending data full report (2024). https://www.equityreleasecouncil.com/documents/q4-fy-2024-lending-data-full-report/
3. MFS Global Retirement Survey, The Road to Better Outcomes, Canadian Results (2024). https://www.mfs.com/en-ca/institutions-and-consultants/insights/retirement-insights/global-retirement-survey-better-outcomes.html
4. Fornero, E., Rossi, M., Brancati, M.C.U.: Explaining why, right or wrong, (Italian) households do not like reverse mortgages. J. Pension Econ. Finance **15**(2), 180–202 (2016)
5. Di Lorenzo, E., Piscopo, G., Roviello, A., Sibillo M.: Exploring reverse mortgage for Italian population: a life-cycle model approach. Annals Operat. Res., 1-19 (2025)
6. Shao, A.W., Hanewald, K., Sherris, M.: Reverse mortgage pricing and risk analysis allowing for idiosyncratic house price risk and longevity risk. Insurance: Math. Econ. **63**, 76–90 (2015)
7. Di Lorenzo, E., Piscopo, G., Sibillo, M., Tizzano, R.: Reverse mortgage and risk profile awareness: proposals for securitization. Appl. Stoch. Model. Bus. Ind. **38**(2), 353–369 (2022)
8. de la Fuente, I., Navarro, E., Serna, G.: Reverse mortgage risks time evolution of VaR in lump-sum solutions. Mathematics **8**(11), 2043 (2020)
9. Di Lorenzo, E., Piscopo, G., Sibillo, M., Tizzano, R.: Reverse mortgages through artificial intelligence: new opportunities for the actuaries. Decisions Econ. Finan. **44**(1), 23–35 (2021)

10. Jahnert, J.R., Schmeiser, H.: The relationship between net promoter score and insurers' profitability: an empirical analysis at the customer level. Geneva Papers Risk Insurance-Issues Pract. **47**(4), 944–972 (2022)
11. Dawes, J.G.: The net promoter score: What should managers know? Int. J. Mark. Res. **66**(2–3), 182–198 (2024)
12. Bertelli B., Di Lorenzo E., Roviello A., Torricelli C.: Who (and why) is still afraid of Reverse Mortgages? Results from survey in Italy, CEFIN Working Papers No 100, ISSN 2282-8168 (2025)

Reverse Mortgages: Exploring the Impact of Risk Factors by Source

Emilia Di Lorenzo[1], Giulia Magni[2(✉)], and Marilena Sibillo[3]

[1] Department of Economic and Statistical Sciences, via Cintia, Complesso Monte S. Angelo, University of Naples Federico II, Naples, Italy
diloremi@unina.it

[2] Department of Statistical Sciences, Piazzale Aldo Moro 5, Sapienza University of Rome, Rome, Italy
giulia.magni@uniroma1.it

[3] Department of Economics and Statistics, via Giovanni Paolo II 132, University of Salerno, Fisciano, Italy
msibillo@unisa.it

Abstract. In recent years, reverse mortgages have gained attention as suitable financial instruments for individuals of retirement age, particularly those classified as "house-rich but cash-poor." Despite growing interest in countries such as the US, the UK, and Australia, the Italian market remains underdeveloped in this area. From a lender's perspective, these contracts are perceived as complex, mainly due to the management of various risk components: longevity risk, financial risk, and house price risk. Our goal is to provide lenders with insights into the impact of each risk factor through a time-dependent profit/loss function. This approach aids in identifying the most critical sources of risk, enabling lenders to take appropriate measures to mitigate potential financial threats. We conduct Monte Carlo simulations of the chosen stochastic risk models and apply a single-component VaR assessment procedure. Our findings indicate that house price risk is the most significant risk factor of a Reverse Mortgage portfolio. Moreover, the lender can control the financial risk through the analysis of the risk premium values. The results also highlight that, beyond a certain time horizon, the lender's exposure stabilizes, with profitability emerging in the long run.

Keywords: Reverse Mortgages · longevity risk · financial risk · house-price risk · single component-VaR

1 Introduction

In a society such as the Italian one, where there is a strong ageing of the population, new financing sources that integrates the current troubled public pension system should be addressed. As a consequence the use of housing as an asset, such us Reverse Mortgage (RM) contracts, are particularly appropriate with the needs of elderly people whose wealth is made up in prevailing percentages by the

M. La Rocca et al. (Eds.): WMAF 2025, *New Perspectives in Mathematical and Statistical Methods for Actuarial Sciences and Finance*, pp. 144–156, 2025.
https://doi.org/10.1007/978-3-032-05551-4_13

value of real estate (the house in which they live) and, to a lesser proportion, by retirement monetary incomes [12].

The Reverse Mortgage is a financial product that allows homeowners to borrow money based on the value of their property, either in form of lump sum, periodic installments, line of credit or any combination of these. In our hypotheses, borrowers retain the right to live in their homes until their death, which marks the contract's termination date. At this point, the loan must be repaid, and there are two ways to do this: in the first, the heirs settle the outstanding debt, including accrued interest, thereby regaining ownership of the property. Alternatively, the lender sells the property, and the proceeds are used to repay the loan. If the value of the house exceeds the outstanding balance, a Non-Negative Equity Guarantee (NNEG) is usually included to prevent the heirs from having to pay the difference. This guarantee implies a reduction in the actual amount disbursed to the borrower to ensure financial equivalence.

The distinctive feature of this contract is that it is issued to people in retirement age, providing them an additional source of pension income, ideally suitable for people who are "house rich but cash poor". It follows that the Italian market context could be particularly appropriate for its spread, where a marked ageing of the population and a widespread ownership of real estate prevail.

According to Word Bank, indeed, the Italian population aged 65 and over has risen from 10 % of the total population in 1960 to 24% in 2023 [17]. Furthermore, people expect to live longer, as the average life expectancy at birth has moved from 69.1 years in 1960 to 79.8 in 2000, reaching 83.1 years in 2023 [17].

It is also well known that for Italians, a population of property owners, the home has always been a source of tranquility and refuge, and it carries a profound significance of transferring a family value to the heirs. According [5], about three quarters (74.3%) of Italians own their homes, and 60.8% of those who own their homes are also at risk of poverty. Regarding the over 65s who live alone, about 88.6% of them have home ownership [4].

Despite the fact all conditions are in place for the propagation, nowadays, the Italian market is not significantly developed in the Reverse Mortgage field. On the borrower side, this is because of the lack of financial education for the groups that might be most interested, and on the lender side because of the complexity of this product. Its complexity lies mainly in managing its different risk components: longevity risk, financial risk, and house price risk.

The purpose of this paper is to provide the lender of a Reverse Mortgage with information about the impact of each risk factor through a time-dependent profit/loss function. This approach helps identify the most critical sources of risk, allowing lenders to take appropriate measures to mitigate potential financial threats.

Many studies have already addressed the problem of calculating risk measures in Reverse Mortgage contracts, primarily focusing on measures such as the Value at Risk (VaR) and the Expected Shortfall (ES) applied simultaneously to all sources of risk.

Hence, [1] calculated VaR and Conditional Value at Risk (CoVaR) for a single borrower without considering longevity risk. [7] assessed VaR and ES for a portfolio of RM in lump-sum solutions without considering stochastic interest rates. [9] estimated the CoVaR using Neural Networks on a portfolio on N contracts in a context of constant risk-free interest rates.

In this paper, we adopt an approach based on the analysis of the impact of each risk component on the VaR of the profit/loss function, considering the worst-case scenario for the single specific component while assuming average values for the remaining risk components. This methodology allows us to identify and quantify the contribution of each risk factor to the VaR. The paper consists of the following sections: Sect. 2 deals with the actuarial model describing the product design, the risk model and the financial outcome from the lender's point of view. In addition, we outline the single-component VaR assessing procedure. Section 3 presents a numerical example illustrating the application of the proposed framework; in Sect. 4 we report some conclusions.

2 Modeling Framework

We consider a RM contract with periodic payments. This means that a bank or an insurer, the lender, issues a RM contract in t=0 to a policyholder, the borrower, who, in turn, provides an initial guarantee represented by the house value or a proportion of it. We assume that the house value, net of the NNEG, is paid out over time in the form of periodic payments at the end of each year, as long as the policyholder is alive.

When the borrower dies, the lender will no longer have to pay instalments; the contract has reached the termination date and, to cover the outstanding balance, the lender will sell the property (or heirs will settle the outstanding debt).

The lender has to manage three main risk factors:

1. The *house price risk*, that is the risk that the property's market value at the time of the borrower's death is not enough to fully repay the remaining outstanding balance of the loan. On the basis of the general assessment in [16], the stochastic value of the house at time t can be expressed as a function of the house appreciation rate, as shown in Eq. 1:

$$\widetilde{H}_t = \delta H_0 \prod_{s=0}^{t-1} \left(1 + \widetilde{r}_s^{HV}\right) \tag{1}$$

where δ is a proportion in the range 0 to 1 of the initial value of the property H_0. $\delta < 1$ means that the borrower receives a loan that is lower than the initial value of the house; $\delta = 1$ equates to the borrower receiving the entire initial value of the house. $\widetilde{r}_s^{HV}$ is the stochastic interest rate of the house appreciation over time.

2. The *financial risk*, that is the volatility risk of interest rates. It is embedded in the rate $\widetilde{r}_s^{RM}$ that the lender applies to the loan. This rate is the sum of the stochastic risk-free interest rate $\widetilde{r}_s$ and the risk premium π. This latter is a strategic variable that guides the lender in potential decisions regarding RM contracts, being π a quantity that, within certain limits, can be controlled by the lender in determining the loan contract details.
The rate $\widetilde{r}_s^{RM}$ is involved in the following valuation:

$$\widetilde{A}_t = \delta H_0 \prod_{s=0}^{t-1} \left(1 + \widetilde{r}_s^{RM}\right) = \delta H_0 \prod_{s=0}^{t-1} \left(1 + \widetilde{r}_s + \pi\right) \tag{2}$$

where $\widetilde{A}_t$ is the uncertain loan value at time t, given by the capitalization of its initial amount at random interest rates $\widetilde{r}_s^{RM}$.

3. The *longevity risk*, that is the risk that the borrower lives longer than expected. As a result, the lender faces a prolonged loan duration, potentially leading to an increased number of periodic payments higher than initially expected. It is represented by the uncertainty in the probability in the following formula:

$$\overline{LS}_0 = \sum_{t=1}^{T} \widetilde{V}_t \left(\prod_{s=1}^{t} (1 + \widetilde{r}_s)^{-1} \right) {}_{t-1/1}\widetilde{q}_x \tag{3}$$

where T is the termination date of the contract, $\widetilde{V}_t$=min($\widetilde{A}_t$,$\widetilde{H}_t$) is the value the heirs or those who take over the contract will pay at time t in case of death of the borrower, ${}_{t-1/1}\widetilde{q}_x$ is the conditional probability that the borrower aged x in t=0 and survived until the age $(x + t - 1)$ dies between the ages $(x + t - 1)$ and $(x + t)$.

$\overline{LS}_0$ represents the expected present value, in $t = 0$, of the payments the lender will made each year, conditioned on the borrower's survival. Equation 3 represents the actuarial-financial balance equation between the lender's and the borrower's obligations at the issue time.

In general, without considering the NNEG, the sum paid to the borrower, corresponding to the actuarial present value of the instalments, has to be perfectly in line with the proportion δ of the initial house value, as in the following equation:

$$\overline{LS}_0 = Ra_x = \delta H_0 \tag{4}$$

where R is the installment the insurer will pay at the end of each year. Inserting the NNEG, Eq. 4 no longer guarantees equality between the quantities: the actuarial present value of the instalments $\overline{LS}_0$ will be lower than the value of the loan δH_0 at $t = 0$. Following [10], the difference between the two values is exactly the NNEG value, which quantifies the cost of the guarantee for the borrower:

$$NNEG = \delta H_0 - \overline{LS}_0 \tag{5}$$

Financially, the NNEG can be viewed as the present value of a set of put options, each maturing at a possible future termination date t. We applied the Black '76 model for price setting [18], with the price of houses as the underlying asset and the accrued loan as strike price of the option.

To assess the risk exposure of the lender in a portfolio of RM contracts, we introduce a time-dependent profit/loss function $\overline{P}_t$ which denotes the expected maximum loss in each year t ($0 \leq t \leq T$) based on the active and ceased contracts in the pool in that year. The approach consists of a portfolio construction with an initial number N_0 of RM contracts and, similar to [8] we assume that the distribution of the number of deaths is binomial $B(N_0, q_x)$, where q_x is the probability that the borrower aged x dies before age $x+1$. The time-dependent profit/loss function is:

$$\overline{P}_t = -R\overline{n}_t^A + \overline{V}_t\overline{n}_t^D \tag{6}$$

where $\overline{n}_t^A$ is the expected number at time t of active contracts, $\overline{n}_t^D$ is the expected number of contracts ceased at time t, $\overline{V}_t$ is the expected value the heirs will pay at time t. In Eq. 6, the first term of the right side is the lender's outflow at time t, the last term is the lender's inflow. The calculation is developed in a year-by-year perspective. If inflows are higher than outflows, $\overline{P}_t$ will be positive and the lender will earn a profit. Conversely, there will be a financial loss for the lender.

Thus, $\overline{P}_t$ is the value of the loss/gain that the lender expects to obtain on average in each t and can be considered the average scenario over all sources of risk.

In the following, we will consider the time-dependent profit/loss function and will set a method to evaluating the impact of each risk factor on it. In this regard, we will consider the Value-at-Risk to measure the risks impacting on the portfolio.

2.1 Single Component VaR Approach

The Value-at-Risk (VaR) measures the worst expected loss under normal market conditions over a specific time interval at a given confidence level [3]. In a nutshell, VaR answers the question: "how much can I lose with x% probability over a given time horizon?" [13].

Following [15], the VaR of a random variable Z is the quantity that verifies the following equation:

$$\mathbb{P}[Z \leq VaR_\alpha(Z)] = \alpha \tag{7}$$

where α is a low probability and $1-\alpha$ is the confidence level. The amount $VaR_\alpha(Z)$ is the α-percentile of the probability distribution of Z. If F is an invertible function, then:

$$VaR_\alpha[Z] = F^{-1}(\alpha) \tag{8}$$

We will consider the VaR of the profit/loss function P_t, $t = 1, 2, \ldots$, at the end of each year during the policy duration identifying the contribution of each source of risk to total profit, in a scenario perspective.

In this way, we will provide the lender not only with a mapping of the main risks on the portfolio, identifying their trend over time, but also highlight some relevant decision variables in order to build contractual control strategies.

The methodology we will follow to study the influence of each risk source on VaR has been structured and implemented in the following three steps:

- We apply to the function P_t the formula:

$$\mathbb{P}\left[P_t^i \leq VaR_t^{\alpha}\left(P_t^i\right)\right] = \alpha \tag{9}$$

  with $i = 1, 2, 3$, referring, respectively, to the house price risk, the longevity risk and the financial risk.
- Once fixed i, we obtain the expected profit or loss under the worst-case scenario for the *i-th* risk, while keeping the other two risk variables at their average values.
- The α-**quantile** (corresponding to the worst-case scenario) is taken for the i-specific source of risk under investigation, while all other sources of risk are kept at their expected values.

The VaR calculation can be considered the VaR referred to each risk component, conditioned on the average values of the others.

From a computational perspective, Monte Carlo simulations are applied to each stochastic process chosen to model the evolution of the corresponding variable involved in the contract. This allows us to obtain, for each instant t, the distribution of each risk component (house prices, risk-free interest rates and longevity). By calculating the average value for all of them and entering it into Eq. 6, we find the loss/gain value that the lender expects to obtain on average in each t. If we analyze the distribution of a single i-source of risk while keeping the other two at their fixed average values, we can calculate P_t^i for each quantile and, setting a low α-value, we obtain $VaR_t^{\alpha}(P_t^i)$.

3 Numerical Results

Aim of this section is to measure the α-quantile of the time-dependent profit/loss function distinguishing the i-specific source of risk, in order to provide a dimensional ordering of the impacts of each risk source on VaR.

Our example focuses on a portfolio of 10.000 homogeneous Reverse Mortgage contracts issued by a financial institution, bank or insurance company to Italian borrowers (both males and females) aged 75 (see for example [6]). Due to data availability, we fix in 2022 the issue year of the contracts; the payment is assumed to be in form of periodic streams. Let us assume that the contract expires either when the individual dies or when the expiry date is reached, once the insured is 105 years old, considered as the extreme age. We set the risk premium π equal to 0,5%. The initial price of the house H_0 is assumed to be €200.000, the lending percentage δ is assumed equal to 30%, so that the maximum loan amount per insured person is €60.000.

To assess the risk exposure, we calibrate models on historical Italian data to evaluate house price risk, financial risk and longevity risk.

To model house price dynamics, we adopt a Geometric Brownian Motion (GBM) model, following the approach of [10]. According to this model the future values of the house prices are given by the Eq. 10:

$$dH_t = \mu H_t dt + \sigma H_t dZ_t \tag{10}$$

where H_t is the house value at time t, μ is the drift of the process, σ is the volatility of housing returns and Z_t is the standard Wiener process.

Calibration is performed using quarterly Italian nominal house price indices from the first quarter Q1 of 1970 to the fourth quarter Q4 of 2021 [14], fitting the model on the logarithmic returns. Based on the estimated parameters, we generate 10.000 Monte Carlo simulations of quarterly house price trajectories spanning the period 20222053. Annualized house prices are obtained by averaging the quarterly values within each simulation. The expected trajectory, representing the average house price evolution, is then computed by averaging across all simulations.

To model the interest rate evolution, we calibrate the Cox-Ingersoll-Ross (CIR) model on monthly 10-year Italian government bond yields observed from April 1991 to December 2021 [2].

The CIR model, ensuring non negative values, is represented by the following Stochastic Differential Equation:

$$dr_t = k(\theta - r_t)dt + s_t \sqrt{r_t}\, dZ_t \tag{11}$$

where r_t is the interest rate value at time t, k is the speed of the mean reversion to the long-term rate θ, s_t is the standard deviation of risk-free rates series and Z_t is the standard Wiener process.

Similar to house price risk, we conduct 10.000 Monte Carlo simulations and determine the expected financial risk trajectory using the mean values of the simulated paths. Longevity risk is assessed through the Lee-Carter mortality model (see Eq. 12), calibrated on Unisex Italian mortality data from 1963 to 2021 [11], according to the well known formula:

$$ln(m_{x,t}) = \alpha_x + b_x k_t \tag{12}$$

where $m_{x,t}$ is the observed central death rate at age x in year t, α_x is the average age-specific pattern of mortality, b_x is the pattern of deviations from the age of profile, k_t is the time index describing mortality trend.

We generate 10.000 stochastic mortality projections, computing the expected longevity risk trajectory by averaging across all simulations over each t.

3.1 Single Component-VaR Results

By exploiting all simulated paths derived in the last paragraph, we can determine the Value-at-Risk (VaR) for each risk factor, averaging out the other risk components.

– *House-VaR*: for house price risk, we set i=1 in Eq. 9. The corresponding house price VaR is computed as:

$$VaR_t^{\alpha}(P_t^1) = -R\overline{n}_t^A + min(\overline{A}_t, VaR_t^{\alpha}(H_t))\overline{n}_t^D \tag{13}$$

where $VaR_t^{\alpha}(P_t^1)$ represents the loss threshold exceeded with probability α, referred to as House-VaR. This metric captures the worst-case evolution of house prices while maintaining the average scenario for financial and longevity risks.

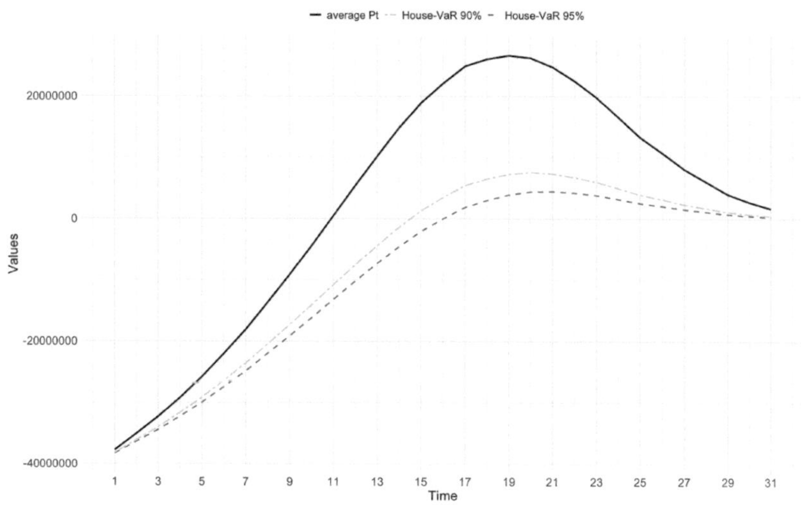

Fig. 1. Average scenario vs House-VaR

The black solid line in Fig. 1 represents the loss/profit function according to Eq. 6. This is the average scenario that the lender expects to face considering all risks simultaneously equal to their corresponding average value. The grey dashed line and the light grey dashed/dotted line respectively represent the VaR at the 95% and 90% confidence level (α=0.05 and α=0.10) considering just house prices in the worst-case scenario.

Under the average scenario the lender will start making profits from the 11th year onward. When considering the House-VaR, losses will cease from the 16th year onward at 95% confidence level and from approximately the 14.5th year onward at 90% confidence level. The minimum profit is significantly lower compared to the average case and converges to zero at maturity ($t = 31$). This highlights the substantial impact that house price risk has on this type of contract.

– *Longevity-VaR*: we set i=2 in Eq. 9, and compute VaR_t^{α} as:

$$VaR_t^{\alpha}(P_t^2) = -RVaR_t^{\alpha}(n_t^A) + min(\overline{A}_t, \overline{H}_t)VaR_t^{\alpha}(n_t^D) \tag{14}$$

Here, the worst-case scenario corresponds to an improvement in mortality rates, which increases the lender's exposure by extending contract durations.
The probability of exceeding $VaR_t^\alpha(P_t^2)$, the Longevity-VaR, is α for each t.

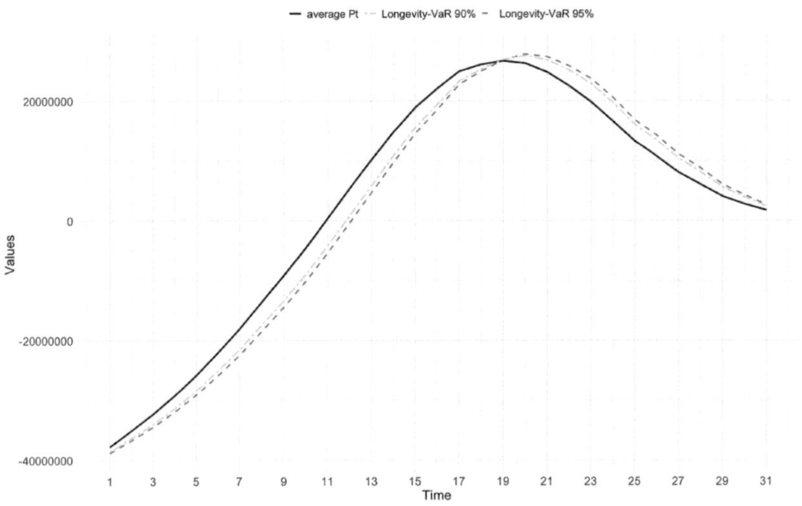

Fig. 2. Average scenario vs Longevity-VaR

By choosing two different alpha values such as $\alpha = 0.10$ and $\alpha = 0.05$, it is clear from Fig. 2 that, for both confidence levels, the maximum loss/minimum gain incurred with a probability of 90% and 95% is very close to the average scenario. Compared to the average scenario, where, as mentioned above, the lender starts to gain from t = 11, with Longevity-VaR the losses will stop from around t = 12 for both confidence levels. Actually, from t = 19 onward, the average scenario even underestimates the minimum gain. This suggests that an error in mortality forecasts is not particularly significant when compared to the average scenario that the lender expects to face.

- *Financial-VaR*: concerning the Financial risk, we set i=3 in Eq. 9, and compute VaR_t^α as:

$$VaR_t^\alpha(P_t^3) = -R\overline{n}_t^A + min(VaR_t^\alpha(A_t), \overline{H}_t)\overline{n}_t^D \tag{15}$$

Again, the worst-case scenario corresponds to low risk-free interest rates and an average scenario for home and longevity risks. The probability of exceeding $VaR_t^\alpha(P_t^3)$, the Financial-VaR, is α for each t.

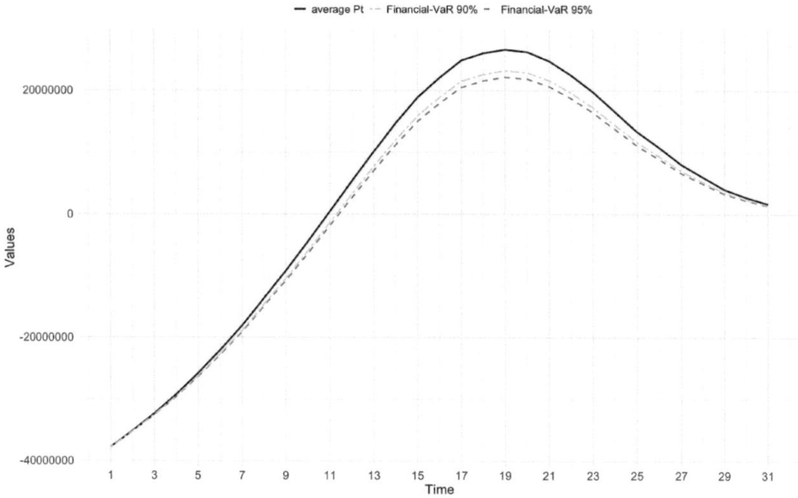

Fig. 3. Average scenario vs Financial-VaR

As for the longevity risk case, Fig. 3 shows that, for both confidence levels, the maximum loss/minimum gain incurred with a probability of 90% and 95% is very close to the average scenario. However, unlike the previous cases, in this context, the lender can manage risk through the analysis of the sensitivity to different risk premiums. With low quantiles of interest rates, indeed, the minimum between $VaR_t^\alpha(A_t)$ and $\overline{H}_t$ is given by $VaR_t^\alpha(A_t)$. The lender can regulate $VaR_t^\alpha(A_t)$ through the risk premium π, adjusting it to mitigate the maximum potential loss and enhance the minimum possible gain. It must be considered that a high risk premium would lead to a situation where the minimum between $VaR_t^\alpha(A_t)$ and $\overline{H}_t$ is determined by $\overline{H}_t$ (i.e., the property value).

We set the initial risk premium equal to 0.5% but the lender can manage financial risk by increasing the risk premium up to a maximum of 1.2%. At this threshold, the maximum loss and minimum gain align precisely with the average scenario P_t. The graphical representation of the sensitivity of the Financial-VaR to changes in the risk premium is shown in Fig. 4. Here, the risk premium varies from 0% to 1.2%, and correspondingly, as can be observed, the Financial-VaR converges towards the average scenario, represented by the black solid line. This behavior stems from the structure of the profit/loss function: as π increases, A_t tends to dominate, but the Financial-VaR is ultimately bounded by H_t, explaining the observed stabilization.

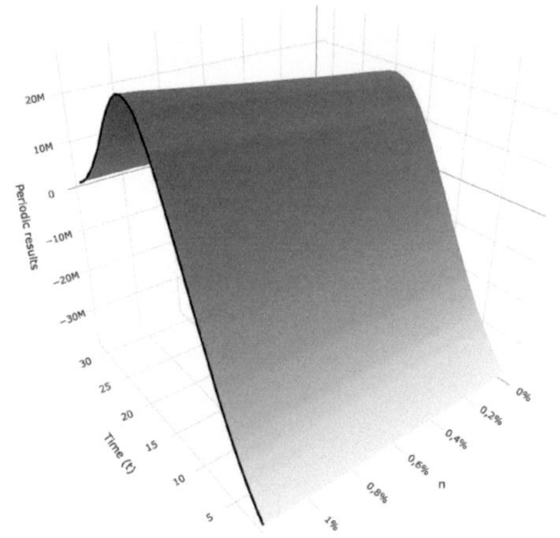

Fig. 4. Behaviour of Financial-VaR when the risk premium changes

4 Conclusions

In this work we provide lenders of Reverse Mortgage contracts with insights into the impact of each of the three main risk factors (House Price Risk, Longevity Risk and Financial Risk) through a specific time-dependent profit/loss function. For this purpose, we will use the Value-at-Risk tool as a measure of risk, considering the impact of each of these components on the annual period result, represented by a profit/loss function. In order to obtain the VaR results for each component, a methodology based on conditional averages was developed to isolate the influence of each component in a general scenario assumed as the expected scenario for the other two. Our findings emphasize that house price risk is the most significant risk factor of a Reverse Mortgage portfolio (RM). Although longevity risk contributes to uncertainty, its effect isn't relevant. The results highlight that, beyond a certain time horizon, the lender's exposure to this specific risk component stabilizes, with profitability emerging in the long run. Furthermore, our analysis demonstrates that the lender can control the financial risk through the analysis of the sensitivity of the profit/loss function to the variability of the risk premiums. This can be considered as a decision variable, quantified in order to mitigate the maximum potential loss and enhance the minimum possible gain. However, an excessive increase in the risk premium may shift the predominant factor in calculating the minimum value of $\overline{V}_t$, which represents the lender's expected inflows, from $\overline{A}_t$ to $\overline{H}_t$, thereby limiting the lender's ability to control financial risk exposure. Within our framework, we find that an optimal risk premium adjustment up to 1.2% allows for a financial risk profile in which the maximum loss and minimum gain converge to the average scenario. These insights are crucial for the development of risk management strategies and actuarial models in RM markets. They suggest that managing house price risk and optimizing the risk premium are key factors in ensuring the financial sustainability of Reverse Mortgages.

References

1. Alai, D.H., Chen, H., Cho, D., Hanewald, K., Sherris, M.: Devoloping equity release markets: risk analysis for reverse mortgages and home reversions. North American Actuarial J. **18**, 217–241 (2014). https://doi.org/10.1080/10920277.2014.882252
2. Banca d'Italia (2024). Rendimenti a scadenza lordi [BMK0100]. https://infostat.bancaditalia.it/inquiry/home?spyglass/taxo:CUBESET=&ITEMSELEZ=&OPEN=false&ep:LC=IT&COMM=BANKITALIA&ENV=LIVE&CTX=DIFF&IDX=1&/view:CUBEIDS=BMK0100 Accessed 01 Feb 2025
3. Benninga, S., Wienwr, Z.: Value-at-risk (VaR). Math. Educ. Res. **7**(4), 39–45 (1998)
4. Ceccarelli, C., Cappiello, A.: Housing as Main Component of the Italian Households' Wealth Importance of the Notarial Data and Reflections on the Role of the Notary for the Protection of Property, Households and Vulnerable Persons. Households and Vulnerable Persons (2023). https://doi.org/10.2139/ssrn.4615882
5. CENSIS e Federproprietà (2022). Gli italiani e la casa: come cambieranno valori e funzioni della casa nell'Italia post-pandemia. Sintesi dei principali risultati. Primo Rapporto Federproprietà - CENSIS. https://www.censis.it/sites/default/files/downloads/Sintesi%20dei%20principali%20risultati%20Federpropriet%C3%A0-Censis%20-%20Gli%20italiani%20e%20la%20casa.pdf Accessed 05 Mar 2025
6. Cho, D., Hanewald, K., Sherris, M.: Risk analysis for reverse mortgages with different payout designs. Asia-Pacific J. Risk Insurance **9**(1), 77–105 (2015). https://doi.org/10.1515/apjri-2014-0012
7. De la Fuente, I., Navarro, E., Serna, G.: Reverse mortgage risks Time evolution of VaR in lump-sum solutions. Mathematics **8**(11), 2043 (2020). https://doi.org/10.3390/math8112043
8. De la Fuente, I., Navarro, E., Serna, G.: Proposal for calculating regulatory capital requirements for reverse mortgages. Socio-Economic Planning Sci. **88**, 101659 (2023). https://doi.org/10.1016/j.seps.2023.101659
9. Di Lorenzo, E., Piscopo, G., Sibillo, M.: Addressing the economic and demographic complexity via a neural network approach: risk measures for reverse mortgages. Comput. Manag. Sci. **21**(1), 11 (2024). https://doi.org/10.1007/s10287-023-00491-x
10. Di Lorenzo, E., Piscopo, G., Sibillo, M., Tizzano, R.: Reverse mortgage and risk profile awareness: proposals for securitization. Appli. Stochastic Models Bus. Indust. **38**(2), 353–369 (2022). https://doi.org/10.1002/asmb.2664
11. HMD 2025. Human Mortality Database. Max Planck Institute for Demographic Research (Germany), University of California, Berkeley (USA), and French Institute for Demographic Studies (France). www.mortality.org, Accessed 01 Feb 2025
12. Itinerari previdenziali. https://www.itinerariprevidenziali.it/site/home/ilpunto/economia-societa/la-ricchezza-degli-over-65-in-italia.html, Accessed 5 Mar 2025
13. Morgan, J.P.: Reuters. RiskMetricsTM—Technical Document (1996)

14. OECD, Housing prices (indicator). https://doi.org/10.1787/63008438-en Accessed 01 Feb 2025
15. Olivieri, A., Pitacco, E.: Introduction to insurance mathematics, 2nd edn. Springer-Verlag, Berlin Heidelberg (2011)
16. Wang, L., Valdez, E.A., Piggott, J.: Securitization of longevity risk in reverse mortgages. North Am. Actuarial J. **12**(4), 345371 (2008). https://doi.org/10.1080/10920277.2008.10597529
17. Wold Bank Group, Data. https://data.worldbank.org/indicator/SP.POP.65UP.TO.ZS?end=2023&locations=IT&start=1960&view=chart, and https://data.worldbank.org/indicator/SP.DYN.LE00.IN?locations=IT, Accessed 5 Mar 2025
18. Dowd, K., Buckner, D., Blake, D., Fry, J.: The valuation of no-negative equity guarantees and equity release mortgages. Econ. Lett. **184**, 108669 (2019). https://doi.org/10.1016/j.econlet.2019.108669

Climate Litigation Risk: Comparing Linear and Non Linear Losses of Insurances

Lorenzo Frattarolo(✉)

Department of Economics, University of Verona, Via Cantarane 24, 37129 Verona, Italy
lorenzo.frattarolo@univr.it

Abstract. We introduce a methodology for quantifying climate litigation risks stemming from insurers' investment in fossil fuels-related assets. It builds on recent advances in liability attribution of climate warming damages to fossil fuel producers, obtaining the Carbon Majors Index (CMI) as a cumulative emission-weighted portfolio of major polluters' stocks. We validate the CDI by showing that the percentage of fossil fuel investments is a significant determinant of the yearly exposure to the CMI. We compare linear losses from linear exposures and non-linear losses from Exposure CoVaR. The size of the large polluters' possible losses from climate litigation is calibrated using losses of Tobacco companies after the US Tobacco Master Settlement Agreement. Our results show that linear and non-linear losses differ in size and ranking of institutions. This suggests the importance of a non-linear long-tail risks approach in assessing climate litigation risk for insurance.

Keywords: Carbon Majors · End-to-end attribution · Fossil Fuel Assets · CoVar

1 Introduction

In, by now, famous speech given at Lloyd's of London in 2015, Mark Carney, then-Governor of the Bank of England and Chairman of the Financial Stability Board, was among the first to suggest that climate litigation poses "long-tail risks" for insurers that may be "significant, uncertain, and non-linear." [5]. The same year, the Financial Stability Board set up a Task Force on Climate-related Financial Disclosures (the "TCFD"). In 2017, the TCFD published a set of recommendations and a model disclosure framework, highlighting climate litigation as a climate-related "transition risk" that affects the entire corporation's structure.

The insurance sector is a large but overlooked segment of institutional investors. Compared to asset managers, insurers did not receive the same level of scrutiny for the climate-related impact of their investments in fossil fuel-related

M. La Rocca et al. (Eds.): WMAF 2025, *New Perspectives in Mathematical and Statistical Methods for Actuarial Sciences and Finance*, pp. 157–168, 2025.
https://doi.org/10.1007/978-3-032-05551-4_14

assets. As the Wall Street Journal reports, in 2023, State Farm and Berkshire Hathaway continued to increase their positions in fossil fuel-related assets [10].

This behavior exposes the insurance investment to the risk of climate litigation, a rising tide that the major polluters face. The total number of climate change cases has more than doubled, from 884 in 2017 to 2,180 in 2022, with most cases brought in the US [7,8] and according to [9], since 2015, there were already 85 "polluter pays" cases filed with 2024 alone contributing with 11 new cases. While climate litigation cases are growing exponentially, their success is still limited. Still, new methodologies, such as the one described in [3], are making climate liability of large polluters scientifically grounded, increasing the probability that those companies will soon bear part of the cost of climate change.

This paper investigates a framework for assessing the risk from climate litigation of large polluters affecting insurers through their investment decisions. It builds on recent advances in "end-to-end" attribution of specific damages from warming to fossil fuel producers [3] by creating a Carbon Majors Index (CMI) as a cumulative emission-weighted portfolio of major polluters' stocks. The paper investigates the percentage of fossil fuel investments reported in the Climate Risk Carbon Initiative of the California Department of Insurance (CDI) as a possible determinant of the yearly exposure to the CMI. Then, following the perspective of a non-linear, long-tail risk, it utilizes the Exposure CoVaR [1] to quantify the climate litigation risk for each insurer. Finally, the size of the large polluters' possible losses from climate litigation is computed by drawing a parallel with the losses incurred by Big Tobacco due to the US Tobacco Master Settlement Agreement [12]. The paper is structured as follows. Section 2 introduces the conceptual framework for assessing climate litigation risk using market data.

Section 3 discusses the different data sources used in the analysis and introduces the CMI. Section 4 compute the climate litigation exposures and 5 summarize and discuss the results.

2 Conceptual Framework

This section describes the simplified model for quantifying climate liability risk using only stock market data. First, we describe how we compute possible losses for large polluters due to climate litigation. We use the methodology outlined in [3] to attribute, for each large polluter, the cumulative emission contribution and damages caused by heatwaves over the period 1991–2020. In [3], a counterfactual simulation by a reduced-complexity climate model, eliminating the cumulative emission contribution for each large polluter, is used to compute the global temperature rise attributable to the polluter's emissions. This rise in global temperature is then converted into local temperature changes. Finally, an empirically derived damage function that generalizes the relationship between extreme heat intensity and economic Growth is used to convert the damage into local GDP losses. In this way, the authors estimate that investor-owned polluters contributed \$ 14 trillion in global heat-driven losses. The authors also found that global losses from extreme heat scale quasi-linearly with cumulative emission contributions. Each extra percentage point generates a further \$ 834 billion in global economic losses from extreme heat in 1991–2020.

Then, in our framework, the losses attributable to the p-th fossil fuel company, in year y, $L_{p,y}$ are the percentage of cumulative emissions $w_{p,y}$ reached in y multiplied by the \$ 14 trillion in global heat-driven losses from investor-owned companies $TOTL_{1991-2020}$. Those losses then would substantiate a climate litigation claim, but in the end, only a fraction of those figures would be paid by polluters. For example, the fraction paid after the litigation by tobacco companies due to the US Tobacco Master Settlement Agreement was $\Delta_{litigation} = 0.27$ [12]. We summarize the model for climate litigation losses of polluters p in year y in the following equation:

$$L_{p,y} = \Delta_{litigation} w_{p,y} TOTL_{1991-2020} \tag{1}$$

The percentage of its potential climate litigation losses is then independent from $\Delta_{litigation}$:

$$\frac{L_{p,y}}{\sum_p L_{p,y}} = \frac{\Delta_{litigation} w_{p,y} TOTL_{1991-2020}}{\Delta_{litigation} TOTL_{1991-2020}} = w_{p,y}$$

To understand how those losses are transferred to the market evaluation of polluters, we build the CMI as a tradable portfolio In which the daily log returns for the polluter p stock p in day t, $r_{p,t}$ are weighted by the percentage of its potential climate litigation losses for that year $w_{p,y}$:

$$r_{CMI,t} = \sum_{p=1}^{P} w_{p,y} r_{p,t} \tag{2}$$

The yearly loss for each insurer is then computed in both linear and non-linear ways.

For the linear loss, the exposure, $\gamma_{i,y}^{CMI}$, for each year y, comes from a linear regression on the daily log returns of insurer i stock $r_{i,t}$ on the CMI, controlling for the factors in the Fama and French 5-factor model [6]:

$$r_{i,t} = \gamma_{i,y}^{CMI} r_{CMI,t} + \sum_{f=1}^{5} \beta_{i,y}^{F_f} r_{F_f,t} + \epsilon_{i,t} \tag{3}$$

We then multiply the estimated $\gamma_{i,y}^{CMI}$ by an estimate of market reaction to litigation using the analogy with Big Tobacco. In particular, [11] reports an $\Delta r = 8\%$ decrease in daily market returns of a value-weighted tobacco companies portfolio after the US Tobacco Master Settlement Agreement. The Master Settlement Agreement also requires tobacco companies to make annual payments proportional to the market share, i.e., to the quantity of tobacco products sold in that year. Analogously, the end-to-end attribution aims to attribute climate change damages based on the cumulative emissions of each Carbon Major. We conjecture that, since the attribution of the cost of long-term negative externalities is comparable, the market reaction could also be similar. Suppose the total damages are a determinant of market response to litigation. In that case, the estimation is conservative because the damages attributed to tobacco companies were around \$ 750 billion, while the conservative estimate of damages from [3] is \$ 14 trillion. If, instead, the reaction is primarily due to reputational damage, it could be appropriate.

The estimated linear daily loss of insurer i in year y is then:

$$LL_{i,y} = \gamma^{\hat{CMI}}{}_{i,y} \times \Delta r \tag{4}$$

We compute the exposure Exposure-CoVaR [1] for the non-linear loss because it measures the "tail-dependency" between two random return variables. We denote by $\mathrm{CoVaR}_q^{i|CMI}$ the VaR of insurer i conditional on the event $-r_{CMI} \leq \mathrm{VaR}_q^{CMI}$. That is, $\mathrm{CoVaR}_q^{i|CMI}$ implicitly defined by the q%-quantile of the conditional probability distribution:

$$\Pr\left(-r_i \,\middle|\{-r_{CMI} \leq \mathrm{VaR}_q^{CMI}\} \leq \mathrm{CoVaR}_q^{i|CMI} = q\%\right) \tag{5}$$

Paralleling the computation done in the linear loss case, the CoVaR level q is computed implicitly by:

$$\mathrm{VaR}_q^{CMI} = \Delta r \tag{6}$$

Exposure-CoVaR measures an individual institution's exposure to CMI distress and is similar to the stress tests performed by individual institutions and regulators. If the CMI is distressed, the insurer will likely face difficulties, resulting in a high Exposure-CoVaR. To estimate the CoVaR, we use quantile regressions that control for market regime, with lagged systematic state variables as in [1].

Non-linear losses are then defined as:

$$NLL_{i,y} = \mathrm{CoVaR}_q^{i|CMI} \tag{7}$$

with q satisfying (6).

In the context of insurers' enterprise risk management (ERM) practices, our linear and non-linear losses estimates can be used as a monitoring tool for understanding and quantifying the financial risk attributable to climate litigation risk. If the risk is relevant, our framework identifies the riskier portfolio positions as those in assets of Carbon Majors with a high percentage of cumulative emissions $w_{p,y}$. Risk mitigation could then be obtained by changing the insurer's portfolio accordingly.

3 Carbon Majors Index

This section describes data sources and introduces and validates the Carbon Majors Index. The computation of the CMI requires stock returns and emissions for large polluters. The list of polluters and their emissions is sourced from the publicly available Carbon Majors database[1], which utilizes publicly available

[1] https://carbonmajors.org/Downloads.

information from company regulatory filings and emissions factors. These data include scope 1 and scope 3 emissions, which include emissions from the production and combustion of the fossil fuels sold by these companies. Stock Market data is sourced from Yahoo Finance and converted to USD using exchange rates provided by the same provider, as needed. For some Carbon Majors, the stock market data was unavailable on Yahoo Finance. The list of the 67 out of 77 carbon majors included in the analysis is reported in the appendix. Additionally, data for some polluters begins later in the sample. We include polluters with stock prices for each year's working days. Cumulative emission weights are computed by dividing by the sum of the cumulative emissions for the companies considered that year. In Fig. 1 we report the daily level of the CMI from 1990 to 2023.

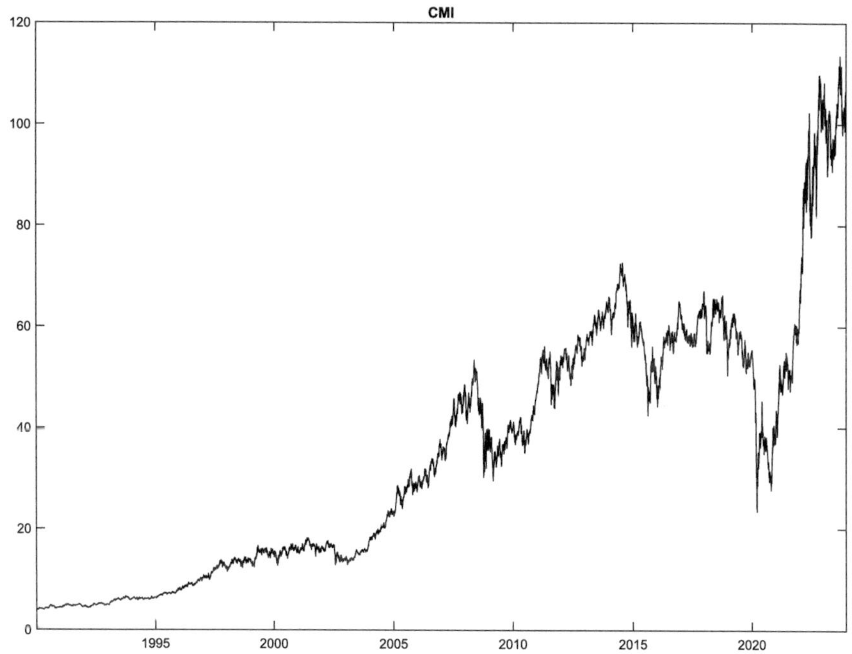

Fig. 1. Carbon Majors Index daily level from 02/01/1990 to 29/12/2023

Low levels of CMI were around the subprime crisis, the Paris Agreement, and the COVID-19 pandemic. The picture also shows a large upward jump at the beginning of the Ukrainian energy crisis. To furtherly validate the CMI, we investigate if investment in fossil fuel-related assets is a determinant of the estimated linear exposure to the index $\gamma^{\hat{CMI}}{}_{i,y}$. In particular, we focus on investment in fossil fuel-related assets by insurance companies active in California. The California Department of Insurance (CDI) required insurers doing business in California with more than \$100 million in annual premiums to publicly disclose their fossil fuel investments from 2015 to 2019[2]. In this study, we selected a subsample of insurances. From the CDI reports, we took the top 100 insurance companies by assets under management in 2019. We selected publicly listed companies or those with a publicly listed parent company, for which data is available on Yahoo Finance. The percentage of investment in fossil fuels for parent companies was calculated as the sum of fossil fuel assets divided by the sum of assets under management, only for the subsidiaries in the original sample. Details of the 29 insurance companies included in the sample are reported in the appendix. For each year y from 2015 to 2019 and each insurer $i = 1, \ldots, 29$ we estimated $\hat{\gamma}_{i,y}^{CMI}$ as in (3).

We then estimated a panel fixed effect model relating $FFA\%_{i,y}$ the percentage of investment in fossil fuel assets to $\hat{\gamma}_{i,y}^{CMI}$:

$$\hat{\gamma}_{i,y}^{CMI} = a_i + bFFA\%_{i,y} + \epsilon_{i,y} \tag{8}$$

The b coefficient estimate using White-Arellano robust standard errors reported in Table 1, shows that the percentage of investment in fossil fuel assets is a determinant of CMI exposure at the significance level of 10%.

Table 1. b estimation from (8) using using White-Arellano robust standard errors

	Estimate	Std. Err.	Stat	P-value
b	-0.02153	0.011505	-1.871	0.06446

We remark that the objective of the analysis is to establish an association between exposures and investment in fossil fuel assets. Investigating a causal link through instrumental variable regression, as suggested by an anonymous referee, is not feasible with the CDI dataset. In fact, we cannot identify a valid instrument with the limited fossil fuel asset information contained in the dataset. Causal analysis will be pursued in future related research.

4 Climate Litigation Losses

In this section, we calculate the potential losses due to climate litigation of carbon majors in both linear and non-linear ways, considering 2023, the last year for which emissions data from the Carbon Majors database are available. The insurance considered is the same as that used in the panel regression of the previous section. Due to limited space, we report the results only for the top ten potential losses. Linear losses are defined in Eq. (4) and reported in Table 2

[2] see https://interactive.web.insurance.ca.gov/apex_extprd/f?p=250:40:0::NO:10,20,30,40:: and https://interactive.web.insurance.ca.gov/apex_extprd/f?p=260:40:15898370807872::NO:10,20,30,40::.

Table 2. Top 10 Potential linear losses in 2023

Insurance Name	Ticker	$LL_{I,2023}$
Lincoln National Life Insurance Co.	LNC	0.0651355
Brighthouse Financial. Inc.	BHF	0.0547344
Equitable Holdings. Inc.	EQH	0.0521064
Metropolitan Life Insurance Co.	MET	0.0454947
Principal Financial Group	PFG	0.0448755
Genworth Financial. Inc.	GNW	0.0447648
American International Group	AIG	0.0446245
CNO Financial Group	CNO	0.0422804
Prudential Insurance Co. of America	PRU	0.0403163
Voya Financial. Inc.	VOYA	0.0366832

In the linear case, potential losses are always lower than $\Delta r = 0.08$, but in any case, sizable.

The non-linear case requires some additional explanation. First, we use the same systematic state variables to control for the market regime as [1], except for the TED spread, which is substituted by the spread between the three-month SOFR rate and the three-month secondary market Treasury bill rate, and equity volatility, which is substituted by the CBOE VIX. Additionally, we excluded the real estate sector return, as it was deemed irrelevant for the current exercise. The level of the VaR corresponding to $\Delta r = 0.08$ is computed considering the unconditional VaR, not including the conditioning on the state variables. We

Table 3. Top 10 Potential non-linear losses in 2023

Insurance Name	Ticker	$NNL_{i,2023}$
Allstate Insurance Co.	ALL	0.1092482
Principal Financial Group	PFG	0.1009361
Munich Re	MURGY	0.0759652
Genworth Financial. Inc.	GNW	0.060881
Primerica. Inc.	PRI	0.0384764
The Travelers Companies. Inc.	TRV	0.0343867
Sun Life Financial Inc.	SLF	0.0279611
CNA Financial Corporation	CNA	0.0261712
Berkshire Hathaway	BRK-A	0.0203527
Equitable Holdings. Inc.	EQH	0.018046

obtained $q = 0.9982$ using the entire history of CMI[3]. The top 10 non-linear losses are reported in Table 3

Concerning the linear case, we note that the top 3 institutions suffer losses bigger than $\Delta r = 0.08$ and that the last 5 positions have lower losses than the last position of the linear case. Principal Financial Group and Genworth Financial are the only two insurance companies reporting top losses in both cases. Berkshire Hathaway, reported to have increased Fossil Fuel-related asset investment in 2023 [10], enters the top 10 only in the non-linear case.

5 Conclusion

In this paper, we propose a conceptual framework to assess investment-related climate litigation risk for insurance companies, using publicly available data. We built the CMI, a portfolio of stocks from major polluters weighted by the percentage of their cumulative emissions. We justify the weighting scheme by recent advances in climate liability attribution reported in [3]. We validate the portfolio by showing that fossil fuel-related asset investment is a significant determinant of CMI exposure. We calibrate potential losses of CDI in the event of Climate litigation by drawing a parallel with Big Tobacco after the US Tobacco Master Settlement Agreement. We compare linear and non-linear potential losses of 29 of the top insurance companies doing business in California. The top 3 non-linear losses exceed the input shock from Carbon Majors. Only two companies appear to have top losses in both cases, leading to entirely different rankings. The methodology outlined in the paper should be considered as a proof of concept for the development of a climate litigation stress test in line with the literature pioneered in [2][4]. We acknowledge, as was pointed out by an anonymous referee, that the sample of insurance companies and polluting firms analyzed is not exhaustive, limiting the generalizability of the results. The choice of using only publicly available data (CDI dataset and Yahoo Finance) ensured replicability. Still, it limited our coverage, excluding most international insurers beyond California and fossil fuel firms from emerging markets, which may be subject to different regulatory and litigation dynamics. Nevertheless, our focus on US companies is also justified by the fact that climate litigations are mostly filed in this country [9]. Our findings, albeit preliminary, highlight the importance of dealing with the "long-tail risks" for insurers that may be "significant, uncertain, and non-linear" [5] when considering the climate litigation. In the context of ERM, we suggest using our non-linear losses for monitoring climate litigation risk. Future research should examine the possibility of applying the proposed framework to assess the climate litigation risk of other institutional investors by leveraging on proprietary ORBIS data as in [2].

---

[3] In Sect. B of the appendix, we report a sensitivity analysis for Δr in the Exposure CoVaR Case.

[4] See Sect. 4 in [4] for additional references.

A Stocks Market Data Included

Table 4. Carbon Majors in the Sample

Company Name	Ticker	Stock Market	Currency
Adani Enterprises	ADANIENT.NS	NSE	INR
Adaro Energy	ADRO.JK	IDX	IDR
Alliance Resource Partners	ARLP	NYSE	USD
Alpha Metallurgical Resources	AMR	NYSE	USD
Anglo American	AAL.L	LSE	GBP
Antero	AR	NYSE	USD
APA Corporation	APA	NASDAQ	USD
Arch Resources	CNR	NYSE	USD
CONSOL Energy	BANPU.BK	NYSE	USD
BASF	BAS.DE	ETR	EUR
BHP	BHP.AX	ASX	AUD
BP	BP.L	LSE	GBP
Bumi Resources	BUMI.JK	IDX	IDR
Canadian Natural Resources	CNQ.TO	TSX	CAD
Cemex	CX	NYSE	USD
Cenovus Energy	CVE.TO	TSX	CAD
Chesapeake Energy	EXE	NASDAQ	USD
Southwestern Energy	CVX	NASDAQ	USD
Chevron	CNX	NYSE	USD
CNX Resources	COP	NYSE	USD
ConocoPhillips	CTRA	NYSE	USD
CRH	CRH.L	LSE	EUR
Devon Energy	DVN	NYSE	USD
Eni	ENI.MI	BIT	EUR
EOG Resources	EOG	NYSE	USD
EQT Corporation	EQT	NYSE	USD
Exxaro Resources Ltd	XOM	JSE	ZAR
ExxonMobil	GLEN.L	NYSE	USD
Glencore	HEI.DE	LSE	GBP
Heidelberg Materials	HES	ETR	EUR
Hess Corporation	HOLN.SW	NYSE	USD
Holcim Group	1605.T	SWX	CHF
Lukoil	LKOH.ME	MCX	RUB
Murphy Oil	MUR	NYSE	USD
Novatek	NVTK.ME	MCX	RUB
Obsidian Energy	OBE.TO	TSX	CAD
Occidental Petroleum	OXY	NYSE	USD
OMV Group	OMV.VI	WBAG	EUR
Orlen	PKN.WA	WSE	PLN
Ovintiv	OVV	NYSE	USD
Peabody Energy	BTU	NYSE	USD
Repsol	REP.MC	BME	EUR
Rio Tinto	RIO.AX	ASX	AUD
RWE	RWE.DE	ETR	EUR
Santos	STO.AX	ASX	AUD
Shell	SHEL.L	LSE	GBP
SM Energy	SM	NYSE	USD
Suncor Energy	SU.TO	TSX	CAD
Surgutneftegas	SNGS.ME	MCX	RUB
Taiheiyo Cement	5233.T	TSE	JPY
Teck Resources	TECK-A.TO	TSX	CAD
TotalEnergies	TTE.PA	EPA	EUR
Tourmaline Oil	TOU.TO	TSX	CAD
Tullow Oil	TLW.L	LSE	GBP
Vale	VALE3.SA	B3	BRL
Vistra	VST	NYSE	USD
Whitehaven Coal	WHC.AX	ASX	AUD
Woodside Energy	WDS.AX	ASX	AUD
EVRAZ	EVR.L	LSE	GBP
Mechel	MTLRP.ME	MCX	RUB
En+ Group	ENPL.L	LSE	GBP
Severstal	CHMF.ME	MCX	RUB
UltraTech Cement	ULTRACEMCO.NS	NSE	INR
Tatneft	TATN.ME	MCX	RUB
Mitsubishi Corporation	8058.T	TSE	JPY
Inner Mongolia Yitai Group	900948.SS	SSE	CNY
ArcelorMittal	MT.AS	EPA	EUR

Table 5. Insurances in the sample

Company Name	Publicly Traded	Parent Company	Name of the Parent Company	Ticker of the Parent Company
National Indemnity Co.	No	Berkshire Hathaway	Berkshire Hathaway	BRK-A
Teachers Insurance and Annuity Association of America (TIAA)	No			
Northwestern Mutual Life Insurance Co.	No			
State Farm Mutual Automobile Insurance Co.	No			
Metropolitan Life Insurance Co.	Yes		Metropolitan Life Insurance Co.	MET
New York Life Insurance Co.	No			
Massachusetts Mutual Life Insurance Co.	No			
Prudential Insurance Co. of America	Yes		Prudential Insurance Co. of America	PRU
American General Life Insurance Co.	No	American International Group	American International Group	AIG
Allianz Life Insurance Co. of North America	No	Allianz SE	Allianz SE	ALIZF
Lincoln National Life Insurance Co.	Yes		Lincoln National Life Insurance Co.	LNC
New York Life Insurance and Annuity Corp.	No			
Columbia Insurance Co.	No			
State Farm Life Insurance Co.	No			
John Hancock Life Insurance Co. (USA)	No	Manulife Financial Corporation	Manulife Financial Corporation	MFC
Principal Life Insurance Co.	Yes		Principal Financial Group	PFG
Pacific Life Insurance Co.	No			
Government Employees Insurance Co. (GEICO)	No	Berkshire Hathaway	Berkshire Hathaway	BRK-A
Jackson National Life Insurance Co.	No	Primerica. Inc.	Primerica. Inc.	PRI
Brighthouse Life Insurance Co.	Yes		Brighthouse Financial. Inc.	BHF
Amer Equity Investment Life Insurance Co.	Yes		American Equity Investment Life Holding Company	not in Yahoo
Equitable Financial Life Insurance Co.	Yes		Equitable Holdings. Inc.	EQH
Guardian Life Insurance Co. of America	No			
Allstate Insurance Co.	Yes		Allstate Insurance Co.	ALL
Midland National Life Insurance Co.	No			
Protective Life Insurance Co.	Yes		Protective Life Corporation	not in Yahoo
Athene Annuity & Life Co.	Yes		Athene Holding Ltd.	not in Yahoo
State Farm Fire & Casualty Co.	No			
Continental Casualty Co.	Yes		CNA Financial Corporation	CNA
Nationwide Life Insurance Co.	No			
United Services Automobile Association (USAA)	No			
Liberty Mutual Insurance Co.	Yes		Liberty Mutual Holding Company Inc.	not in Yahoo
Variable Annuity Life Insurance Co.	No			
Great American Life Insurance Co. (OH)	No	American Financial Group	American Financial Group	AFG
Genworth Life Insurance Co.	Yes		Genworth Financial. Inc.	GNW
Transamerica Life Insurance Co.	No	Aegon N.V.	Aegon N.V.	AEG
Symetra Life Insurance Co.	Yes		Symetra Financial Corporation	not in Yahoo
Voya Retirement Insurance and Annuity Co.	Yes		Voya Financial. Inc.	VOYA
North American Co. for Life and Health Insurance	No	Samaritan's Purse	Samaritan's Purse	not in Yahoo
RGA Reinsurance Co.	Yes		Reinsurance Group of America. Inc.	RGA
Allstate Life Insurance Co.	Yes		Allstate Insurance Co.	ALL
Fidelity & Guaranty Life Insurance Co. (IA)	Yes		Fidelity & Guaranty Life	not in Yahoo
USAA Life Insurance Co.	No			
Security Benefit Life Insurance Co.	No			
Nationwide Life & Annuity Insurance Co.	No			
Nationwide Mutual Insurance Co.	No			
Forethought Life Insurance Co.	No			
Unum Life Insurance Co. of America	Yes		Umm Group	UNM
Union Fidelity Life Insurance Co.	No			
U.S. Life Insurance Co. in the City of New York	No			
Transamerica Premier Life Insurance Co.	No	Aegon N.V.	Aegon N.V.	AEG
Zurich American Insurance Co.	No	Zurich Insurance Group	Zurich Insurance Group	not in Yahoo
General Reinsurance Corp.	No	Berkshire Hathaway	Berkshire Hathaway	BRK-A
Athene Annuity & Life Assurance Co.	Yes		Athene Holding Ltd.	not in Yahoo
Minnesota Life Insurance Co.	No			
State Compensation Insurance Fund	No			
Venerable Insurance & Annuity Co.	No			
Commonwealth Annuity and Life Insurance Co.	No			
Prudential Retirement Insurance and Annuities Co.	Yes		Prudential Insurance Co. of America	PRU
Berkley Insurance Co.	No	W. R. Berkley Corporation	W. R. Berkley Corporation	WRB
RiverSource Life Insurance Co.	No	Ameriprise Financial	Ameriprise Financial	AMP
Factory Mutual Insurance Co.	No			
National Fire & Marine Insurance Co.	No			
GEICO Indemnity Co.	No	Berkshire Hathaway	Berkshire Hathaway	BRK-A
United of Omaha Life Insurance Co.	No	Mutual of Omaha	Mutual of Omaha	MUT
Life Insurance Co. of the Southwest	No			
Wilton Reassurance Co.	No			
Travelers Indemnity Co.	Yes		The Travelers Companies. Inc.	TRV
Hartford Fire Insurance Co.	Yes		The Hartford Financial Services Group. Inc.	HIG
American Home Assurance Co.	No	American International Group	American International Group	AIG
ACE American Insurance Co.	No	ACE Limited	ACE Limited	CB
St. Paul Fire & Marine Insurance Co.	Yes		The Travelers Companies. Inc.	TRV
Employers Reassurance Corp.	No			
Bankers Life & Casualty Co.	No	CNO Financial Group	CNO Financial Group	CNO
Berkshire Hathaway Life Insurance Co.	No	Berkshire Hathaway	Berkshire Hathaway	BRK-A
National Union Fire Insurance Co.	No	American International Group	American International Group	AIG
ReliaStar Life Insurance Co.	Yes		Voya Financial. Inc.	VOYA
Travelers Casualty & Surety Co.	Yes		The Travelers Companies. Inc.	TRV
Great-West Life & Annuity Insurance Co.	Yes		Great-West Lifeco Inc.	GWL
EquiTrust Life Insurance Co.	No			
Metropolitan Tower Life Insurance Co.	Yes		Metropolitan Life Insurance Co.	MET
Cincinnati Insurance Co.	Yes		Cincinnati Financial Corporation	CINF
Penn Mutual Life Insurance Co.	No			
American National Insurance Co.	Yes		American National Insurance Company	not in Yahoo
Munich Reinsurance America. Inc.	No	Munich Re	Munich Re	MURGY
UnitedHealthcare Insurance Co. (CT)	Yes		UnitedHealth Group Incorporated	UNH
Sun Life Assurance Co. of Canada	Yes		Sun Life Financial Inc.	SLF
Delaware Life Insurance Co.	No			
Prudential Annuities Life Assurance Corp.	Yes		Prudential Insurance Co. of America	PRU
Life Interinsurance Exchange	No			
Swiss Reinsurance America Corp.	No	Swiss Re AG	Swiss Re AG	not in Yahoo
Western-Southern Life Assurance Co.	No			
Peerless Insurance Co.	No			
Genworth Life & Annuity Insurance Co.	Yes		Genworth Financial. Inc.	GNW
Ameritas Life Insurance Corp.	No			
CMFG Life Insurance Co.	No			
Federal Insurance Co.	No	Chubb Limited	Chubb Limited	CB
Transatlantic Reinsurance Co.	No			
USAA Casualty Insurance Co.	No			
American United Life Insurance Co.	No			

B Exposure CoVaR Sensitivity to Δr

We report the Exposure CoVaR Sensitivity to Δr. We note that for the linear losses case, changing Δr does not affect the ranking, and the losses would remain lower than the input shock due to linearity (Table 4 and 5).

Table 6 reports results for $\Delta r = 0.04$ (VaR level $q = 0.99$) and $\Delta r = 0.0232$ (VaR level $q = 0.95$). For $\Delta r = 0.04$, results are almost indistinguishable from the benchmark case. For $\Delta r = 0.232$, there are changes in the ranking. Still, the main findings are reinforced: the losses for 7 insurances, instead of 2 in the benchmark case, exceed the input shock, and the ranking is, in any case, different from the one implied by linear losses.

Table 6. Exposure CoVaR Sensitivity to Δr

$\Delta r = 0.04$	$q = 0.99$		$\Delta r = 0.232$	$q = 0.95$	
Insurance Name	ticker	$\text{CoVaR}_q^{i\mid CMI}$	Insurance Name	ticker	$\text{CoVaR}_q^{i\mid CMI}$
Allstate Insurance Co.	ALL	0.109248	Principal Financial Group	PFG	0.071378
Principal Financial Group	PFG	0.100936	Protective Life Corporation	PL	0.046282
Munich Re	MURGY	0.075965	Primerica. Inc.	PRI	0.041342
Genworth Financial. Inc.	GNW	0.060881	CNA Financial Corporation	CNA	0.03334
Primerica. Inc.	PRI	0.038476	Voya Financial. Inc.	VOYA	0.032465
The Travelers Companies. Inc.	TRV	0.034387	Sun Life Financial Inc.	SLF	0.030337
Sun Life Financial Inc.	SLF	0.027961	Chubb Limited	CB	0.025828
CNA Financial Corporation	CNA	0.026171	UnitedHealth Group Incorporated	UNH	0.020703
Berkshire Hathaway	BRK-A	0.020353	Unum Group	UNM	0.019513
Equitable Holdings. Inc.	EQH	0.018046	American Financial Group	AFG	0.017889

References

1. Adrian, T., Brunnermeier, M.K.: Covar. American Economic Review **106**(7), 1705–41 (2016)
2. Battiston, S., Mandel, A., Monasterolo, I., Schütze, F., Visentin, G.: A climate stress-test of the financial system. Nat. Clim. Chang. **7**(4), 283–288 (2017)
3. Callahan, C.W., Mankin, J.S.: Carbon majors and the scientific case for climate liability. Nature **640**(8060), 893–901 (2025)
4. Campiglio, E., Daumas, L., Monnin, P., von Jagow, A.: Climate-related risks in financial assets. J. Econ. Surv. **37**(3), 950–992 (2023)
5. Carney, M.: Breaking the tragedy of the horizon–climate change and financial stability. Speech given at Lloyd's of London https://www.bis.org/review/r151009a.pdf, 29, 220–230 (2015)
6. Fama, E.F., French, K.R.: A five-factor asset pricing model. J. Financ. Econ. **116**(1), 1–22 (2015)
7. United Nations Environment Programme. Global climate litigation report: 2023 status review (2023)
8. Setzer, J., Higham, C.: Global trends in climate change litigation: 2023 snapshot (2022)

9. Setzer, J., Higham, C.: Global trends in climate change litigation: 2025 snapshot (2025)
10. Shifflett, S., Eaglesham, J.: The two big insurers still betting on fossil fuels. The Wall Street Journal (2025). Accessed 21 May 2025
11. Sloan, F.A., Trogdon, J.G., Mathews, C.A.: Litigation and the value of tobacco companies. J. Health Econ. **24**(3), 427–447 (2005)
12. Thomä, J., Kastl, J., Bayer, C., Cooke, D.: A burden they will carry: the potential economic & financial cost of climate liabilities to companies and investors. Technical report, Sustainable Finance Observatory (2021)

Option Hedging Through Reinforcement Learning

Federico Giorgi, Stefano Herzel(✉), and Paolo Pigato

Department of Economics and Finance, University of Rome "Tor Vergata", Rome, Italy
stefano.herzel@uniroma2.it

Abstract. We propose a Reinforcement Learning algorithm to hedge the payoff of a European call option. The algorithm is first tested on the Black-Scholes-Merton model, where the problem has a well known solution, so that we can compare the strategy obtained by the algorithm to the theoretical optimal one. Then, in a more realistic case that includes transaction costs, the algorithm outperforms the standard delta hedging strategy.

Keywords: Reinforcement Learning · Dynamic Strategies · Risk management

1 Introduction

The hedging of European option payoffs is an extensively studied problem in quantitative finance. This problem arises when a trader sells an option and seeks to hedge the risk of the option payoff by using a dynamic strategy involving the underlying asset. In the Black-Scholes-Merton (BSM) framework, this can be done through the "Delta Hedging" strategy. It involves holding an amount of the underlying asset equal to the option's "Delta," which measures the sensitivity of the option's price to changes in the underlying asset. Mathematically, Delta is computed as the first derivative of the option price with respect to the price of the underlying asset.

The BSM model assumes that the price of the underlying follows a geometric Brownian motion and that trading occurs continuously and without transaction costs. Under these assumptions one can derive a closed-form solution for option prices and hedging strategies. However, in real financial markets, these assumptions often do not hold. Market frictions such as transaction costs, discrete trading intervals, and deviations from log-normal price distributions complicate the implementation of Delta Hedging. Despite these limitations, Delta Hedging remains a cornerstone of practical financial risk management due to its simplicity and effectiveness in many scenarios.

In this paper, we explore the hedging problem in a realistic setting: we address, with an approach based on Reinforcement Learning (RL), the challenges of Delta Hedging in markets where trading occurs at discrete intervals and

M. La Rocca et al. (Eds.): WMAF 2025, *New Perspectives in Mathematical and Statistical Methods for Actuarial Sciences and Finance*, pp. 169–178, 2025.
https://doi.org/10.1007/978-3-032-05551-4_15

transaction costs are significant, which are practical constraints faced by traders. RL is a branch of machine learning where an agent interacts with an environment over time, observing its state and taking actions to maximize cumulative rewards. At each time step, the agent observes a new state and receives feedback in the form of a reward, which guides its future actions. RL algorithms train agents by simulating a large number of "episodes," enabling them to improve their understanding of the environment and refine their strategies through trial and error. Seminal works on RL, such as [Sutton and Barto, 2018], provide the theoretical foundation for applying these algorithms to a wide range of dynamic decision-making problems.

Recent applications of RL in finance have demonstrated its potential to address complex optimization problems. For instance, [Kolm and Ritter, 2019] and [Giorgi et al., 2024], apply RL to dynamic portfolio optimization under transaction costs, while [Buehler et al., 2019] and [Cao et al., 2019] leverage RL to develop optimal hedging strategies for derivatives. [Vittori et al., 2020] further extend these approaches by combining RL with neural networks to handle high-dimensional state spaces.

We present a RL algorithm for Delta Hedging in discrete-time settings with transaction costs. The proposed algorithm combines the SARSA (State-Action-Reward-State-Action) algorithm with a neural network to estimate the value function. This combination allows the model to handle a continuous set of states (e.g., current holdings, market realizations) and actions (e.g., the quantity of shares to trade). The value function is iteratively refined through batches of simulated episodes. Initially, the agent follows its current policy to make decisions, but its performance improves over time as it incorporates new information from simulated rewards. It is important to note that RL algorithms, while powerful, are not guaranteed to perform optimally in all instances. To validate the effectiveness of our approach, we conduct comparative tests in scenarios where the optimal solution is known.

The paper is organized as follows. Section 2 provides a detailed description of the hedging problem, including the key challenges associated with discrete-time trading and transaction costs. Section 3 introduces the fundamental concepts of RL relevant to our approach. In Sect. 4, we present the algorithm and discuss its implementation. In Sect. 5 we evaluate its performance through numerical experiments. Finally, Sect. 6 concludes with a summary of findings and directions for future research.

2 The Hedging Problem

We analyze the problem faced by a trader who sells a contingent claim at time t_0 for a price P_0 and subsequently adopts a dynamic strategy to cover the exposure to the payoff of the claim P_T at time T. We consider a market where it is possible to trade a single risky asset, whose price at time t is denoted by S_t. Let n_k represent the units of the underlying held in the portfolio at time t_k, and let c_k denote the cost incurred at time t_k to adjust the portfolio from n_{k-1} to n_k

(this cost can be negative, reflecting a gain). Assuming that hedging occurs at discrete times $t_0, t_1, \ldots, t_{N-1}$, with $t_{N-1} < T$, the total cost of the strategy is given by $\sum_{k=0}^{N-1} c_k$, while the liquidation value of the final position in the asset is $n_{N-1} S_T$. Thus, the final profit and loss (P&L) of the trader's position at time T is:

$$H_T = -P_T - \sum_{k=0}^{N-1} c_k + n_{N-1} S_T + P_0.$$

The goal of the hedging strategy is to minimize the risk of H_T, specifically by minimizing its variance. A contingent claim is said to be "replicable" if a strategy exists that reduces the variance of H_T to zero. Markets in which any contingent claim can be perfectly replicated are referred to as "complete."

The completeness of a market model depends on the dynamics of the price process S_t and the cost functions c_k. In the BSM model [Black and Scholes, 1973], [Merton, 1973], the dynamics of the underlying asset are described by:

$$dS_t = \mu S_t \, dt + \sigma S_t \, dW_t,$$

where μ is the drift, σ is the volatility, and W_t is a standard Brownian motion. Additionally, the model assumes the existence of a risk-free asset with continuously compounded interest rate r, continuous-time trading, and the absence of transaction costs. Under these conditions, any contingent claim is replicable.

We focus on a European Call option, a contingent claim with a payoff:

$$P_T = \max(S_T - K, 0), \tag{1}$$

where K is the strike price and T is the maturity. The objective is to determine a dynamic strategy that adjusts n_k at each time step to replicate the payoff in (1). Since P_T is an increasing function of S_T, the hedging strategy typically involves holding positive amounts of the underlying asset.

In the BSM model, the optimal hedging strategy is determined by the option's "Delta," the first derivative of the option's no-arbitrage price with respect to the underlying asset. The Delta is computed as:

$$n_t = \Phi \left(\frac{\ln(S_t/K) + \tau \left(r + \frac{\sigma^2}{2} \right)}{\sigma \sqrt{\tau}} \right), \tag{2}$$

where $\Phi(\cdot)$ is the cumulative distribution function of the standard normal distribution, and $\tau = T - t$ is the time to maturity. This Delta hedging strategy achieves perfect replication in the BSM framework, ensuring $H_T = 0$ almost surely, provided that P_0 is the no-arbitrage price and the strategy is continuously executed.

However, in practice, perfect hedging is impossible when trading is restricted to discrete times $t_k \in \{t_0, \ldots, t_{N-1}\}$, since the Delta hedging strategy introduces replication errors, and the variance of the error depends on the time discretization and the "Gamma" of the option, i.e., the second derivative of the option price with respect to the underlying asset (see [Angelini et al., 2009] and [Angelini and Herzel, 2015] for details).

Additionally, the assumption of no transaction costs in the BSM model is unrealistic. When transaction costs are present, the Delta hedging strategy becomes suboptimal and alternative strategies are needed. The remainder of this work explores how RL can provide an effective alternative to the classical Delta hedging approach.

3 Reinforcement Learning

In RL problems, an agent interacts with an environment over time. The *environment* refers to the system outside the agent's direct control, and the agent's goal is to maximize cumulative rewards through a sequence of actions. At each time step t, the agent observes the environment's *state*, represented by $s_t \in \mathcal{S}$, where $\mathcal{S}$ denotes the *state space.* Based on this observation, the agent selects an *action* $a_t \in \mathcal{A}(s_t)$, where $\mathcal{A}(s_t)$ represents the set of feasible actions, which may depend on the current state s_t. After taking the action, at time $t+1$, the agent receives a *reward* $R_{t+1} = R(s_t, a_t)$, which depends, possibly randomly, on the current state and the action taken. Additionally, the agent transitions to a new state s_{t+1}, completing the interaction cycle.

The sequence of states, actions, and rewards forms a stochastic process. The mathematical framework commonly used to model this process is the Markov Decision Process (MDP). In an MDP, the state and reward at time $t+1$ depend only on the current state and action, not on the full history:

$$\mathcal{L}\left(s_{t+1}, R_{t+1} | s_0, a_0, R_1, \ldots, s_t, a_t\right) = \mathcal{L}\left(s_{t+1}, R_{t+1} | s_t, a_t\right), \tag{3}$$

where $\mathcal{L}(X|Y)$ denotes the conditional probability law of X given Y. In most RL problems, the transition dynamics $\mathcal{L}$ are not explicitly known and the agent learns by trial and error, interacting with the environment.

A *policy* defines the agent's behavior, specifying the action a to take for a given state s. A stationary Markovian policy depends only on the current state:

$$\begin{aligned} \pi : \mathcal{S} &\to \mathcal{A}(s), \\ s &\mapsto a = \pi(s). \end{aligned} \tag{4}$$

The combination of an initial state distribution, the MDP dynamics (3), and a policy π determines the evolution of the system as a homogeneous Markov chain (see [Csáji and Monostori, 2008] for extensions to time-varying environments).

To evaluate policies, we define the cumulative discounted reward:

$$G_t = \sum_{k=0}^{\infty} \gamma^k R_{t+k+1}, \tag{5}$$

where $\gamma \in (0, 1]$ is the discount factor, which weights immediate rewards more heavily than distant ones. The state-value function for a policy π is:

$$v_\pi(s) = \mathbb{E}_\pi\left[G_t | s_t = s\right], \tag{6}$$

where $\mathbb{E}_\pi$ denotes the expectation assuming the agent follows policy π. Similarly, the action-value function is:

$$q_\pi(s,a) = \mathbb{E}_\pi \left[G_t | s_t = s, a_t = a \right], \tag{7}$$

which quantifies the expected cumulative reward starting from state s, taking action a, and subsequently following π. An *optimal policy* π^* maximizes the expected cumulative reward for all states:

$$\pi^* = \underset{\pi}{\operatorname{argmax}}\, v_\pi(s) \quad \forall s \in \mathcal{S}. \tag{8}$$

All optimal policies share the same optimal value functions:

$$v_*(s) = \max_\pi v_\pi(s), \tag{9}$$

$$q_*(s,a) = \max_\pi q_\pi(s,a). \tag{10}$$

These satisfy the *Bellman optimality equations* (see [Sutton and Barto, 2018]):

$$v_*(s) = \max_{a \in \mathcal{A}(s)} \mathbb{E}\left[R_{t+1} + \gamma v_*(s_{t+1}) | s_t = s, a_t = a \right], \tag{11}$$

$$q_*(s,a) = \mathbb{E}\left[R_{t+1} + \gamma \max_{a_{t+1} \in \mathcal{A}(s_{t+1})} q_*(s_{t+1}, a_{t+1}) | s_t = s, a_t = a \right]. \tag{12}$$

In practice, RL algorithms alternate between *exploration* and *exploitation*. During exploitation, the agent selects actions based on the current estimate of the optimal policy, while exploration involves trying alternative actions to improve the policy. A common approach is the ϵ-greedy policy

$$\pi_{\epsilon\text{-greedy}}(s) = \begin{cases} \tilde{a}, & \text{if } u < \epsilon, \\ \operatorname{argmax}_a q(s,a), & \text{if } u \geq \epsilon, \end{cases} \tag{13}$$

where $\epsilon \in (0,1)$, $u \sim U(0,1)$, and $\tilde{a}$ is a random action. Over time, ϵ decreases to favor exploitation as the policy improves.

Under appropriate conditions, RL algorithms converge to the optimal value functions (10) through repeated training and interaction with the environment (see [Sutton and Barto, 2018]). This process is a hallmark of RL, distinguishing it from classical dynamic programming by its reliance on experience rather than explicit knowledge of the MDP dynamics.

4 The Algorithm

The hedging problem must be formulated in a manner suitable for applying RL. Define the change in the units of the asset S at time t_k as

$$\Delta n_k := n_k - n_{k-1}.$$

The cost of updating the portfolio at time t_k is given by

$$c_k := S_{t_k} \Delta n_k + c(\Delta n_k),$$

where the first term reflects the cost of trading in a perfectly liquid market, and the second term, $c(\Delta n_k)$, represents transaction costs, which are always positive and become zero only in a perfectly liquid market.

The hedging target H_T can then be expressed as

$$H_T = -P_T - \sum_{k=0}^{N-1} (S_{t_k} \Delta n_k + c(\Delta n_k)) + n_{N-1} S_T + P_0. \tag{14}$$

Let us define the "profit and loss" (P&L) of the portfolio position at time t_k as

$$\varrho_k := (-P_{t_k} + P_{t_{k-1}}) - n_{k-1}(S_{t_k} - S_{t_{k-1}}).$$

Substituting this into H_T, we obtain

$$H_T = \sum_{k=1}^{N} \varrho_k - c(\Delta n_{k-1}), \tag{15}$$

where the hedging target is expressed as the sum of "local costs", defined as the differences between the P&L and transaction costs. The objective is to optimize the mean-variance utility of the hedging target H_T

$$\mathbb{MV}[H] = \mathbb{E}[H] - \frac{\kappa}{2} \mathbb{V}[H], \tag{16}$$

where κ is a constant representing the trader's risk aversion.

As a reward function we use

$$R_t = (\varrho_t - c(\Delta n_{t-1})) - \frac{\kappa}{2} \left(\varrho_t - c(\Delta n_{t-1})\right)^2, \tag{17}$$

which encourages the agent to optimize the mean-variance utility of the local costs, and consequently, the overall hedging target. This is validated under the assumptions

$$\mathbb{V}[H_T] = \sum_{k=1}^{N} \mathbb{V}[\varrho_k - c(\Delta n_{k-1})], \quad \mathbb{E}[\varrho_k] = 0, \; k = 1, \ldots, N.$$

We define the state at time t as

$$s_t := (S_t, \tau, n_{t-1}) \in \mathbb{R}^3,$$

where S_t is the current value of the underlying, $\tau = T - t$ is the time to maturity, and n_{t-1} is the current position in the underlying asset.

The action variable is defined as

$$a_t := \Delta n_t \in \mathbb{R},$$

where the action space is constrained by

$$a_t \in (-n_{t-1}, L - n_{t-1}) = \mathcal{A}(s_t), \tag{18}$$

ensuring that the trader neither short-sells nor exceeds a lot size L. This state-dependent action space is also bounded, which facilitates optimization during the greedy policy computation.

We compute the trading costs as

$$c(a) = m \times tick \times (|a| + 0.01a^2), \tag{19}$$

where $tick > 0$ is the tick size, and $m > 0$ determines market friction. The term $tick \times |a|$ reflects the cost of crossing a bid-offer spread, while the quadratic term models the market impact of the trade. This function is often used in the literature to quantify trading costs, but the flexibility of the algorithm allows to change it to reflect specific market conditions.

Although the reward function depends on the option price P_t, we do not include P_t in the state space. Instead, P_t is simulated using the BSM formula $P_t = p^{BSM}(S_t)$, ensuring compatibility with the RL framework without relying on a specific pricing model. Importantly, the intermediate option prices cancel out, as shown

$$\sum_{t=1}^{T} (p^{BSM}(S_t) - p^{BSM}(S_{t-1})) = p^{BSM}(S_T) - p^{BSM}(S_0) = (S_T - K)^+ - P_0.$$

Our algorithm is based on the SARSA approach, which was introduced in [Rummery and Niranjan, 1994] and refined in [Sutton and Barto, 2018]. SARSA estimates the optimal action-value function q_* through simulations of State-Action-Reward-State-Action (SARSA) sequences

$$s_0^{(j)}, a_0^{(j)}, r_1^{(j)}, s_1^{(j)}, a_1^{(j)}, \ldots. \tag{20}$$

The value function is updated iteratively using the SARSA formula:

$$q_{k+1}(s_{t-1}, a_{t-1}) = q_k(s_{t-1}, a_{t-1}) + \alpha \left(R_t + \gamma q_k(s_t, a_t) - q_k(s_{t-1}, a_{t-1})\right), \tag{21}$$

where $\alpha \in [0, 1]$ is the learning rate. Setting $\alpha = 1$ simplifies this to

$$q_{k+1}(s_{t-1}, a_{t-1}) = R_t + \gamma q_k(s_t, a_t). \tag{22}$$

The algorithm combines the SARSA method with supervised regression for value function approximation. Training proceeds in batches, starting with an initial value function $\hat{q}^{(0)}(s, a)$. Exploration is managed using the ϵ-greedy policy (13), where ϵ is gradually reduced during training.

To optimize the greedy policy we tested several global optimization , including basin-hopping [Wales and Doye, 1997], brute force, differential evolution [Storn and Price, 1997], dual annealing [Xiang et al., 1997], and we finally opted for SHGO [Endres et al., 2018] that, in our experience, ensured more stable and robust performances even in complex state-action spaces.

In the next section, we present numerical results demonstrating the algorithm's effectiveness.

5 Applications

In our first experiment, we compare the RL policy to the standard delta hedging strategy under the assumption of no transaction costs.

We model the underlying dynamics using a geometric Brownian motion (2), with daily drift and volatility parameters set to $\mu = 0\%$ and $\sigma = 1\%$, respectively. The initial price is $S_0 = \$100$. The scenario involves a lot of $L = 100$ plain vanilla European call options on the underlying S, traded at-the-money with a strike price $K = \$100$. The options have a maturity of 10 days, during which the trader rebalances their position 5 times a day, resulting in $T = 50$ total rebalancing times. The tick size (minimum price change) is $tick = \$0.1$, and the risk-free rate is assumed to be $r = 0\%$. Transaction costs are set to zero by applying $m = 0$ in the general cost formula (19).

The RL agent is trained over $N_B = 10$ consecutive batches, each generating $J = 75,000$ episodes using the MDP (2). The epsilon-greedy policy begins with $\epsilon = 50\%$ and decreases at each batch iteration as $\epsilon \leftarrow \epsilon/3$. During training, the SHGO global optimization algorithm [Endres et al., 2018] is used to compute the policy's maximization.

For out-of-sample validation, $J = 10,000$ MDP paths are simulated. The optimal delta strategy (2) and the RL-based strategy achieve similar outcomes concerning the hedging target, with the delta strategy showing less variance due to its deterministic nature.

We introduce transaction costs by setting the multiplier $m = 1$ in the cost formula (19). The RL agent is trained in this new environment using the same setup as in the no-cost case. After the training we compare the outcome of the RL strategy to those of the standard delta-hedging. The variance of the hedging target is smaller for the RL agent. The RL agent effectively accounts for transaction costs, resulting in more conservative trading. In Fig. 1 we see that the trading costs for the RL strategy are significantly lower, because the RL agent learned that with transaction costs it is more convenient to limit the trades.

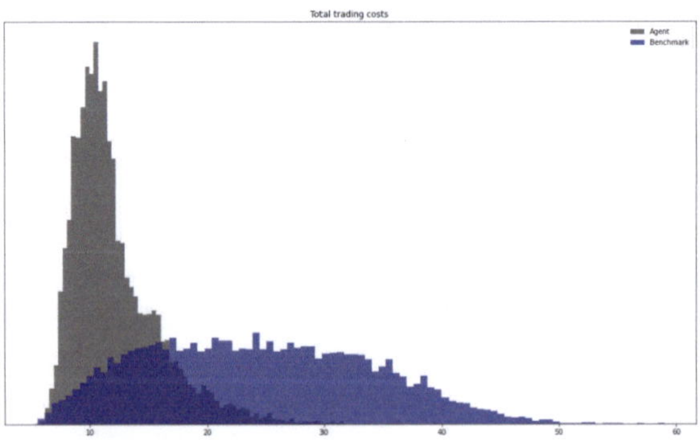

Fig. 1. Total trading costs for the delta hedging strategy (blue) and RL agent (black). (Color figure online)

6 Conclusion

We have proposed a RL algorithm for hedging European call options in market settings that reflect real-world constraints, such as discrete-time trading and transaction costs. Our approach combines the SARSA algorithm with SHGO global optimization algorithm [Endres et al., 2018] to compute the policy's maximization. The use of RL provides a flexible framework that allows the algorithm to learn and adapt to complex market dynamics, overcoming many of the limitations inherent in traditional Delta Hedging strategies.

While the results are promising, we believe that for an effective practical use of RL in finance, one should aim to some more model-free approach, where the data for training are provided directly from empirical observations, in a non-parametric way. Research directions along this avenue are exemplified by [Yang et al., 2024] and [Wiese et al., 2020].

References

Angelini, F., Herzel, S.: Evaluating discrete dynamic strategies in affine models. Quant. Finan. **15**(2), 313–326 (2015)

Angelini, F., Herzel, S., et al.: Measuring the error of dynamic hedging: a laplace transform approach. J. Comput. Finan. **13**(2), 47 (2009)

Black, F., Scholes, M.: The pricing of options and corporate liabilities. J. Polit. Econ. **81**(3), 637–654 (1973)

Buehler, H., Gonon, L., Teichmann, J., Wood, B.: Deep hedging. Quant. Finan. **19**(8), 1271–1291 (2019)

Cao, J., Chen, J., Hull, J.C., Poulos, Z.: Deep hedging of derivatives using reinforcement learning. SSRN 3514586 (2019)

Csáji, B.C., Monostori, L.: Value function based reinforcement learning in changing markovian environments. J. Mach. Learn. Res. **9**(8), 1–31 (2008)

Endres, S.C., Sandrock, C., Focke, W.W.: A simplicial homology algorithm for lipschitz optimisation. J. Global Optim. **72**(2), 181–217 (2018)

Giorgi, F., Herzel, S., Pigato, P.: A reinforcement learning algorithm for trading commodities. Appl. Stoch. Model. Bus. Ind. **40**(2), 373–388 (2024)

Kolm, P.N., Ritter, G.: Dynamic replication and hedging: a reinforcement learning approach. J. Finan. Data Sci. **1**(1), 159–171 (2019)

Merton, R.C.: Theory of rational option pricing. Bell J. Econ. Manag. Sci. 141–183 (1973)

Rummery, G.A., Niranjan, M.: On-line Q-learning using connectionist systems, vol. 37. Department of Engineering, University of Cambridge Cambridge (1994)

Storn, R., Price, K.: Differential evolution-a simple and efficient heuristic for global optimization over continuous spaces. J. Global Optim. **11**(4), 341–359 (1997)

Sutton, R.S., Barto, A.G.: Reinforcement Learning: An Introduction. The MIT Press, Cambridge (2018)

Vittori, E., Trapletti, M., Restelli, M.:Option hedging with risk averse reinforcement learning. arXiv preprint arXiv:2010.12245 (2020)

Wales, D.J., Doye, J.P.: Global optimization by basin-hopping and the lowest energy structures of lennard-jones clusters containing up to 110 atoms. J. Phys. Chem. A **101**(28), 5111–5116 (1997)

Wiese, M., Knobloch, R., Korn, R., Kretschmer, P.: Quant gans: deep generation of financial time series. Quant. Finan. **20**(9), 1419–1440 (2020)
Xiang, Y., Sun, D., Fan, W., Gong, X.: Generalized simulated annealing algorithm and its application to the Thomson model. Phys. Lett. A **233**(3), 216–220 (1997)
Yang, Y., et al.: A survey on diffusion models for time series and spatio-temporal data. arXiv preprint arXiv:2404.18886 (2024)

Parameter Stability in Yield Curve Fitting

Lorenzo Mercuri[1], Andrea Perchiazzo[1], Edit Rroji[2], and Ilaria Stefani[3](✉)

[1] Department of Economics and Quantitative Methods, University of Milan, Milan, Italy
{lorenzo.mercuri,andrea.perchiazzo}@unimi.it
[2] Department of Statistics and Quantitative Methods, University of Milano-Bicocca, Milan, Italy
edit.rroji@unimib.it
[3] Department of Economics and Business Sciences, University of Parma, Parma, Italy
ilaria.stefani@unipr.it

Abstract. We investigate the yield term structure of different countries in order to assess the impact of recent interventions from central banks. We adopt a static approach through the use of the Nelson-Siegel and its extended version named Nelson-Siegel-Svensson model. Empirical results suggest that fitted parameters in the restricted model are more stable across all countries for the period 2020–2025.

Keywords: Yield Term Structure · Nelson-Siegel · Nelson-Siegel-Svensson · Sovereign bonds

1 Introduction

The last few years have been characterized by extraordinary macroeconomic turbulence. The COVID-19 pandemic triggered an economic shock, leading to aggressive monetary interventions by central banks, both in the European Union and the United States. Subsequent to the pandemic, the outbreak of the Russia-Ukraine war in 2022 further exacerbated economic instability, fueling inflationary pressures and inducing further monetary tightening. This sequence of events led to a period of rapid interest rate adjustments and a substantial transformation in the dynamics of sovereign bond markets.

Frequent monetary policy interventions caused significant shifts in the shape of the yield to maturity (YTM) term structures across various countries. Our goal is to understand which nodes of the yield term structure have experienced the greatest impact. Additionally, we aim to identify which countries observed the greatest variability in part of the nodes in the yield term structure. Along with a rolling window analysis based on historical volatility of nodes in the yield term structure, we use the Nelson-Siegel model [2] and its extended version, i.e., Nelson-Siegel-Svensson model [4], to fit the yield term structure extracted from sovereign bonds. A similar comparison for these two parsimonious models is

M. La Rocca et al. (Eds.): WMAF 2025, *New Perspectives in Mathematical and Statistical Methods for Actuarial Sciences and Finance*, pp. 179–189, 2025.
https://doi.org/10.1007/978-3-032-05551-4_16

performed in [5] using only US Treasuries for the period 2000–2019. We compare the fitting results using daily data from July 2020 to January 2025. The countries under investigation are France, Germany, Italy, Japan, and the USA.

2 Volatility Analysis

We first perform a non-parametric analysis of the volatility of nodes in the yield term structure over time using rolling windows. Specifically, we analyze the 30-day rolling volatility of yields for different countries to assess the areas of the term structure that exhibited the greatest variability. This analysis enables us to identify the maturities most impacted by macroeconomic shocks. Figure 1 presents for different nodes the 30-day rolling volatility for Italy, Germany, France, Japan, and the USA.

Several important insights emerge from this analysis. In particular, except for the case of Japan, the highest levels of rolling volatility were observed at the short end of the yield curve, particularly concentrated between 1 and 3 years. The most volatile period for the European markets coincides with the beginning of the Russian-Ukraine war. This pattern suggests that, in the European and US economies, short-term market expectations were extremely sensitive to the rapid and frequent adjustments in monetary policy undertaken in response to the evolving macroeconomic environment. Investors and market participants seemed to react promptly to policy announcements and shifts in economic outlook, resulting in heightened volatility at the shorter maturities.

In contrast, Japan showed a slightly different behavior, with greater variability for the range of maturities between 20 and 30 years. This outcome reflects the unique characteristics of the Japanese monetary framework, where the Bank of Japan's yield curve control (YCC) policy adjustments, aimed at managing both short and long-term interest rates, played a crucial role. Consequently, volatility was persistent for long-term maturities, highlighting the specific challenges of the Japanese context.

Regarding the USA, the rolling volatility takes large values starting from 2022. This indicates that both short and medium-term segments of the yield curve were significantly impacted, consistent with the Federal Reserve's aggressive and sustained interest rate hikes aimed at containing high inflation rates. The broad span of volatility suggests that monetary tightening influenced a wide range of market expectations, affecting borrowing costs and investment decisions across multiple time horizons.

Overall, these observations are coherent with the different monetary policy strategies. They emphasize how differences in central banking approaches, economic resilience, and market dynamics shape the behavior of yield curves under conditions of macroeconomic stress. In the next section, we will look not at single nodes in the term structure but at the parameters of parsimonious models. The idea is to compact the information in few elements in order to grasp the impact of recent events in the market.

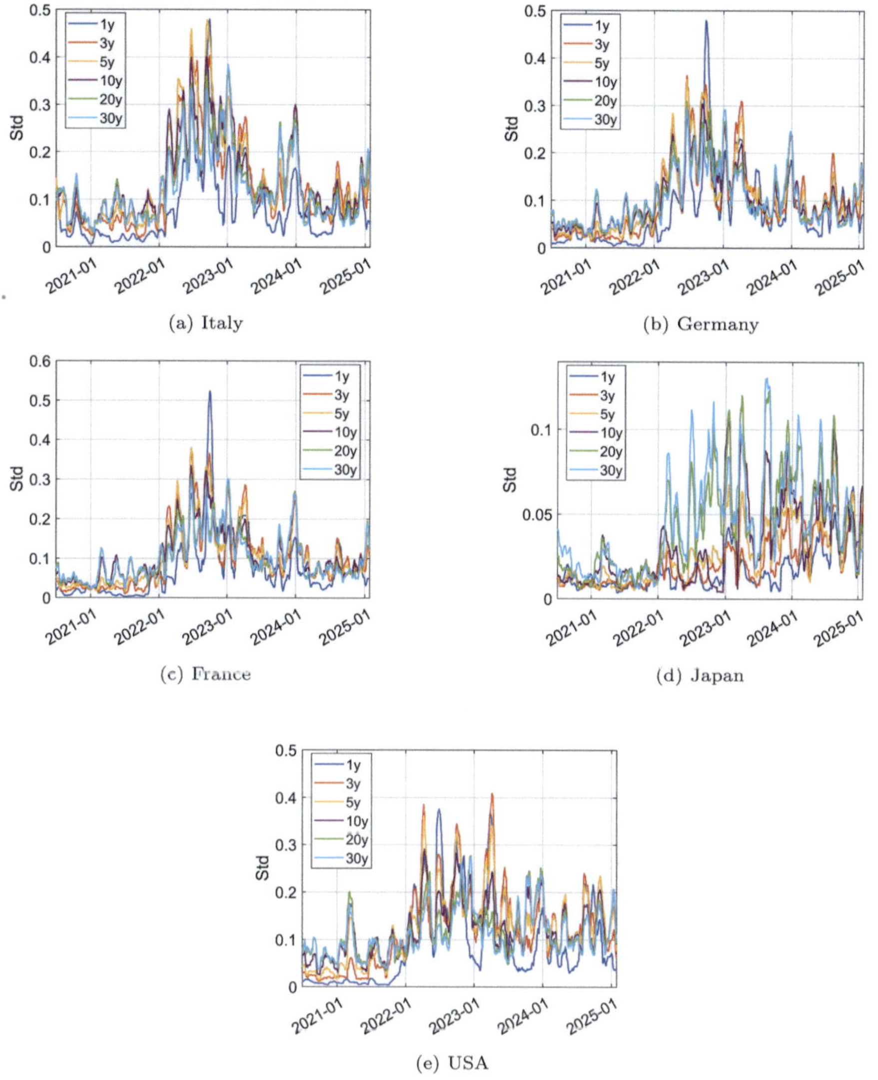

(a) Italy

(b) Germany

(c) France

(d) Japan

(e) USA

Fig. 1. Rolling 30-day volatility, respectively for Italy, Germany, France, Japan and the USA. The standard deviation is computed for the nodes 1y, 3y, 5y, 10y, 20y, and 30y.

2.1 Parsimonious Models

We consider two models to fit the yield curves. The first is the Nelson-Siegel (NS) model proposed in [2] while the second is its extension, i.e., Nelson-Siegel-Svensson (NSS) model [4]. For practical implementation, we use the `R` package *YieldCurve* based on [1].

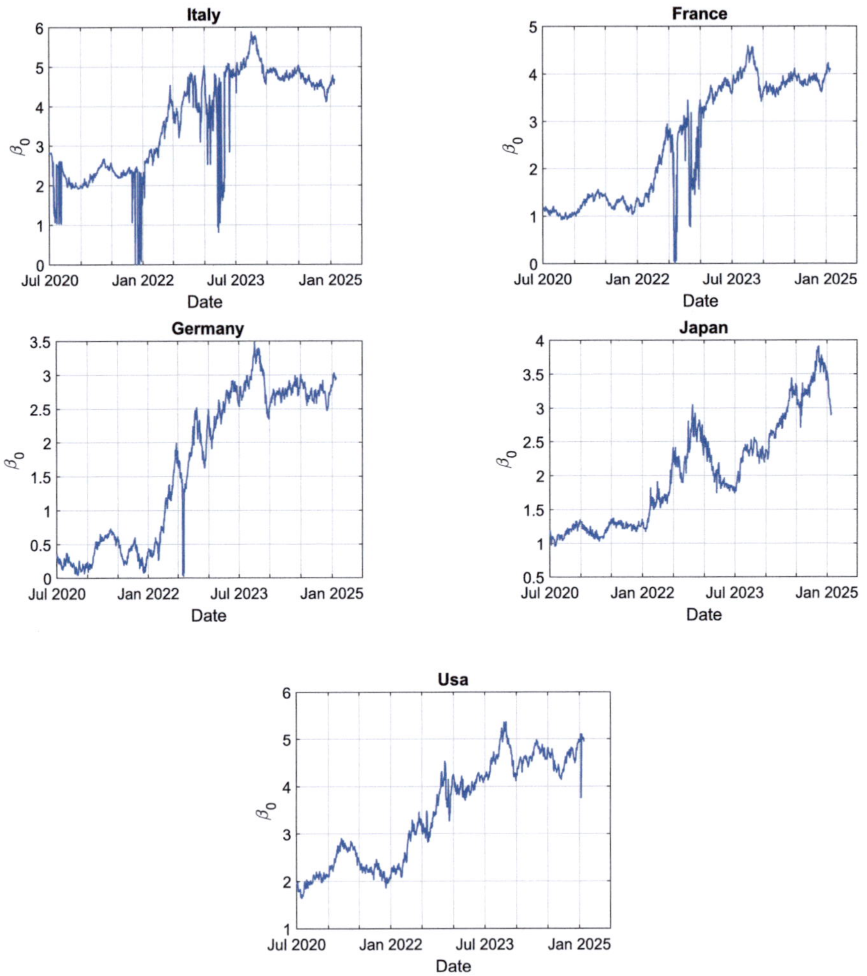

Fig. 2. Estimation of the parameter β_0 in the Nelson-Siegel model for different countries.

In the Nelson-Siegel model, the yield for a given maturity m is expressed as:

$$y\left(m\right) = \beta_0 + \beta_1\left(\frac{1 - e^{-\frac{m}{\tau}}}{\frac{m}{\tau}}\right) + \beta_2\left(\frac{1 - e^{-\frac{m}{\tau}}}{\frac{m}{\tau}} - e^{-\frac{m}{\tau}}\right), \tag{1}$$

where β_0 can be interpreted as the long-term interest rate level; β_1 captures the short-term rate sensitivity, reflecting how the yield curve changes in the near term; and β_2 captures the curvature or the hump in the yield curve, which can reflect the medium-term yield behavior.

The term $\left(\frac{1-e^{-\frac{m}{\tau}}}{\frac{m}{\tau}}\right)$ is the "exponential decay" component that controls the near-term yields (shorter maturities). In Figs. 2, 3, and 4 the estimated model

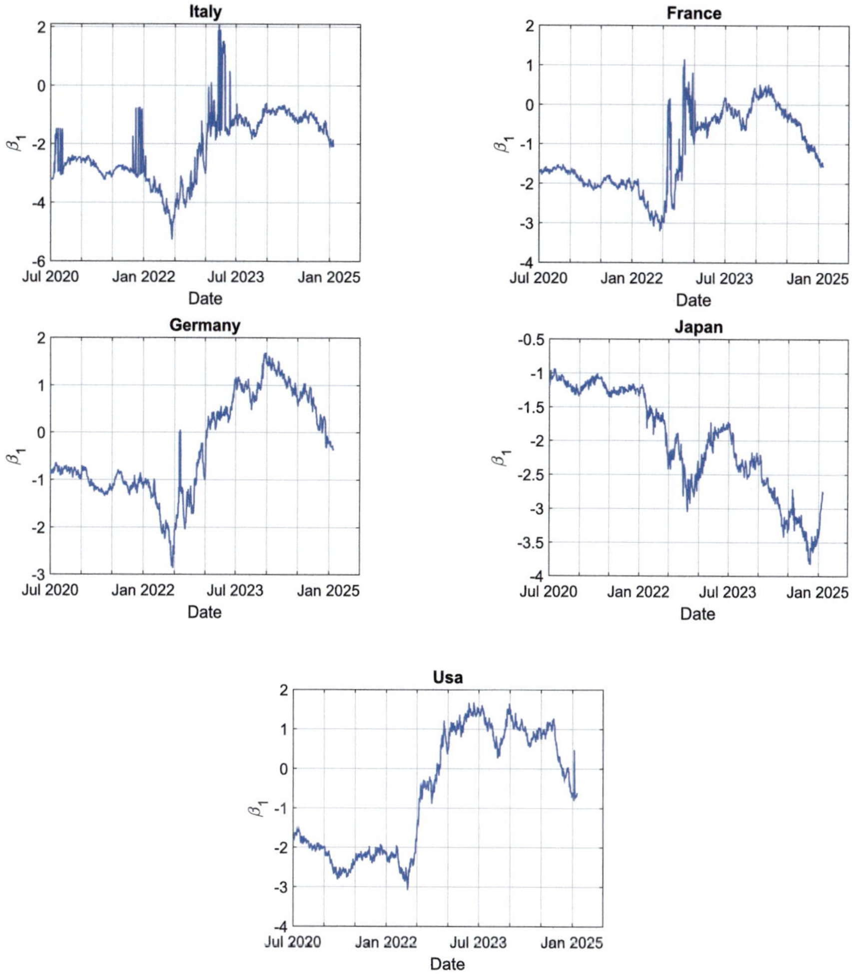

Fig. 3. Estimated β_1 parameter in the Nelson Siegel model for different countries.

parameters for different countries are shown. For the long-term yields, we observe very few spikes in the sequences of β_0. The country with greater instability seems to be Italy. Similar results are confirmed also for β_1 that collects information on the near term. The fitted values for Japan and the USA do not contain spikes. Some instabilities in fitting the curvature, i.e., β_2, emerge for all countries. It is interesting to notice that for Japan spikes appear only during the last year, whereas European countries exhibit significant instability in the fitted parameter throughout 2022. Furthermore, it is worth to notice that during periods of turbulence β_0 and β_1 tend to compensate each other, in the sense that they jump in opposite directions. For β_2, jumps seem to be clustered in most of the cases, except for the last jump for the USA and Japan.

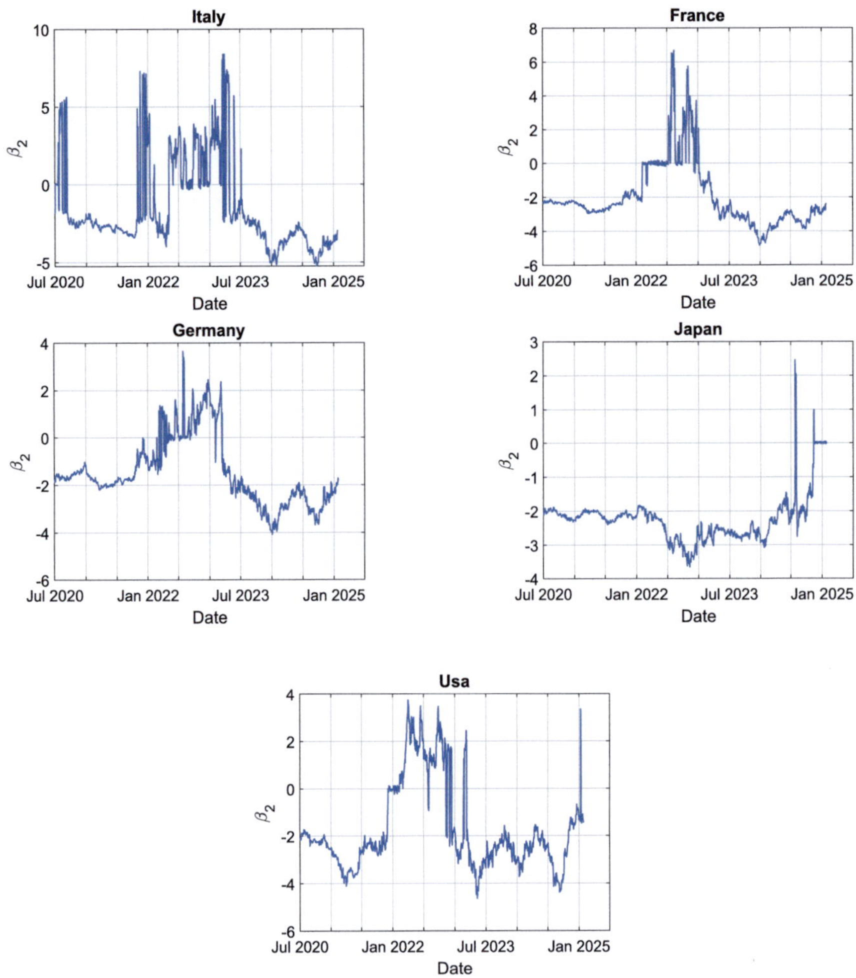

Fig. 4. Estimation of the β_2 parameter in the Nelson-Siegel model for different countries.

The Nelson-Siegel model, while useful, can sometimes struggle to capture more complex behaviors of the yield curve, especially when there are significant inflections (or "humps") in the curve, which is where the Svensson extension comes in.

The second model we use to fit the yield curves is the Nelson-Siegel-Svensson (NSS) model that improves the flexibility and accuracy of modelling yield curves. The NSS model adds two key components: the parameter β_3 which accounts for the second curvature in the yield curve and the additional decay factor τ_2, which enhances the model's ability to capture more complex shapes in the yield curve, especially for longer maturities. In particular, the yield in the Nelson-Siegel-

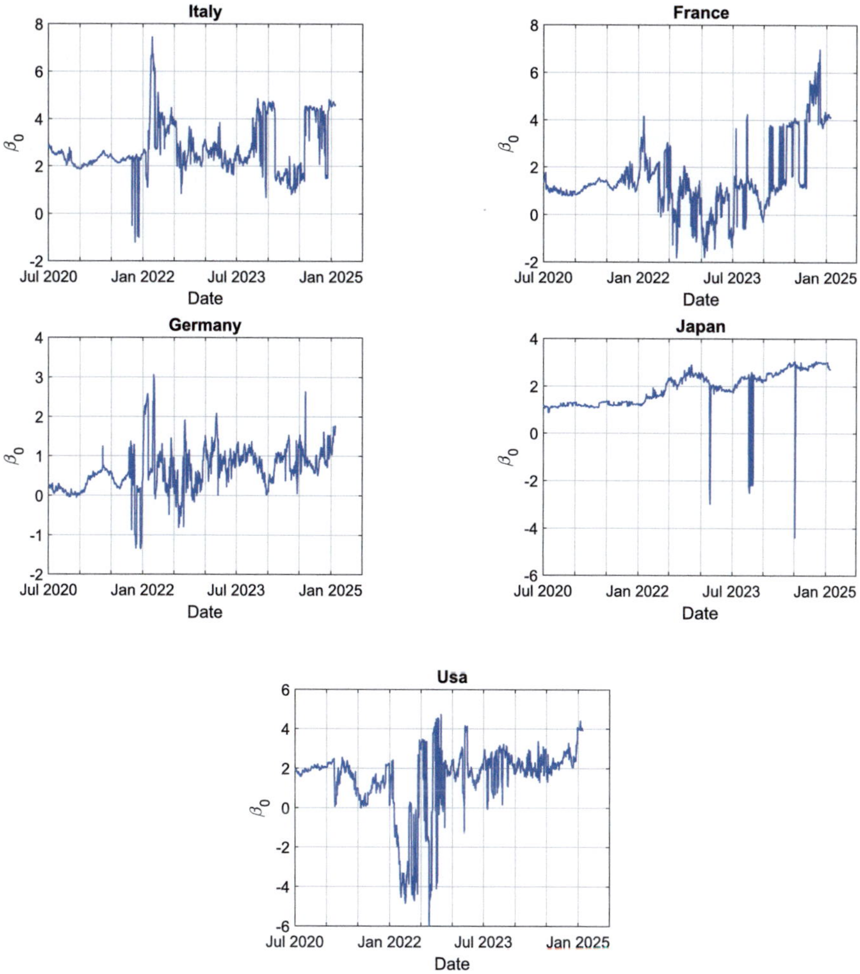

Fig. 5. Estimation of the β_0 parameter of the Nelson-Siegel-Svensson model for different countries.

Svensson model writes

$$y(m) = \beta_0 + \beta_1 \left(\frac{1 - e^{-\frac{m}{\tau_1}}}{\frac{m}{\tau_1}} \right) + \beta_2 \left(\frac{1 - e^{-\frac{m}{\tau_1}}}{\frac{m}{\tau_1}} - e^{-\frac{m}{\tau_1}} \right) + \beta_3 \left(\frac{1 - e^{-\frac{m}{\tau_2}}}{\frac{m}{\tau_2}} - e^{-\frac{m}{\tau_2}} \right).$$

In addition to the Nelson-Siegel parametrization, the NSS model introduces:

- β_3 that measures the second curvature. The addition of this term provides greater flexibility, allowing to capture more complex yield curve shapes, such as multiple humps or changes in curvature over time.

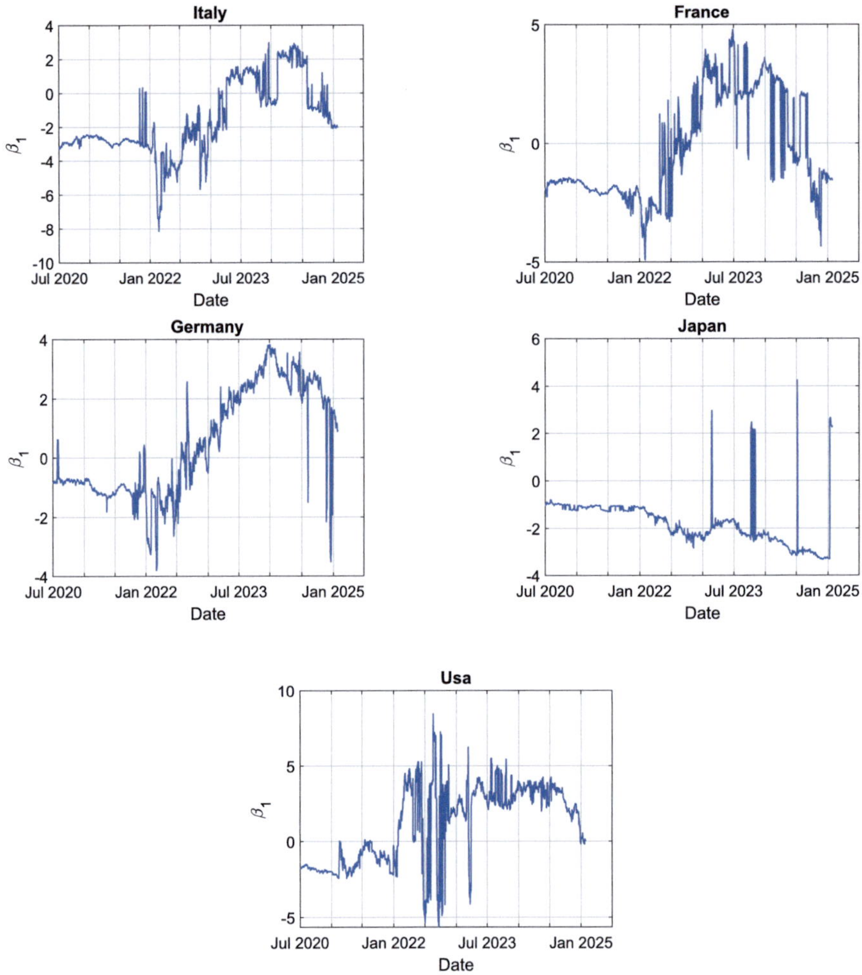

Fig. 6. Estimated β_1 parameter in the Nelson-Siegel-Svensson model for different countries.

- Two decay factors, τ_1 and τ_2, which permit to capture a wider range of shapes for the yield curve, especially at long maturities. Specifically, τ_2 determines the location of the second hump (if $\beta_3 > 0$) or trough (if $\beta_3 < 0$).

Figures 5, 6, 7, and 8 display the parameters estimated with this model. The initial observation is that, starting in 2023, the NSS model yields unstable results for Japan, characterized by frequent spikes in all parameters. In contrast, the preceding period under investigation did not exhibit any stability concerns. Despite the reduced magnitude, analogous spikes emerge in the analyses conducted for

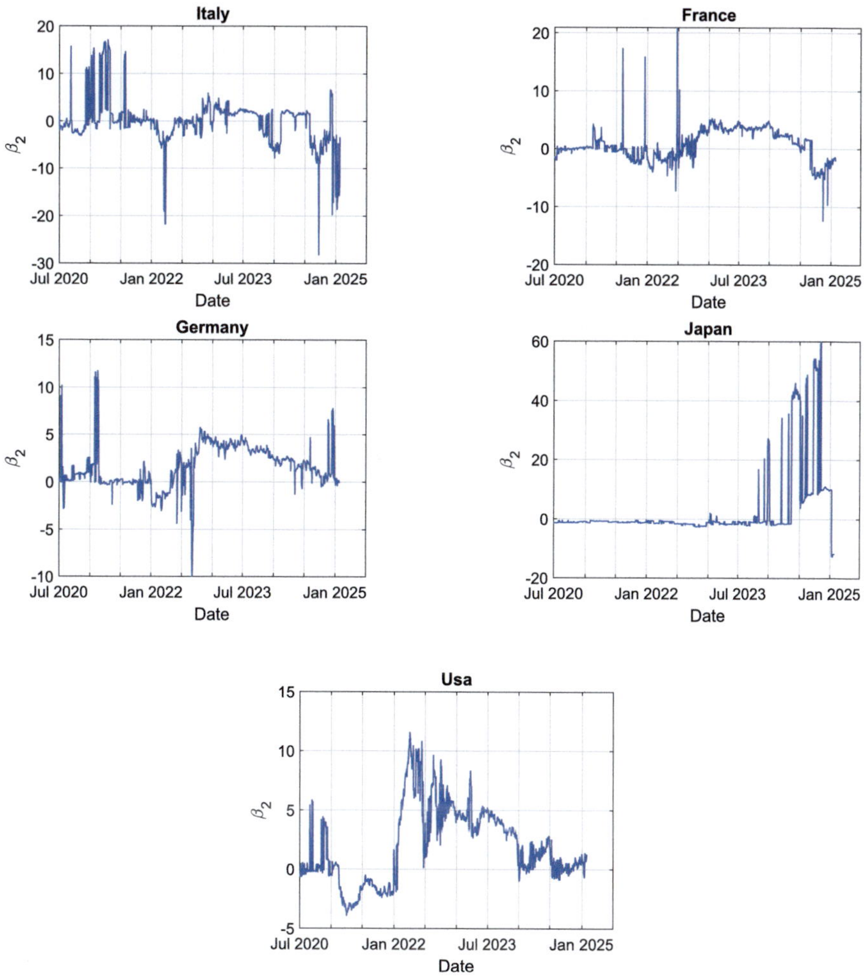

Fig. 7. Estimated β_2 parameter in the Nelson-Siegel-Svensson model for different countries.

the other countries. A direct comparison indicates that the Nelson-Siegel model should be preferred due to its versatility, allowing for application in various contexts.[1]

[1] The findings, which are available upon request, demonstrate that the Nelson-Siegel model yields to results that are analogous to those of NSS in terms of Root Mean Square Error.

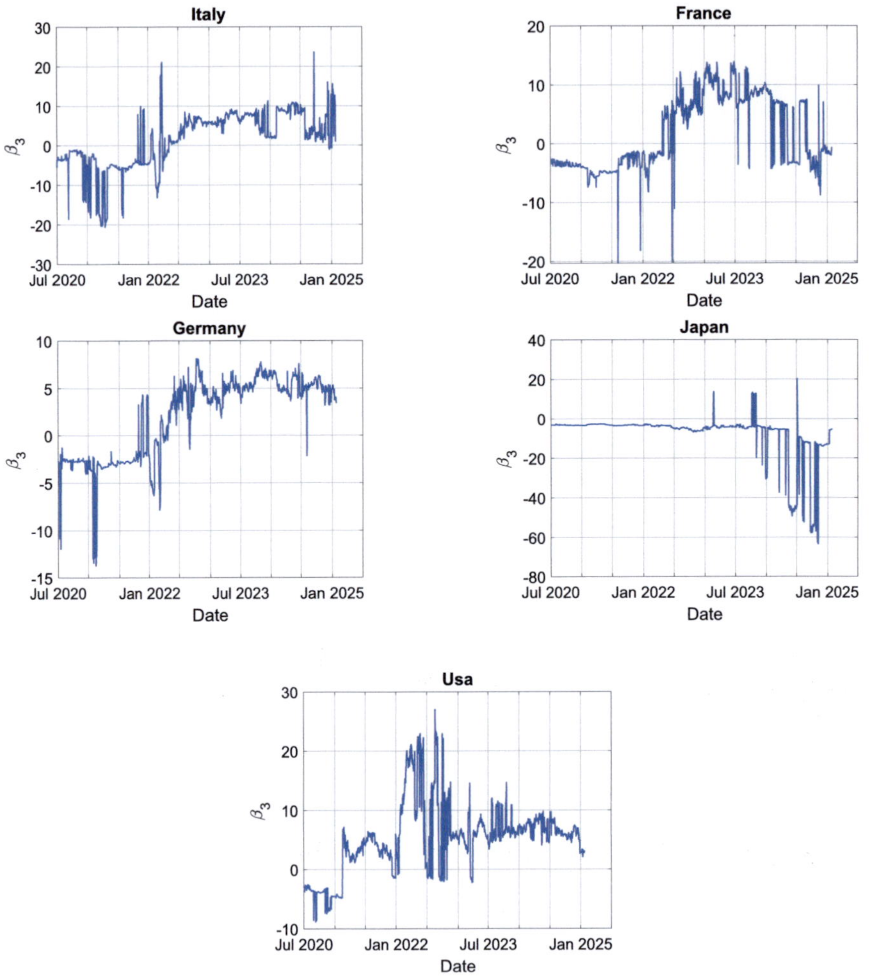

Fig. 8. Estimated β_3 parameter in the Nelson-Siegel-Svensson model for different countries.

References

1. Consiglio, A., Guirreri, S.: Simulating term structure of interest rates with arbitrary marginals. Int. J. Risk Assess. Manag. **15**(4), 299–313 (2011)
2. Nelson, C.R., Siegel, A.F.: Parsimonious modeling of yield curves. J. Bus. **60**(4), 473–489 (1987)
3. Nymand-Andersen, P.: Yield curve modelling and a conceptual framework for estimating yield curves: evidence from the European Central Bank's yield curves. In: ECB Statistics Paper, 27 (2018)

4. Svensson, L.E.: Estimating and interpreting forward interest rates: Sweden 1992–1994. In: NBER Working Paper 4871 (1994). https://doi.org/10.3386/w4871
5. Wahlstrøm, R.R., Paraschiv, F., Schürle, M.: A comparative analysis of parsimonious yield curve models with focus on the nelson-siegel, svensson and bliss versions. Comput. Econ. **59**(3), 967–1004 (2022)

Deep Learning for Tabular Data: Application to Credit Risk Modeling

Steven Mphaya[1,2(✉)], Marialuisa Restaino[2], and Michele La Rocca[2]

[1] University School for Advanced Studies IUSS Pavia, Piazza della Vittoria 15, Pavia 27100, Italy
[2] University of Salerno, Via Giovanni Paolo II, 132, Fisciano, SA 84084, Italy
smphaya@unisa.it

Abstract. Credit risk is one of the primary risks that banks face, deserving special attention. Previously, modelling credit risk data using parametric models required significant labour, which was time-consuming. Despite the dominance of machine learning (ML) models, deep learning (DL) models for tabular data have emerged to address their drawbacks, including interpretability issues. We seek to determine whether the TabNet model is worth paying the price of its sophisticated computation and interpretability abilities. We used 37991 Italian manufacturing companies to determine their default likelihood. We adopted the Boruta method and sequential attention mechanism for feature selection, SMOTEENN for data balance, and SHAP values to quantify features' contribution toward model output. A comparative analysis revealed that XGBoost remains a state-of-the-art model in balanced and imbalanced data cases. Thus, leveraging XGBoost can assist lenders in predicting and classifying potential defaulters. Data limitations and feature exclusions set the stage for further exploration of TabNet's performance in default prediction tasks.

Keywords: Credit Risk · Deep Learning · Tabular Data · TabNet

1 Introduction

Credit risk describes the likelihood that a debtor will fail or unwilling to repay the obligor due to financial crisis, bankruptcy, or other reasons. Its occurrence exposes lending entities to direct and indirect losses, affects the utilization of companies' funds, and increases management costs [30]. Failure to repay a loan can be seen as an early sign of financial distress, leading to the loss of financial opportunities. Thus, credit risk assessment is crucial to stakeholders and scholars worldwide since careless approval of loans may result in acute effects such as the 2008 financial crisis. Increased attention to credit risk prediction has triggered the proposition of comprehensive and proactive models to understand borrowers' default probability.

Parametric statistical models, especially the logistic regression (LR), have been used in credit risk prediction and classification to determine and quantify

M. La Rocca et al. (Eds.): WMAF 2025, *New Perspectives in Mathematical and Statistical Methods for Actuarial Sciences and Finance*, pp. 190–203, 2025.
https://doi.org/10.1007/978-3-032-05551-4_17

the borrower's default risk [25]. However, their strict parametric assumptions might expose the modelling workflow to severe misspecification problems, giving rise to the need for more advanced and innovative models. Machine learning (ML) and deep learning (DL) models have emerged to address the drawbacks of parametric statistical models in various domains, including credit risk, mainly when the analysis focuses on prediction. In this latter case, both single and ensemble ML models like decision trees (DT) and XGBoost have been proven to outperform parametric models due to their ability to learn complex data structures, minimizing the need for extensive feature engineering while achieving high performance [14].

However, in many applications, classical ML models have been outperformed by DL models, which can better model complex, nonlinear structural data with minimal or no effort in pre-processing and feature engineering [18]. On the one hand, DL models have excelled in learning complex data structures such as images and videos. On the other hand, they tend to perform poorly on tabular data, despite being the common form of real-world data. Heterogeneity of tabular data, lack of inherent positional information, and their limited sample size are among the factors contributing to the DL models' poor performance, hence their under-utilisation [6]. See also Shwartz-Ziv and Armon [24] for a detailed comparison among ML and DL models.

Moreover, the black-box nature of both ML and DL strongly limits their interpretability. It may be risky when such models are entrusted with making critical decisions that impact institutions' and individuals' welfare. Therefore, we perform a comparative analysis of highly performing ML models against the TabNet model [3], one of the most innovative DL models, specifically developed to address the limitations of DL models when applied to tabular data. In this paper, we aim to determine if the TabNet model is worth paying the price of its complex computations as compared to the classical ML models when the focus is on Credit Risk Modelling, with a specific attention to interpretability. To the best of our knowledge, this paper marks the inaugural adoption of the TabNet model with SHAP values for interpretability in the default prediction task.

The rest of this paper is organized as follows: Sect. 2 provides a review of related work on DL models in credit risk prediction. Section 3 describes the data and methodology used, including the models. The empirical results are presented in Sect. 4, whereas Sect. 5 discusses the conclusion and future research projections.

2 Related Work

The LR model has been widely adopted in credit risk prediction and classification tasks. Fernando et al. [13] applied the panel LR model in a corporate governance and financial distress analysis, achieving average performance. Abdou et al. [1] used the LR model in default prediction on the Indian banking sector, whereas Agrawal and Maheshwari [2] proposed LR-based multiple discriminant analysis (MDA) in Indian firms, achieving promising predictive performance than the

MDA model. ML methods are used to overcome the LR model's drawbacks in estimating default probability. Yuan et al. [31] applied the Support Vector Machine (SVM) model in enterprise credit evaluation, achieving better performance than the logit model. Chen [10] combined a DT-based model with boosting techniques in individual credit evaluation of banks, achieving high performance. However, DT-based models are prone to overfitting and feature selection bias; hence, the use of ensemble DT-based models [22].

Arora and Kaur [4] used the random forest (RF) model in the debtors' behavior prediction, achieving higher performance than LR and single DT-based models. Xia et al. [27] proposed a cost-sensitive DT-boosted loan evaluation model incorporating XGBoost to enhance the discrimination capabilities for potential defaulters, achieving the best performance in unbalanced data. Chang et al. [9] applied the XGBoost model in the risk assessment of financial institutions, and the results confirmed its performance advantages. Thus, it can be concluded that gradient-boosted models achieve excellent performance, hence their use as benchmark models, as they are the current novel models in tabular data tasks.

However, ML models struggle to handle large-scale and high-dimensional data, with high reliance on feature engineering and distributional drift. Additionally, the "no free lunch" theorem indicates that no single method consistently outperforms others across all tabular data tasks, necessitating comprehensive evaluations to understand DL models' performance in real-world applications. Consequently, the continuous evolution in the credit risk domain has surfaced DL models as the potential solution. Xu et al. [28] combined the Long Short-Term Memory (LSTM) network and the Adaptive Boosting (AdaBoost) algorithm to deal with imbalanced credit risk data, with the model outperforming other ML models. McDonnell et al. [18] applied the TabNet model in insurance risk classification and bankruptcy prediction, respectively, achieving similar accuracy performance as the XGBoost, whilst outperforming the weighted XGBoost model by providing highly interpretable results. Thus, given this framework, we adopt the TabNet model to assess the default likelihood of Italian manufacturing companies and compare its predictive and interpretability performance against the highly performing gradient-boosted ML models.

3 Methodology

3.1 Data

We used data from the Organization of Economic Cooperation and Development (OECD) data stream for company-level microdata based on the Commercial ORBIS data stream of the Bureau Van Dijk [23]. Our sample comprises 37991 Italian manufacturing companies from 2000 to 2018. The set of companies is divided into two groups, based on their status: i) active or non-default class, and ii) inactive or default class, which includes bankrupt, liquidated, and dissolved companies. Thus, the variable of interest **"Status"** takes values 1 if the companies are active and 0 otherwise.

For each company, using the information from the financial statements, a list of financial ratios is computed. Since the number of possible predictors that may affect the default probability is sufficiently high, we choose the features following the criteria as shown in Caroni et al. [7]. Then, we clean the data by checking the missingness rate and correlation among variables. As suggested in Caroni et al. [7], we retain only the variables for which the correlation is less than 0.70. Thus, the final set of features consists of 28 financial ratios considered as potential predictors of predicting the default, as listed in Table 1.

The presence of imbalanced data in credit risk datasets, whereby the number of non-defaulters and defaulters is not equal, negatively affects the performance of DL models as they assume balanced data classes by default [8]. Thus, we used the SMOTEENN algorithm, combining the oversampling technique of Synthetic Minority Oversampling Technique (SMOTE) with the under-sampling approach of Edited Nearest Neighbour (ENN) [16] to balance the data. Furthermore, the presence of irrelevant features deteriorates the model's performance. Thus, we applied the Boruta method and built-in sequential attention mechanism for feature selection for benchmark models (since it works perfectly with DT-based models) and the TabNet model, respectively. Boruta method replicates original features to create shadow features and randomizes their values systematically to remove any relationships with the outcome variable [21]. Each model extracts a specific set of important features, and the final feature set is obtained through the voting system.

3.2 Models

The TabNet is a novel DL model designed specifically for tabular data learning tasks. It uses a sequential attention mechanism for feature selection and an encoder-decoder framework for self-supervised learning [3]. Contrary to DT-based models, the model provides several benefits in terms of generalization, performance, and interpretability. The model's central idea involves sequentially utilizing an Attention Transformer for feature selection and a Feature Transformer for changing selected features into meaningful and useful representations [30]. This iterative process mimics decision-making in DT-based models. The model also integrates feature reuse and sparsity to reduce overfitting and boost performance, prior scales to control the feature selection rate, and a cross-entropy-based loss function to make attention masks sparser [30]. Furthermore, the model offers both local and global interpretability, allowing visualization of feature importance at each decision step and feature contribution to the overall model output, respectively. These mechanisms help TabNet to construct a sophisticated nonlinear transformation in the feature space, enhancing predictive power while maintaining interpretability.

The Logistic Regression model has demonstrated effective performance in most classification tasks. Specifically, the LR model is a type of regression used when the response variable is binary, and the explanatory variables are categorical or continuous [12]. The model outputs the probability values ranging between 0 and 1, illustrating whether a particular subject belongs to a specific class. Using

Table 1. Financial indices considered as potential predictors in the analysis.

ID	Formula	ID	Formula
ind004	Operating revenue/Stock	ind064	Net income/Current assets
ind008	Defensive assets - current liabilities to fund expenditures for operations	ind065	Net income/Fixed assets
ind009	Cash flow/Net worth	ind066	Net income/Net worth
ind014	Cash/Current liabilities	ind070	Net worth/Sales
ind017	Cash/Total assets	ind073	shareholders fund/ Total assets
ind020	Ln(Total assets)	ind077	Current assets - Inventory
ind029	Current liabilities/Total liabilities	ind078	Quick assets/Current liabilities
ind044	Equity/Fixed assets	ind079	Quick assets/Sales
ind048	Fixed assets/Total assets	ind084	Net income/Shareholder's equity
ind051	Inventory/Total assets	ind089	Sales/EBIT*
ind053	Loans/Total assets	ind092	Sales/Total assets
ind057	Long term debts/Total assets	ind101	Total debts/Equity
ind062	(Profit (Loss) for period + interest paid)/ (Shareholders funds + Non-current liabilities)	ind118	Net worth/Total liabilities
ind063	Net income/Cash flow	ind122	Current assets/Operating income

*Earnings before interest and tax

the debtor-specific features, the model computes the default likelihood by estimating the probability that a borrower will fail to meet their obligations. The model is extensively used due to its simplicity and interpretability, hence its broad adoption in Credit Risk Modelling and use as a benchmark model.

The Random Forest model is an ensemble randomized learning algorithm that uses the construction of an assemblage of basic DTs during model training [29]. The model's underlying principle is to create a batch of trees using various subsets of the input samples and features and then aggregate the outputs of individual DTs to produce overall results. The model minimizes the overfitting of individual DTs through bootstrap sampling and feature randomization techniques, thereby improving the accuracy and robustness of the model [29]. Thus, the final model output is obtained by considering the class receiving the most "votes" from the individual DTs in the classification task or the average of all predictions from individual DTs in the regression task, hence its adoption as a benchmark model in this paper.

The XGBoost algorithm is a distributed and efficient boosting integrated classification technique based on DTs [11]. This algorithm combines several low-accuracy and precision-weak models into an overall high-accuracy and precision model. It is known for its ability to handle sparse datasets, hence high speed, parallelism, and robust performance without imposing any distributional assumption [11]. It improves the traditional gradient-boosting method by introducing regularization in its loss function to avoid overfitting. The model has proven to perform excellently, evidenced by many Kaggle competition prizes, hence its adoption.

The LightGBM model focuses on improving overall model accuracy by training various weak models sequentially and optimizing the prediction residuals from the previous weak models by introducing two novel techniques, namely, Gradient-based One-Sided Sampling (GOSS) and Exclusive Feature Bundling (EFB) [19]. The GOSS technique filters out data features to focus on those with higher gradients by calculating variance gain, whereas the EFB technique reduces the number of features in the sparse dataset by bundling exclusive features into fewer features. Unlike XGBoost, the LightGBM model utilizes leaf-wise tree growth as opposed to level-wise tree growth [19]. Thus, trees are grown vertically, focusing on expanding the leaves that most contribute to the overall loss reduction [19]. This approach enables the model to prioritize relevant features, hence achieving rapid convergence. Thus, we use the LightGBM model as a benchmark due to its high customization and ability to handle categorical features natively.

3.3 SHapley Additive ExPlanation (SHAP)

Due to the complexity of both ML and DL models and their "black-box" nature, interpretability is challenging. The SHAP method describes the decision-making process of complex models, clarifying individual features' contributions to the model's overall output [5]. The method further applies game theory ideas to assign importance scores to each feature for a particular prediction. This method is preferred as it unifies the model's local and global interpretability [5]. Thus, it exhibits each feature's relevance and quantifies how each feature affects model outputs. A SHAP score is a weighted sum of features' marginal contributions across all possible feature subsets, indicating the model prediction when a particular feature is included and excluded. We apply the SHAP technique to the highest-performing model after the comparative analysis.

3.4 Model Evaluation Metrics

To evaluate the performance of the models, we computed some metrics based on the confusion matrix, widely used to assess the performance of the models [5]. The table consists of information about the actual and predicted classifications obtained by the model, thus enhancing the visualization of the model's performance. In particular, the terms in the confusion matrix used for computing the metrics are:

- *True Positives (TP):* The number of cases correctly classified as "default".
- *True Negatives (TN):* The number of cases correctly classified as "non-default".
- *False Positives (FP):* The number of cases incorrectly classified as "default".

– *False Negatives (FN):* The number of cases incorrectly classified as "non-default".

Thus, starting from these quantities, some metrics are computed using a default threshold of 0.5 (all predicted probabilities greater than 0.5 are assigned to class 1 and 0 otherwise). In this paper, we considered accuracy, precision, recall, specificity, and F1-measure.
Accuracy measures the proportion of the observations correctly predicted by the model, as shown in Eq. (1).

$$Accuracy = \frac{TP + TN}{TP + FP + TN + FN} \tag{1}$$

Precision defines the proportion of positive cases that are correctly predicted out of all cases that were predicted as positive as shown in Eq. (2). A higher precision score implies the model's ability to minimize FPs effectively.

$$Precision = \frac{TP}{TP + FP} \tag{2}$$

Recall measures the proportion of correctly predicted positive observations out of all actual positive cases in the dataset [20], as shown in Eq. (3). The higher the recall value, the more effective the model is in identifying the majority of TP observations.

$$Recall = \frac{TP}{TP + FN} \tag{3}$$

Specificity measures the model's ability to correctly recognize negative observations among all actual negatives in the dataset, as shown in Eq. (4).

$$Specificity = \frac{TN}{TN + FP} \tag{4}$$

Higher specificity values indicate the model's ability to correctly recognize negative instances and minimize FP cases.
F1 measure defines a harmonic mean of precision and recall metrics, balancing the model's accuracy and coverage, given by Eq. (5).

$$F1\ measure = 2 \times \frac{\text{Precision} \times \text{Recall}}{\text{Precision} + \text{Recall}} \tag{5}$$

Furthermore, we also computed the *AUC score*, which measures the area under the receiver operating characteristic (ROC) curve (AUC) that determines the model's ability to discriminate between classes as shown in Eq. (6), with a higher AUC value indicating better performance.

$$AUC\ score = \int_0^1 \text{ROC}(t)dt \tag{6}$$

4 Results and Discussion

We discovered highly imbalanced classes of the response variable "Status" with 10.1% and 89.9% of inactive and active companies, respectively. To prevent bias in model performance, we balanced the data using SMOTEEN as shown in Fig. 1.

Fig. 1. Distribution of the "Status" before and after SMOTEENN. Thus, the SMOTEENN algorithm leveled the class imbalance, achieving 49.6% and 50.4% class distribution, respectively.

We randomly partitioned the data into 80% and 20% for training and testing. Each model's performance is enhanced by tuning the hyperparameters using Random-Search-Cross-Validation with 5 folds to optimize model performance. Since each model has a different format, we fine-tuned each model separately (although some models have some common hyperparameters). The metrics shown in Tables 2 and 3 are computed on the test dataset, which is widely used for evaluating the predictive abilities of the models. The models' performance summary when fitted to the unbalanced data is recorded in Table 2. The bolded metrics represent the highest-performing value.

Table 2. Model performance using imbalanced data.

	Models				
Evaluation metric	Logistic Regression	Random Forest	LightGBM	XGBoost	TabNet
Accuracy	0.8078	0.8587	0.8480	**0.9162**	0.8922
Precision	0.9715	0.9077	**0.9731**	0.9453	0.9140
Recall	0.8091	0.7123	0.8538	0.9625	**0.9715**
Specificity	0.2930	0.3452	0.1040	**0.5026**	0.1845
F1 score	0.8829	0.7982	0.9096	**0.9538**	0.9419
AUC ROC	0.8612	0.8992	0.8970	**0.9021**	0.8060

As shown in Table 2, on average, the XGBoost model outperformed other benchmark models, including the TabNet model, in almost all evaluation metrics.

The TabNet model detected the positive class better (active companies) (Recall = 0.9715) than other models. However, the specificity metric is too low in all the models, implying the models' struggle to identify the minority class (inactive companies) correctly. Thus, Table 2 may not be the true representation of the models' performance since the models are likely to be biased toward predicting the majority class due to the data imbalance. This is in line with the literature by Williams [26] and Lunque et al. [17], who found that recall and precision metrics of a binary classifier depend on the ratio of positive to negative cases in the test dataset.

The models' performance results after applying SMOTEENN are consistent and unbiased, with the models exhibiting novel performance, as shown in Table 3. The bolded metrics represent the highest-performing value.

Table 3. Model performance using balanced data.

Evaluation metric	Models				
	Logistic Regression	Random	LightGBM	XGBoost	TabNet
Accuracy	0.8405	0.8504	0.8624	**0.9342**	0.8562
Precision	0.8429	0.9118	0.8410	**0.9580**	0.8628
Recall	0.8408	0.7870	0.8964	**0.9154**	0.8496
Specificity	0.8401	0.8152	0.8278	**0.9552**	0.8629
F1 score	0.8419	0.8448	0.8678	**0.9362**	0.8561
AUC ROC	0.9117	0.9376	0.9380	**0.9848**	0.9296

As evident in Table 3, the XGBoost model consistently outperforms all benchmark models, including the TabNet model, as indicated by the highest evaluation metric values. Thus, the XGBoost model has achieved state-of-the-art performance in predicting and classifying firms as active or inactive. This is in line with Chen and Guestrin [11] and Ke et al. [15], who disclosed the outperformance of the XGBoost in tabular data tasks, but contrary to McDonnell et al. [18]. Furthermore, the TabNet model achieves a comparable performance, with the RF and LightGBM models in its ability to discriminate (AUC ROC) between default and non-default companies. The LR model remains the least-performing model, signifying its unreliability. Thus, in terms of performance, the TabNet model may not be the best model worth paying the price of its computational complexities in default prediction in balanced and unbalanced data.

TabNet model provided instance-wise features' importance to the model output at each decision step, as illustrated in Fig. 2. The brighter the color, the more relevant the feature is to the model output at a particular step. It is recommended to use moderate decision steps to allow the model to learn all features thoroughly while preventing data memorization. Therefore, we used 6 decision steps.

As evident in Fig. 2, feature colors are mixed and scattered from steps 1 to 3, denoting model training difficulties due to the presence of noise and feature

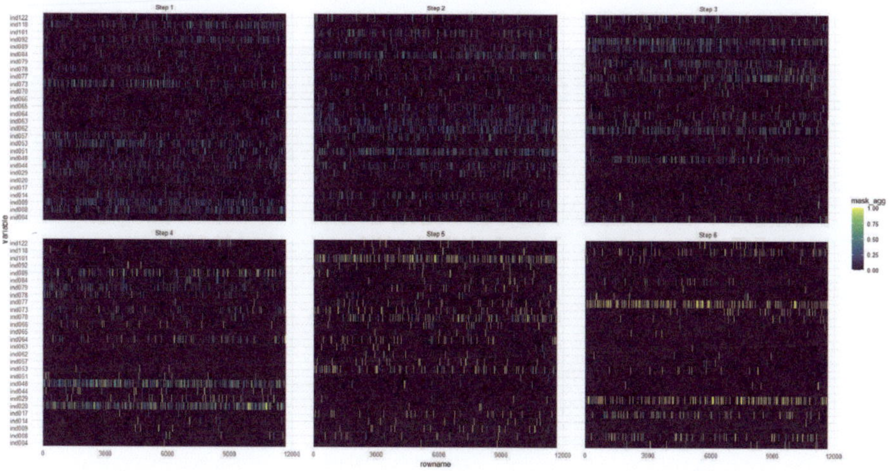

Fig. 2. Instance-wise mask plot illustrating step-wise feature importance to the model's output (Local interpretability). The brighter the color, the more relevant the feature.

interaction in the dataset. From steps 4 to 6, the feature colors are clearer. Step 6 shows that features such as no credit interval (ind008), cash flow to total assets ratio (ind017), current liabilities to total liabilities ratio (ind029), and quick asset (ind077) were considered relevant than other features at this decision step.

The TabNet model also demonstrated relevant features to the overall model decision.

As illustrated in Fig. 3, features like no credit interval (ind008), cash flow to current debts ratio (ind014), net income to fixed assets ratio (ind065), return on equity (ind084), and total debts to shareholders' funds ratio (ind101) are more relevant to the model's overall output than other features, implying that such features are crucial in predicting and classifying firms' default likelihood.

Lastly, we use SHAP values to quantify the impact of each feature on the model output using the summary beeswarm plot shown in Fig. 4.

The y-axis denotes the feature's names in order of importance based on the XGBoost (outperforming model) from top to bottom. The x-axis indicates the SHAP value describing the change in the log-odds of a feature. The gradient color represents the original value for that feature. Overlapping points are scattered slightly on the y-axis to indicate the distribution of the SHAP values. As evident in Fig. 4, high values of no credit interval record (ind008), net income to current assets ratio (ind064), net income to net worth ratio (ind066), sales to EBIT ratio (ind089), and shareholders fund to total assets ratio (ind073) have high positive contribution to the prediction, implying that firms with such high ratios are highly likely to be non-defaulters. High values of net income to cash flow ratio (ind063) and net income plus interest paid to shareholders fund plus non-current liabilities ratio (ind062) have high negative contributions to the prediction, implying a high likelihood of default. Furthermore, low values of the

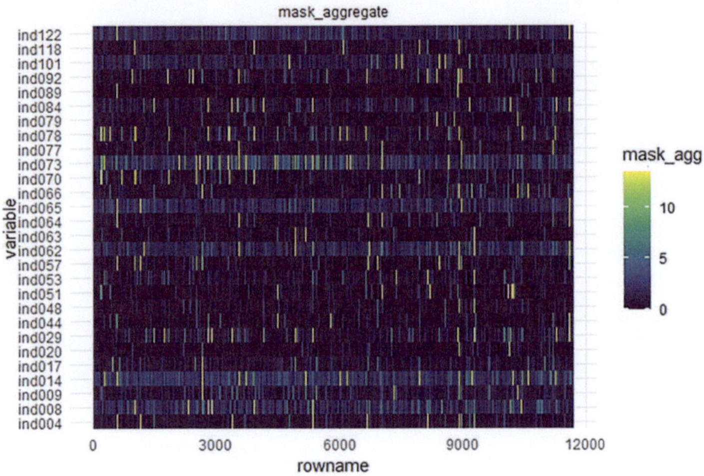

Fig. 3. Aggregated-mask plot showing features' importance to the model's overall output. The brighter the color, the more relevant the feature (Global interpretability).

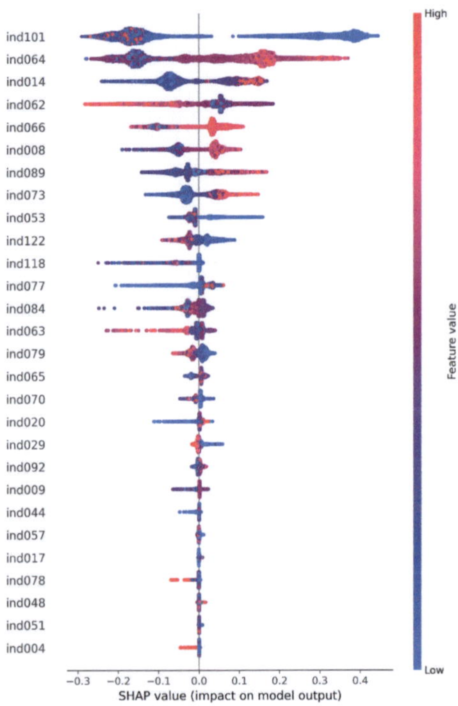

Fig. 4. SHAP values summary beeswarm plot quantifying the contribution of each feature on the XGBoost model output.

liquidity ratio (ind053) have a low positive contribution, whereas low values of the natural logarithm of total assets (ind020) and quick assets (ind077) have a low negative contribution. Low values of debts to equity ratio (ind101) have both low positive and negative contributions, implying its insensitiveness to the firm's default likelihood. Financial indices like sales to total assets ratio (ind092), cash to total assets ratio (ind017), fixed assets ratio (ind048), inventory to total assets ratio (ind051), and long-term debts to total assets ratio (ind057) have almost no contribution, whether their values are high or low.

5 Conclusion and Future Work

Studies on credit risk prediction have been a top priority in the financial domain. Our study aimed at comparing the performance of the TabNet against the boosted ML models to determine whether the TabNet model is worth paying the price of its sophisticated computational capabilities in firms' default prediction. We discovered that the XGBoost model performs well in predicting and classifying a firm's credit risk, showcasing its vast potential applications in this field compared to TabNet. We found that the TabNet model has a comparable performance with the LR, RF, and LightGBM models in predicting firms' credit risk. The good performance of XGBoost in all metrics entails its robustness and suitability for default prediction tasks. Our findings underscore the crucial role of data pre-processing and feature engineering in developing powerful models. Boruta and attention mechanism techniques helped important feature extraction, while SMOTEENN balanced the class distribution, resulting in robust model performance.

The TabNet model provides insight into the relevance of each feature to the model's overall prediction. Additionally, SHAP values addressed the "black-box" nature of the XGBoost by quantifying the contribution of each feature to the overall output. Thus, continuous monitoring and updating of the TabNet model is essential to improve its performance over time. Acknowledging data limitations and feature exclusions sets the stage for further exploration of the predictive performance of TabNet in default prediction. Thus, while our study demonstrates the effectiveness of the XGBoost in predicting a firm's default risk in both balanced and unbalanced data cases, further research is needed to assess its performance against TabNet in various industrial contexts. Our findings have significant implications for financial institutions involved in credit risk assessment. By leveraging ML models like XGBoost together with appropriate data preprocessing techniques and interpretation tools, firms can enhance their risk management strategies and make more informed decisions.

Acknowledgments. This paper and related research have been conducted during and with the support of the Italian national inter-university PhD course in Sustainable Development and Climate Change (link: https://www.phd-sdc.it) with the support of a scholarship financed by the Ministerial Decree no. 351 of 9th April 2022, based on the NRRP - funded by the European Union - NextGenerationEU - Mission 4 11Education and Research", Component 1 "Enhancement of the offer of educational services:

from nurseries to universities" Investment 4.1 "Extension of the number of research doctorates and innovative doctorates for public administration and cultural heritage".

References

1. Abdou, H.A., Mitra, S., Fry, J., Elamer, A.A.: Would two-stage Scoring models alleviate bank exposure to bad debt? Expert Syst. Appl. **128**, 1–13 (2019)
2. Agrawal, K., Maheshwari, Y.: Efficacy of industry factors for corporate default prediction. IIMB Manag. Rev. **31**(1), 71–77 (2019)
3. Arik, S.O., Pfister, T.: TabNet: attentive interpretable tabular learning. In: Proceedings of the AAAI conference on Artificial Intelligence, vol. 35(8), pp. 6679–6687 (2021)
4. Arora, N., Kaur, P.D.: A bolasso based consistent feature selection enabled random forest classification algorithm: an application to credit risk assessment. Appli. Soft Comput. **86**, 105936 (2020)
5. Aruleba, I., Sun, Y.: Effective credit risk prediction using ensemble classifiers with model explanation. IEEE Access **12**, 115015–115025 (2024). https://doi.org/10.1109/ACCESS.2024.3445308
6. Baecke, P., Bocca, L.: The value of vehicle telematics data in insurance risk selection processes. Dec. Support Syst. **98**, 69–79 (2017)
7. Caroni, C., Pierri, F., Restaino, M.: A comparison of variable selection methods in competing risks models for business failures. . Statistica & Applicazioni, **21**(1), 1–17 (2023). https://doi.org/10.26350/999999_000053
8. Chang, Y.C., Chang, K.H., Chu, H.H., Tong, L.I.: Establishing decision tree-based short-term default credit risk assessment models. Commun/ Stat. Theory Methods **45**(23), 6803–6815 (2016)
9. Chang, Y.C., Chang, K.H., Wu, G.J.: Application of extreme gradient boosting trees in the construction of credit risk assessment models for financial institutions. Appl. Soft Comput. **73**, 914–920 (2018)
10. Chen, M.Y.: A hybrid ANFIS model for business failure prediction utilizing particle swarm optimization and subtractive clustering. Inf. Sci. **220**, 180–195 (2013)
11. Chen, T., Guestrin, C.: XGBoost: a scalable tree boosting system. In: Proceedings of the 22nd ACM SIGKDD International Conference on Knowledge Discovery and Data Mining, pp. 785–794 (2016)
12. Chow, S.M., Zu, J., Shifren, K., Zhang, G.: Dynamic factor analysis models with time-varying parameters. Multivar. Behav. Res. **46**(2), 303–339 (2011)
13. Fernando, J.M.R., Li, L., Hou, G.: Financial versus non-financial information for default prediction: Evidence from Sri Lanka and the USA. Emerg. Mark. Financ. Trade **56**(3), 673–692 (2020)
14. Guelman, L.: Gradient boosting trees for auto insurance loss cost modeling and prediction. Expert Syst. Appl. **39**(3), 3659–3667 (2012)
15. Ke, G., Meng, Q., Finley, T., et al.: Lightgbm: a highly efficient gradient boosting decision tree. In Advances in Neural Information Processing Systems, vol. 30, pp. 3146–3154. Curran Associates Inc (2017)
16. Lin, M., Zhu, X., Hua, T., et al.: Detection of ionospheric scintillation based on xgboost model improved by SMOTE-ENN technique. Remote Sensing **13**(13), 2577 (2021)

17. Luque, A., Carrasco, A., Mart'ın, A., de Las Heras, A.: The impact of class imbalance in classification performance metrics based on the binary confusion matrix. Pattern Recogn. **91**, 216–231 (2019)
18. McDonnell, K., Murphy, F., Sheehan, B., et al.: Deep learning in insurance: accuracy and model interpretability using TabNet. Expert Syst. Appli. **217**, 119543 (2023)
19. Nguyen, H.V., Byeon, H.: Predicting depression during the COVID-19 pandemic using interpretable TabNet: A case study in South Korea. Mathematics **11**(14), 3145 (2023)
20. Obaido, G., Ogbuokiri, B., Swart, T.G., et al.: An interpretable machine learning approach for hepatitis B diagnosis. Appli. Sci. **12**(21), 11127 (2022)
21. Prasad, P., Loveson, V.J., Das, B., Kotha, M.: Novel ensemble machine learning models in flood susceptibility mapping. Geocarto Int. **37**(16), 4571–4593 (2022)
22. Quan, Z., Valdez, E.A.: Predictive analytics of insurance claims using multivariate decision trees. Dependence Model **6**(1), 377–407 (2018)
23. Ribeiro, S.P., Menghinello, S., De Backer, K.: The OECD ORBIS database: responding to the need for firm-level micro-data in the OECD. OECD Statistics Working Papers, vol. 2010/01, pp. 1–22 (2010). https://doi.org/10.1787/5kmhds8mzj8w-en
24. Shwartz-Ziv, R., Armon, A.: Tabular data: deep learning is not all you need. Inform. Fusion **81**, 84–90 (2022)
25. Wang, W., Xi, J.: A rapid pattern-recognition method for driving styles using clustering-based support vector machines. In: 2016 American Control Conference (ACC), pp. 5270–5275. IEEE (2016)
26. Williams, C.K.I.: The effect of class imbalance on precision-recall curves. Neural Comput. **33**(4), 853–857 (2021)
27. Xia, Y., Liu, C., Li, Y., Liu, N.: A boosted decision tree approach using Bayesian hyper-parameter optimization for credit scoring. Expert Syst. Appl. **78**, 225–241 (2017)
28. Xu, Z., Shen, D., Nie, T., Kou, Y.: A hybrid sampling algorithm combining M-SMOTE and ENN based on random forest for medical imbalanced data. J. Biomed. Inform. **107**, 103465 (2020)
29. Younas, N., Ali, A., Hina, H., et al.: Optimal causal decision trees ensemble for improved prediction and causal inference. IEEE Access **10**, 13000–13011 (2022)
30. Yu, C., Jin, Y., Xing, Q., et al.: Advanced user credit risk prediction model using LightGBM, XGBoost, and TabNet with SMOTEENN. In: 2024 IEEE 6th International Conference on Power, Intelligent Computing and Systems (ICPICS), pp. 876–883. IEEE (2024)
31. Yuan, K., Chi, G., Zhou, Y., Yin, H.: A novel two-stage hybrid default prediction model with K-Means clustering and support vector domain description. Res. Inter. Bus. Finance **59**, 101536 (2022)

Modeling Economic Recovery via Diffusion Processes with Multisigmoidal Logistic Mean Subject to Random Catastrophes

Sabina Musto(✉) and Paola Paraggio

Dipartimento di Matematica, Università degli Studi di Salerno, Fisciano, SA, Italy
{smusto,pparaggio}@unisa.it

Abstract. This work analyzes lognormal diffusion processes with a multisigmoidal logistic mean, subject to random catastrophic events. The occurrence of such catastrophes is governed by a counting process $N(t)$, and upon each catastrophic event, the process restarts from a random state. The resulting process is employed as a model to describe phenomena of an economic and financial nature. In particular, we conduct a simulation study in which the process replicates the dynamics of bank capital, with catastrophes governed by a Poisson process and restart points fixed at the bailout level. Furthermore, we propose an application to real data related to Industrial Production Index (IPI) in which the time arrivals of the catastrophes are fixed and the restart points are binomially distributed. Such application confirms the relevance of the proposed model.

Keywords: lognormal diffusion processes · simulation study · financial crisis · Industrial Production Index (IPI)

1 Introduction

Stochastic processes exhibiting sigmoidal behavior are widely used in various fields (see, for instance, some recent works as [1,2] and [3]), since it is reasonable to assume that the growth of any population is subject to reaching a plateau and evolving under the influence of random fluctuations. However, the sample paths of such processes are usually continuous, which may not be entirely realistic, especially in situations where growth is affected by external catastrophic events. These phenomena induce abrupt changes in the system's dynamics, resulting in discontinuities. In mathematical modeling, such events are referred to as catastrophes (or jumps) and have been the subject of several studies over the years (see, for example, [4,5] and [6]).

Typically, catastrophes are modeled with a twofold stochastic component: randomness in time, meaning that the occurrence of the catastrophe is considered a random event, and randomness in space, implying that the process restarts from a random value after the catastrophe. Our research fits within this

M. La Rocca et al. (Eds.): WMAF 2025, *New Perspectives in Mathematical and Statistical Methods for Actuarial Sciences and Finance*, pp. 204–215, 2025.
https://doi.org/10.1007/978-3-032-05551-4_18

framework, focusing on lognormal diffusion processes subject to random catastrophes, with particular attention to processes characterized by a multisigmoidal logistic mean. This choice is motivated by the flexibility of such models, as they are capable of representing various phenomena, including growth patterns with multiple phases.

The paper is organized as follows. In Sect. 2, we provide a brief theoretical description of lognormal diffusion processes with multisigmoidal logistic mean subject to random catastrophes. Section 3 is devoted to applications of the proposed stochastic model in economic and financial contexts. Specifically, we present a simulation study modeling the evolution of a bank's capital under financial shocks, alongside a detailed analysis of a real-world dataset concerning Industrial Production Index (IPI). The first study also includes an empirical analysis of the first-passage time through a critical threshold, examining its dependence on the intensity of a Poisson counting process governing the occurrence of catastrophes. The second application presents maximum likelihood estimates and compares sample means with estimated means, considering random catastrophes which arrive at deterministic times and relocate the process to randomly chosen points (described by a binomial distribution). The best fitting model is then selected by considering two different criteria: the minimization of the Mean Square Error (MSE) and the maximization of posterior probability.

2 General Background

We consider a lognormal diffusion process $\{X(t); t \in I\}$ with $I = [t_0, +\infty)$ and $t_0 \geq 0$ having the following infinitesimal moments

$$A_1(x,t) := h(t)x, \qquad A_2(x) := \sigma^2 x^2, \tag{1}$$

where $h(t)$ is a positive and integrable function in any set (t_0, t) with $t \in I$. The corresponding state space of $X(t)$ is the set of positive real numbers $\mathbb{R}^+$. The function $h(t)$ represents the time-dependent growth rate. Clearly, different choices of $h(t)$ lead to different lognormal diffusion processes. From now on, we will focus on a special case in which the mean of $X(t)$ is a multi sigmoidal logistic function, i.e.

$$h(t) := h_\theta(t) = \frac{P_\beta(t)\, e^{-Q_\beta(t)}}{\eta + e^{-Q_\beta(t)}}, \qquad t \in I,$$

where $\eta > 0$, $\theta = (\eta, \beta^\mathsf{T})^\mathsf{T}$, $\beta^\mathsf{T} = (\beta_1, \ldots, \beta_p)$, $p \in \mathbb{N}$, and $Q_\beta(t)$ and $P_\beta(t)$ are polynomials defined as

$$Q_\beta(t) := \sum_{i=1}^{p} \beta_i t^i, \quad \beta_p > 0, \qquad P_\beta(t) := \frac{\mathrm{d}}{\mathrm{d}t} Q_\beta(t).$$

For such choice of $h(t)$, the mean of the process $X(t)$ is represented by a multisigmoidal logistic growth function, i.e.

$$\mathsf{E}\,(X(t)) := x(t) = x_0\, \frac{\eta + e^{-Q_\beta(t_0)}}{\eta + e^{-Q_\beta(t)}}, \qquad t \in I, \tag{2}$$

being $x_0 := x(t_0) > 0$. The curve given in Eq. (2) is known as multi-sigmoidal logistic function since it may exhibit multiple inflections and since for $p = 1$ it reduces to the classical logistic function (for further details see [7] and [8]). The diffusion process $X(t)$ having the infinitesimal moments given in Eq. (1) is the solution of the following Stochastic Differential Equation (SDE)

$$\mathrm{d}X(t) = h_\theta(t)X(t)\mathrm{d}t + \sigma X(t)\mathrm{d}W(t), \qquad X(t_0) = X_0, \tag{3}$$

where $W(t)$ denotes a standard Wiener process independent on the initial state X_0 for any $t \in I$. The SDE given in (3) can be solved by means of Itô's formula. The corresponding expression of the resulting process is the following

$$X(t) = X_0 \exp\left(H_\xi(t_0, t) + \sigma\left(W(t) - W(t_0)\right)\right), \qquad t \in I \tag{4}$$

with

$$H_\xi(s,t) := \int_s^t h_\theta(u)\mathrm{d}u - \frac{\sigma^2}{2}(t-s) = \log\left(\frac{\eta + e^{-Q_\beta(s)}}{\eta + e^{-Q_\beta(t)}}\right) - \frac{\sigma^2}{2}(t-s),$$

for $t_0 < s < t$ and $\xi = (\theta^\intercal, \sigma^2)^\intercal$. It is very well known in literature that if X_0 follows a lognormal distribution, i.e. $X_0 \sim \Lambda_1(\mu_0, \sigma_0^2)$, or if X_0 is degenerate, i.e. $\mathsf{P}(X_0 = x_0) = 1$, then all the finite dimensional joint distributions of the process $X(t)$ are lognormal and in particular the conditional distribution of $X(t)$ given $X(s) = y$ with $t_0 < s < t$ and $y > 0$ is also lognormal, i.e.

$$[X(t) \mid X(s) = y] \sim \Lambda_1\left(\log y + \log\left(\frac{\eta + e^{-Q_\beta(s)}}{\eta + e^{-Q_\beta(t)}}\right) - \frac{\sigma^2}{2}(t-s), \sigma^2(t-s)\right).$$

The use of this diffusion process as a model for growth phenomena characterized by multiple inflection points is quite widespread. However, such a process does not account for the possibility of discontinuities in growth trajectories caused by external catastrophic events. Given that such disruptions are highly plausible in real-world scenarios, a jump-diffusion counterpart of the lognormal diffusion process with a multisigmoidal logistic mean has been recently introduced (see [4]). In particular, let $N(t)$ be the counting process that describes the number of jumps occurred until the time $t \in I$, i.e.

$$N(t) := \max\left\{n \in \mathbb{N} \mid \Theta_n \leq t\right\}, \qquad t \in I,$$

where Θ_n, for $n \in \mathbb{N}$, represents the arrival time of the n-th catastrophe. We denote the probability density function (PDF) of the absolutely continuous random variable Θ_n with $\gamma_n(t)$, i.e.

$$\gamma_n(t) := \frac{\mathrm{d}}{\mathrm{d}t}\mathsf{P}\left(\Theta_n \leq t\right), \qquad t \in I, \quad n \in \mathbb{N}.$$

Moreover, by setting $\Theta_0 = t_0$ a.s., let us denote with $I_n := \Theta_{n+1} - \Theta_n$ the interarrival time between two consecutive catastrophes and by $\psi_n(t)$ the corresponding PDF, i.e.

$$\psi_n(t) := \frac{\mathrm{d}}{\mathrm{d}t}\mathsf{P}(I_n \leq t), \qquad t \in I, \quad n \in \mathbb{N}_0.$$

Hence, we define the process $X(t)$ subject to random catastrophes as

$$X(t) := \sum_{n=0}^{+\infty} X_n(t)\mathbb{I}\left(N(t) = n\right), \qquad t \in I, \tag{5}$$

where $X_n(t)$ is a lognormal diffusion process as (4) and $\mathbb{I}(A)$ denotes the indicator function related to A, so it is equal to 1 when A is true, otherwise it is equal to 0. It is clear from Eq. (5) that the process $X(t)$ evolves as $X_n(t)$ in the time interval between the n-th catastrophe and the following one, for any $n \in \mathbb{N}_0$. Assuming that the restart point of the process immediately after a catastrophic event is represented by an absolutely continuous random variable with PDF given by $\phi_n(s)$, it is then possible to derive the transition PDF of the process $X(t)$ as follows (see also [4] and [5]):

$$\begin{aligned} f(x,t \mid x_0,t_0) =& \ \mathsf{P}\left(I_0 > t - t_0\right) f_0(x,t \mid x_0,t_0) \\ &+ \sum_{n=1}^{+\infty} \int_{t_0}^{t} \mathsf{P}\left(I_n > t - \tau\right) \left(\int_0^{+\infty} \phi_n(s) f_n(x,t \mid s,\tau) \mathrm{d}s \right) \psi^{[n]}(\tau) \mathrm{d}\tau. \end{aligned}$$

It should be noted that, in the expression above, the function $f_n(x,t \mid s,\tau)$ denotes the transition PDF of the process $[X_n(t) \mid X_n(\tau) = s]$, i.e.

$$f_n(x,t \mid s,\tau) = \frac{1}{x\sigma\sqrt{2\pi(t-\tau)}} \exp\left(- \frac{\left(\log\left(\frac{x}{s}\right) - \log\left(\frac{\eta + e^{-Q_\beta(\tau)}}{\eta + e^{-Q_\beta(t)}}\right) + \frac{\sigma^2}{2}(t-\tau)\right)}{2\sigma^2(t-\tau)} \right),$$

with $t_0 < \tau < t$ and $x, s > 0$.

3 Case Studies

In the following sections, we focus our attention on two case studies of particular interest. More specifically, in Sect. 3.1, we analyze a lognormal diffusion process, whose mean follows a classical logistic profile, subject to random jumps. In this setting, jumps occur according to a Poisson process with fixed intensity $\lambda > 0$ and reset the process to a predetermined deterministic value (the bailout level). On the contrary, in Sect. 3.2, we model real-world data regarding the IPI by means of a lognormal diffusion process subject to random jumps with a multisigmoidal logistic mean. In this latter case, jumps occur at predetermined times but reset the process to a random value governed by a binomial distribution.

3.1 Application to Simulated Data

In this section, we assume that the arrival of catastrophic events $N(t)$ is governed by a classical Poisson process with intensity $\lambda > 0$. The occurrence of a catastrophe resets the process to a predetermined state, so that in this case the associated PDF $\phi_n(s)$ is degenerate and centered at $r_C > 0$, i.e.

$$\phi_n(s) = \delta(s - r_C), \qquad s > 0,$$

where $\delta(x)$ denotes the Dirac-delta function. In this framework, the process $X(t)$ can effectively represent the capital of a bank (in millions of dollars) evolving according to a diffusion driven by profits and losses. In the event of a catastrophic crisis, such as the default of major borrowers, liquidity shortages, and similar disruptions, the government or central bank typically intervene through bailouts, recapitalizations, or public guarantees, restoring the bank's capital to a sufficient level that allows it to continue its operations. It is reasonable to assume that the timing of such catastrophic events is random and, specifically, follows a Poisson process (a model widely used for catastrophic events), while the bailout level from which the process resumes is taken to be deterministic. It should also be noted that a bank's capital cannot grow indefinitely, as it is subject to market, regulatory, and business constraints. The classical logistic model, i.e. the one obtained for the following choice of the growth rate

$$h_\theta(t) = \frac{\beta e^{-\beta t}}{\eta + e^{-\beta t}}, \qquad t \in I,$$

describes perfectly growth phenomena that are initially rapid but gradually slow down until reaching a plateau known as carrying capacity. This behavior closely mirrors the dynamics of a bank's capital, which tends to expand during economic cycles and eventually stabilizes around a target level. Therefore, it is reasonable to model the evolution of bank capital between successive crises by means of a lognormal diffusion process with a logistic mean.

In the simulation study, conducted using the software R, we performed Monte Carlo simulations of the lognormal diffusion process with logistic mean, integrated with randomly timed jumps that reset the process to the bailout deterministic level (see Fig. 1(a)).

We use the simulated sample paths to construct the empirical estimate of the PDF of the first-passage-time of the process $X(t)$ through critical thresholds represented by percentage increases in the level of bailout. In studies involving diffusion processes, the analysis of first-passage-times through critical thresholds is of particular interest, as such times provide a way to quantify the vulnerability of a system and to estimate the critical moment when intervention becomes necessary (such as the implementation of risk mitigation measures). The chosen thresholds correspond to values 10%, 20% and 50% above the bailout level. A failure to cross this threshold indicates that a critical situation has not yet been resolved. Similarly, the mean first passage time can be interpreted as the expected time before overcoming the critical condition. Figure 1(b) presents an empirical estimate of the density of the first-passage-time of the process through the critical thresholds. We point out that, when the percentage increase of the threshold increases, the tail of the densities is heavier. Moreover, Fig. 2 shows the mean of the first-passage-time as well as the percentage of sample paths which

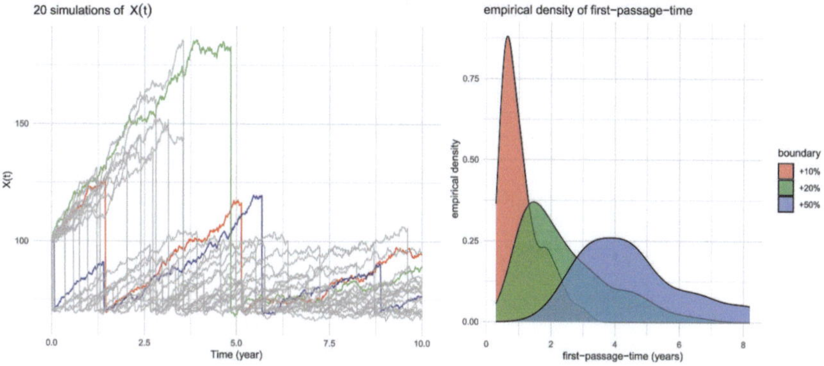

(a) (b)

Fig. 1. (a) Twenty simulated sample paths of the process $X(t)$ subjected to random catastrophes governed by a Poisson process with intensity $\lambda = 0.5$ for $\eta = 0.5$, $\beta = 0.25$, $\sigma = 0.05$, $\mathsf{P}(X_0 = 100) = 1$ and $r_C = 70$. (b) The empirical density of the first-passage-time of the process $X(t)$ through the boundary $S_1 = 1.10\ r_C$ (red), $S_2 = 1.20\ r_C$ (green) and $S_3 = 1.50\ r_C$ (blue) obtained using 500 simulated sample paths. (Color figure online)

do not cross the boundary as a function of the Poisson process intensity λ. The observed increase in the mean first-passage time and in the proportion of sample paths that do not cross the critical threshold as λ increases can be explained as follows. When events occur too frequently, the process does not have sufficient time to recover, and as a result, it never crosses the threshold. Consequently, the mean time can become significantly longer, due to the absence of any threshold crossings. A similar reasoning can be adapted also for the percentage of non-crossing sample paths. Moreover, the higher the threshold is, the less frequently the process manages to exceed it after a restart (see the heavy tails observed in Fig. 1(b)). As a result, the mean is computed over fewer valid crossings and is therefore more sensitive to variability. This explains why, for higher thresholds, the trend in the mean or in the percentage of non-crossing paths appears less smooth.

3.2 Application to Real Data

This section presents a real data study concerning the IPI across six European countries, namely Austria, Belgium, Denmark, Germany, Italy and Sweden. The IPI is an economic indicator that measures the real output of the industrial sector relative to a designated base year, expressed as a percentage. The IPI encompasses production in mining, manufacturing, and public utilities (including electricity, gas, and water supply), while excluding activities related to construction. The dataset covers the period from 2000 to 2023 and includes two major catastrophic events: the 2009 financial crisis (following the collapse of

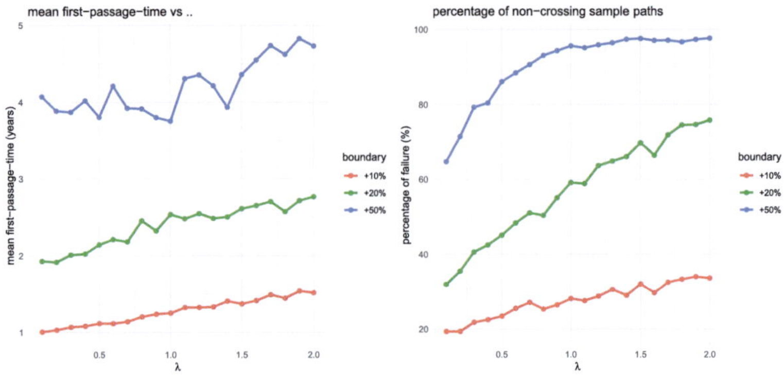

(a) (b)

Fig. 2. (a) The mean of the first-passage-time of the process $X(t)$ through the boundary $S_1 = 1.10\, r_C$ (red), $S_2 = 1.20\, r_C$ (green) and $S_3 = 1.50\, r_C$ (blue). (b) The percentage of sample paths which do not cross the boundary S. The values of the parameters are the same of Fig. 1. (Color figure online)

Lehman Brothers) and the 2020 COVID-19 pandemic. The data are publicly available at [?] and are displayed in Fig. 3.

With the aim of describing the observed growth patterns through a stochastic multisigmoidal logistic model subject to catastrophes, we assume that the catastrophe arrival times are known a priori. Consequently, the associated density function $\gamma_n(t)$ is degenerate, i.e.

$$\gamma_n(t) = \delta(t - \theta_n), \qquad n \in 1, 2.$$

Note that $n \in \{1, 2\}$ since the catastrophes occurred in the selected time interval are only two. The randomness in the model is instead introduced through the restart point, which is assumed to follow a binomial distribution (a scheme recently adopted in a similar application context [4]). Thus, the probability density function of the restart point is given by

$$\phi_n(s) = \binom{N_n}{s} p_n^s \,(1 - p_n)^{N_n - s}, \qquad s = 1, 2, \ldots, N_n$$

The objective of this study is to identify the most suitable model by selecting among lognormal diffusion processes with multisigmoidal logistic mean with polynomials $Q_\beta(t)$ of increasing degree p within each time interval between two consecutive catastrophes. The maximum likelihood estimates of the model parameters, denoted with $\xi = \left(\eta, \beta^\intercal, \sigma^2\right)^\intercal$, for each time interval are obtained using the `maxLik` function of `R`.

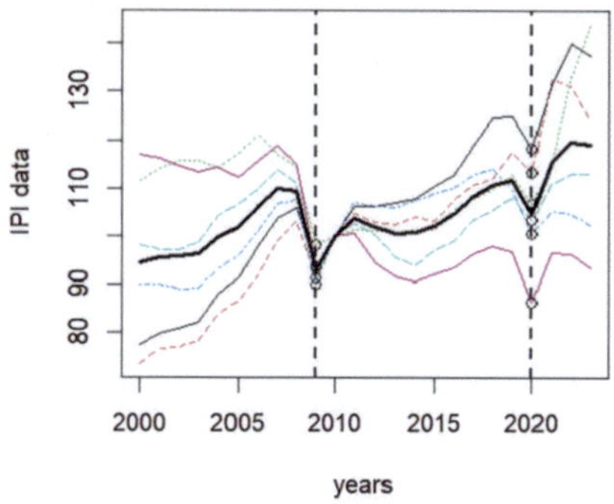

Fig. 3. IPI of Austria, Belgium, Denmark, Sweden, Germany and Italy (from top to bottom near 2020) from 2000 to 2023. The solid black line represents the mean of the data. The vertical dashed lines represent the time arrival of the catastrophes (the first in $\theta_1 = 2009$ and the second in $\theta_2 = 2020$).

The best-fitting model is then selected as the one minimizing the mean square error (MSE) between the estimated mean and the empirical sample mean, i.e.

$$MSE = \frac{1}{n}\sum_{i=1}^{n}(y_i - \hat{y}_i)^2,$$

being n the total size of the data set, y_i the real data and $\hat{y}_i$ the estimated value. The results are summarized in Table 1 and illustrated in Fig. 4. It is evident that the optimal model corresponds to the selection of a polynomial of degree 3 for the first interval and 4 for the second. This selection is further supported by a Bayesian model selection approach commonly employed in normal regression, using intrinsic priors for the model parameters and uniform priors over models, as proposed in [4] and [9]. Note that for the third time interval, no posterior probability was computed, since only a first-degree polynomial could be used as an approximation due to a limited amount of data. The selection of the polynomial degree for $Q_\beta(t)$ follows a two-step procedure: first, we fix a maximum degree of 6; then, among the candidate polynomial regression models in the class $\mathfrak{M}_j$, $j = 1, \ldots, 6$, we choose the one with the highest posterior probability.

Given the dataset (t, x^j) from model class $\mathfrak{M}_j$, the posterior probability of model M_j is defined as

$$\mathsf{P}(M_j \mid t, x) = \frac{B_{j,1}(t,x)\pi(M_j)}{\sum_{i=1}^{6} B_{i,1}(t,x)\pi(M_i)}, \qquad j = 1, \ldots, 6$$

Table 1. The MSE and the posterior probability for polynomial degrees from 1 to 6 across three time periods. Best models per period are underlined. The value of the posterior probability in first time interval, corresponding to $p = 3$, has been approximated to 1 since the first decimal digit different from 9 appears in the 10th position.

Degree	2000–2008		2009–2019		2020–2023	
	MSE	Posterior Probability	MSE	Posterior Probability	MSE	Posterior Probability
1	1.97656	9.04936×10^{-18}	7.82247	3.60587×10^{-39}	0.29052	—
2	0.81799	4.72039×10^{-10}	7.73599	6.93663×10^{-40}	—	—
3	<u>0.27251</u>	<u>$\simeq 1$</u>	0.28653	0.46194	—	—
4	3.92124	2.95813×10^{-26}	<u>0.25284</u>	<u>0.53806</u>	—	—
5	3.88989	1.65832×10^{-26}	3.20832	1.90487×10^{-31}	—	—
6	17.2527	1.734565×10^{-41}	1.75889	1.38008×10^{-25}	—	—

where $\pi(M_j) = \frac{1}{6}$ is the prior probability related to the selection of the model M_j (assumed to be uniform),

$$B_{i,1}(t,x) = \frac{2}{\pi}(i+1)^{(i-1)/2} \int_0^{\pi/2} \frac{\sin^{i-1}(\varphi)\left(n + (i+1)\sin^2(\varphi)\right)^{(n-i)/2}}{\left(n\mathcal{B}_{i,1} + (i+1)\sin^2(\varphi)\right)^{(n-1)/2}} d\varphi,$$

is the Bayes factor for the comparison between the model M_i and the model M_1, and

$$\mathcal{B}_{i,1} = \frac{\sum_{k=1}^{n}\left(m_k - \widehat{m}_k^{(i)}\right)^2}{\sum_{k=1}^{n}\left(m_k - \widehat{m}_k^{(1)}\right)^2},$$

being $n = \sum_{i=1}^{d}(n_i - 1)$ the total number of data, m_k the sample mean at the k-the time instant and $m_k^{(j)}$ with $j = 1, \ldots, 6$ the estimated mean at the k-th time instant obtained by using the model M_j.

After obtaining the maximum likelihood estimates of the parameters ξ, the focus shifts to estimating the parameters $N_n \in \mathbb{N}$ and $p_n \in (0,1)$ of the binomial distribution representing the restart point. Specifically, the binomial size N_n is chosen as the largest integer not exceeding the estimated mean in the year immediately preceding the catastrophe. Whereas, the binomial probability parameter p_n is then estimated as usual using the following formula:

$$p_n = \frac{m_{\theta_n}}{\widehat{N}_n},$$

being $\widehat{N}_n$ the estimate of N_n and m_{θ_n} the value of the sample mean in the year θ_n. The parameter estimates of the binomial distribution of the catastrophe are given by $\widehat{p}_B^{(1)} = 0.84530$, $\widehat{N}_1 = 110$ for the first intertime and by $\widehat{p}_B^{(2)} = 0.93500$, $\widehat{N}_2 = 112$ for the second one (as reported in Table 2). The average IPI

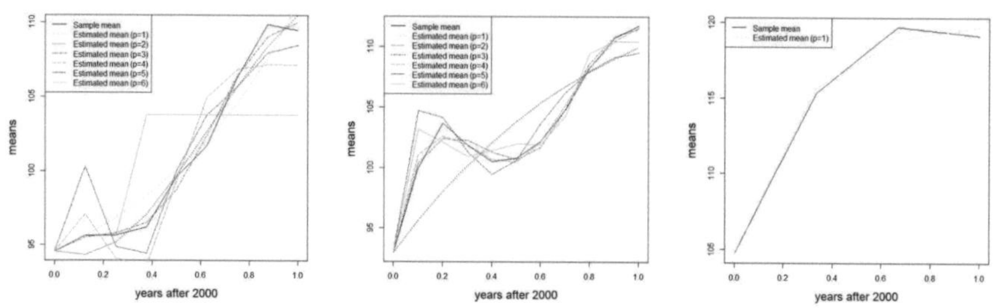

(a) First time interval (b) Second time interval (c) Third time interval

Fig. 4. Comparison between sample means and the estimated means for each choice of the degree of the polynomial (p).

Table 2. Estimates of the parameters of the binomial distribution.

arrival time of the catastrophe	$\widehat{p}_B$	$\widehat{N}$
$\theta_1 = 2009$	$\widehat{p}_B^{(1)} = 0.84530$	$\widehat{N}_1 = 110$
$\theta_2 = 2020$	$\widehat{p}_B^{(2)} = 0.93500$	$\widehat{N}_2 = 112$

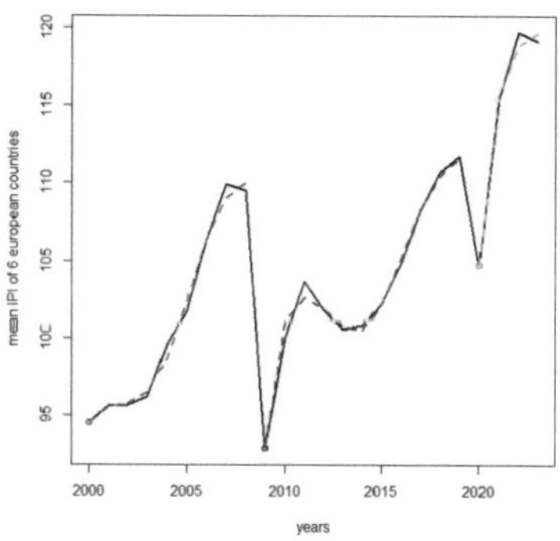

Fig. 5. Mean IPI of Austria, Belgium, Denmark, Germany, Italy and Sweden (solid line) and estimated curve (dashed line) from 2000 to 2023.

data along with the estimated mean are shown in Fig. 5, from which the good correspondence between the model and the actual data is evident.

4 Conclusion

This work has presented a stochastic framework for modeling diffusion processes with multisigmoidal logistic mean subject to random catastrophes, with particular emphasis on applications in economic and financial domains. Through simulation studies, we demonstrated how random jumps, modeled via Poisson processes, affect the first-passage-time dynamics of systems subject to external interventions, such as bank bailouts. The analysis of real-world data, specifically the IPI of several European countries, further confirmed the suitability of the proposed model in the particular case in which the restart points are distributed according to a binomial distribution. The results highlight the relevance of using jump-diffusion models with flexible mean structures to account for sudden disruptions and suggest directions for future research in both methodological refinement and broader applications.

Acknowledgments. S.M. and P.P. are members of the research group GNCS of INdAM (Istituto Nazionale di Alta Matematica) and acknowledge support received from the "European Union – Next Generation EU" through MUR-PRIN 2022, project 2022XZSAFN "Anomalous Phenomena on Regular and Irregular Domains: Approximating Complexity for the Applied Sciences", and MUR-PRIN 2022 PNRR, project P2022XSF5H "Stochastic Models in Biomathematics and Applications".

References

1. Albano, G., Barrera, A., Giorno, V., Romàn-Romàn, P., Torres-Ruiz, F.: First passage and first exit times for diffusion processes related to a general growth curve. Commun. Nonlinear Sci. Numer. Simul. **126**, 107494 (2023). https://doi.org/10.1016/j.cnsns.2023.107494
2. Barrera, A., De la Peña Cuao, A., Serrano Pérez, J.J., Torres-Ruiz, F.: Estimating the consumer price index using the lognormal diffusion process with exogenous factors: the Colombian case. AIMS Math. **10**(2), 3334–3380 (2025). https://doi.org/10.3934/math.2025155
3. García-Burgos, A., Paraggio, P., Romero-Molina, D., Castro, N.: Inference with non-homogeneous lognormal diffusion processes conditioned on nearest neighbor. Mathematics **12**(23), 3703 (2024). https://doi.org/10.3390/math12233703
4. Di Crescenzo, A., Musto, S., Paraggio, P., Torres-Ruiz, F.: Special lognormal diffusion processes with binomial random catastrophes and applications to economic data. Appl. Math. Model. **145**, 116146 (2025). https://doi.org/10.1016/j.apm.2025.116146
5. Giorno, V., Romàn-Romàn, P., Spina, S., Torres-Ruiz, F.: Estimating a non-homogeneous Gompertz process with jumps as model of tumor dynamics. Comput. Stat. Data Anal. **107**, 18–31 (2017). https://doi.org/10.1016/j.csda.2016.10.005
6. Schlomann, B.H.: Stationary moments, diffusion limits, and extinction times for logistic growth with random catastrophes. J. Theor. Biol. **454**, 154–163 (2018). https://doi.org/10.1016/j.jtbi.2018.06.007
7. Di Crescenzo, A., Paraggio, P., Román-Román, P., Torres-Ruiz, F.: Applications of the multi-sigmoidal deterministic and stochastic logistic models for plant dynamics. Appl. Math. Model. **92**, 884–904 (2021). https://doi.org/10.1016/j.apm.2020.11.046
8. Di Crescenzo, A., Paraggio, P., Román-Román, P., Torres-Ruiz, F.: Statistical analysis and first-passage-time applications of a lognormal diffusion process with multi-sigmoidal logistic mean. Stat. Papers **64**, 1391–1438 (2023). https://doi.org/10.1007/s00362-022-01349-1

9. Moreno, E., Serrano-Pérez, J.J., Torres-Ruiz, F.: Consistency of Bayes factors for linear models. Rev. R. Acad. Cienc. Exactas, Fis. Nat. Ser. A Mat. **119**, 1–18 (2025). https://doi.org/10.1007/s13398-024-01685-x
10. UNECE Statistical Database. https://w3.unece.org/PXWeb2015/pxweb/en/STAT/STAT__20-ME__5-MEPW/0_en_MECCIndProdY_r.px/. Accessed 15 May 2025

Scaling the Tails: Intraday Quantiles for Forecasting Value-at-Risk and Expected Shortfall

Antonio Naimoli[1(✉)], Ostap Okhrin[2], and Giuseppe Storti[1]

[1] Dipartimento di Scienze Economiche e Statistiche (DISES), Università degli Studi di Salerno, Fisciano, Italy
{anaimoli,storti}@unisa.it

[2] Institute of Transport and Economics, TUD Dresden University of Technology, Dresden, Germany
ostap.okhrin@tu-dresden.de

Abstract. Incorporating realized volatility measures into risk forecasting models can lead to more accurate forecasts. This paper introduces innovative risk forecasting models that replace realized volatility measures with observable risk proxies derived from high-frequency data through the scaling of intraday quantiles. Specifically, we present a flexible approach for Value-at-Risk and Expected Shortfall forecasting by proposing novel scaling factor estimation methods based on consistent loss functions combined with Multiplicative Error Models using the Generalized F distribution. The empirical analysis across 27 Dow Jones Industrial Average stocks reveals that our proposed approach can achieve significant accuracy improvements in tail risk forecasting.

Keywords: Value-at-Risk · Expected Shortfall · high-frequency quantiles · consistent loss functions · Multiplicative Error Models · Generalized F distribution

1 Introduction

The stability of financial institutions and economic systems increasingly depends on effective management of extreme or tail risks. Value-at-Risk (VaR) and Expected Shortfall (ES) have become essential quantitative risk measures, forming the foundation for regulatory capital requirements, strategic decision-making, and comprehensive risk management frameworks under the Basel III regulatory standards.

Recent research has developed multiple approaches for forecasting daily risk measures with increasing precision. A prominent strand of literature provides evidence that high-frequency measures significantly enhance the accuracy of volatility and tail risk forecasts at daily scales, as evidenced in numerous studies employing Realized Volatility (RV) measures [7,14,16,20,25,28]. A parallel approach involves directly estimating quantiles of daily financial returns from high-frequency data, leveraging distributional scaling laws [5,10,18,19]. In particular,

M. La Rocca et al. (Eds.): WMAF 2025, *New Perspectives in Mathematical and Statistical Methods for Actuarial Sciences and Finance*, pp. 216–225, 2025.
https://doi.org/10.1007/978-3-032-05551-4_19

assuming self-similarity of the return process, quantiles are first estimated at the intraday frequency and then rescaled to the daily frequency using the Hurst exponent H as scaling factor.

However, determining appropriate scaling factors for transforming intraday quantiles into daily risk forecasts remains challenging. Empirical studies suggest that using fixed scaling factors often outperforms existing estimation techniques, including the Detrended Fluctuation Analysis (DFA) method of [31], the DMA estimator of [1], and Fractal Dimension (FD) estimators of [32]. This solution underscores the need for alternative methodologies that can accurately estimate scaling factors across different market conditions and distributional characteristics.

This paper tackles these challenges by proposing a novel quantile regression framework that estimates the Hurst coefficient H for each quantile order α by minimizing strictly consistent loss functions for the quantiles [23] or for the pair (VaR, ES) [15]. This approach allows the scaling factor to vary across quantiles, accommodating distributional heterogeneity in financial returns [3,23,24,33], while remaining robust to potential violations of scaling assumptions.

Second, we develop a forecasting framework applying the Multiplicative Error Models (MEM) [12] to directly predict quantile-based risk measures, departing from conventional MEM applications that typically model tail-risk measures indirectly through RV dynamics. We enhance this framework by employing the Generalized F (GF) distribution, which offers greater flexibility in representing the error term compared to the standard Gamma distribution [13], encompassing a wider range of distributional shapes including Exponential, Weibull, Generalized Gamma, and Log-Logistic [22,29].

An additional advantage of our approach is the ability to specify separate models for daily VaR and ES series using high-frequency data to estimate observable proxies for tail risk measures. This circumvents the non-elicitability limitation of ES [17] while avoiding computational challenges associated with minimizing joint loss functions in the Fissler-Ziegel class [15] due to the sensitivity to the choice of initial values, requiring appropriate initialization strategies [30,34]. Specifically, our approach enables the direct generation of ES forecasts independent of VaR forecasts, which proves particularly valuable when conditional return distributions undergo structural changes, necessitating distinct dynamics for different risk measures.

We evaluate our methodology using 1-min intraday returns for 27 constituents of the Dow Jones Industrial Average (DJIA) over the period 2001–2016. The results demonstrate significant improvements in forecast accuracy compared to existing approaches using fixed quantile rescaling and heterogeneous autoregressive (HAR) modeling structures [7].

The remainder of this paper is organized as follows. Section 2 presents the theoretical framework for scaling variables. Section 3 details our methodology for deriving ex-post estimates of daily VaR and ES from intraday quantiles and introduces our consistent loss function approach to scaling factor estimation. Section 4 presents the MEM framework for VaR and ES forecasting. Section 5 provides a comprehensive description of the financial dataset and forecasting design implemented in our study. Section 6 evaluates empirical forecasting performance. Section 7 summarizes findings and implications.

2 Theoretical Background

This section presents the theoretical foundations underlying our methodology for estimating and forecasting daily VaR and ES using intraday data, with particular emphasis on *self-affine* processes [11,26].

A random process $X(t)$ is defined as *self-affine* or *self-similar* [6] if, for some $H > 0$ and $(c; k; t_1, ..., t_k) \geq 0$, it satisfies:

$$X(ct_1), \ldots, X(ct_k) \stackrel{d}{=} c^H X(t_1), \ldots, c^H X(t_k),$$

where $\stackrel{d}{=}$ denotes equality in distribution. This relationship can be expressed more concisely as:

$$X(ct) \stackrel{d}{=} c^H X(t)\,, \tag{1}$$

where H represents the scaling exponent or Hurst exponent of $X(t)$, while c denotes the scaling factor governing the temporal transformation.

For self-similar processes with stationary increments and $0 < H < 1$, this distributional scaling property holds locally:

$$X(t + c\Delta) - X(t) \stackrel{d}{=} c^H \left(X(t + \Delta) - X(t)\right). \tag{2}$$

In financial modeling, two prominent examples of self-affine processes are L-stable processes and Fractional Brownian Motion (fBM). While L-stable processes feature independent and stable increments, fBM is characterized by continuous sample paths and Gaussian increments. The coefficient H determines the fundamental properties of fBM: $H = 0.5$ corresponds to standard Brownian Motion (BM); $0 < H < 0.5$ implies anti-persistent behavior with negative autocorrelation; $0.5 < H < 1$ indicates persistence with long memory properties. In long memory analysis, H is commonly identified with the *Hurst* coefficient [9].

Time scaling plays a critical role in financial applications, particularly in risk analysis and derivatives pricing. The widely adopted *square root of time rule* is frequently used for scaling quantiles of return distributions to lower frequencies [8,35]. This rule typically applies in volatility scaling within the Black-Scholes framework, where the d-period volatility is expressed as $\sigma\sqrt{d}$. Under the Basel regulatory framework [2], this scaling principle extends to VaR, suggesting that d-day VaR can be derived by scaling daily VaR by $\sqrt{d}$.

This relationship is valid when returns follow a Normal distribution with stationarity and independence of increments, properties characteristic of Brownian Motion. However, empirical evidence consistently reveals departures from these assumptions in financial time series. For processes exhibiting self-similarity with Hurst exponent H, the appropriate scaling factor becomes d^H for d-day horizons [27], accounting for long-range dependencies and heavy-tailed distributions in financial returns.

For a financial asset with price process $P(t)$ where $t \in \mathcal{R}, t \geq 0$, we define the log-return process as:

$$X(t) = \log P(t) - \log P(0).$$

Consequently, from (2) and for any probability level $\alpha \in (0,1)$, quantiles exhibit the following scalar behavior:

$$Q_\alpha (X(t + c\Delta) - X(t)) \stackrel{d}{=} c^H Q_\alpha (X(t+\Delta) - X(t)). \tag{3}$$

This scaling property enables the computation of VaR and ES at daily frequency through appropriate scaling of their high-frequency counterparts, forming the theoretical foundation of our methodological approach.

3 Scaling Intraday Risk Measures Using Consistent Loss Functions

To estimate daily financial return quantiles from high-frequency data, we sample $c \in \mathbb{N}$ intraday returns each day, denoted by $\mathbf{r}_{c,t} = \{r_{1,t}, \ldots, r_{c,t}\}$, over the period $t = 1, \ldots, T$. The corresponding daily log-return is denoted by r_t. To compute intraday VaR, denoted as $\widehat{Q}_\alpha(\mathbf{r}_{c,t})$, and ES, denoted as $\widehat{ES}_\alpha(\mathbf{r}_{c,t})$, for a given risk level, α, the quantiles are derived using the empirical sample-quantile estimator, while the ES is calculated via the empirical sample-ES estimator:

$$\widehat{ES}_\alpha(\mathbf{r}_{c,t}) = \frac{1}{\alpha\, c} \sum_{i=1}^{c} r_{i,t} \times 1\left\{ r_{i,t} \leq \widehat{Q}_\alpha(\mathbf{r}_{c,t}) \right\}.$$

Finally, to allow intraday measures to be rescaled to the daily level, the scaling index H is fixed or estimated from the intraday log-return data. This rescaling transforms the quantiles and ES as follows:

$$\widehat{RQ}_{c,\alpha,t} = c^H \widehat{Q}_\alpha(\mathbf{r}_{c,t}), \tag{4}$$

$$\widehat{RES}_{c,\alpha,t} = c^H \widehat{ES}_\alpha(\mathbf{r}_{c,t}). \tag{5}$$

These Realized Quantile (RQ) and Realized ES (RES) measures [10] are inputs for time series modeling to generate VaR and ES forecasts.

After comparing various methods, [10] found that a fixed $H = 0.5$ yields better forecasting results than existing estimation techniques. This finding drives the development of new estimation methods based on VaR and ES loss functions.

In particular, we estimate the scaling index H via the Quantile Loss (QL) function:

$$QL_{\alpha,t} = (\alpha - 1\{r_t \leq VaR_{\alpha,t}\})(r_t - VaR_{\alpha,t}). \tag{6}$$

Since $c^H \widehat{Q}_\alpha(\mathbf{r}_{c,t})$ represents the scaled quantile estimate corresponding to the daily VaR, $VaR_{\alpha,t}$, we obtain H by minimizing:

$$\underset{H}{\operatorname{argmin}}\, QL_{\alpha,t}(H) = \left(\alpha - 1\left\{r_t \leq c^H \widehat{Q}_\alpha(\mathbf{r}_{c,t})\right\}\right) \left(r_t - c^H \widehat{Q}_\alpha(\mathbf{r}_{c,t})\right). \tag{7}$$

The rescaled risk measures are:

$$\widehat{RQ}_{c,\alpha,t}(QL) = c^{\hat{H}_{QL}} \widehat{Q}_\alpha(\mathbf{r}_{c,t}) \tag{8}$$

$$\widehat{RES}_{c,\alpha,t}(QL) = c^{\hat{H}_{QL}} \widehat{ES}_\alpha(\mathbf{r}_{c,t}). \tag{9}$$

An alternative approach relies on the zero-degree homogeneous loss function (FZ0) by [30]:

$$\begin{aligned} FZ0_{\alpha,t} =& \frac{1}{\alpha ES_{\alpha,t}} 1\{r_t \leq VaR_{\alpha,t}\}\left(r_t - VaR_{\alpha,t}\right) \\ &+ \frac{VaR_{\alpha,t}}{ES_{\alpha,t}} + \log(-ES_{\alpha,t}) - 1\,. \end{aligned} \tag{10}$$

The scaling index H is obtained by minimizing:

$$\begin{aligned} \underset{H}{\operatorname{argmin}}\, FZ0_{\alpha,t}(H) =& \frac{1}{\alpha\, c^H \widehat{ES}_\alpha(\mathbf{r}_{c,t})} 1\left\{r_t \leq c^H \widehat{Q}_\alpha(\mathbf{r}_{c,t})\right\}\left(r_t - c^H \widehat{Q}_\alpha(\mathbf{r}_{c,t})\right) \\ &+ \frac{c^H \widehat{Q}_\alpha(\mathbf{r}_{c,t})}{c^H \widehat{ES}_\alpha(\mathbf{r}_{c,t})} + \log\left(-c^H \widehat{ES}_\alpha(\mathbf{r}_{c,t})\right) - 1\,. \end{aligned} \tag{11}$$

The corresponding risk measures are:

$$\widehat{RQ}_{c,\alpha,t}(FZ0) = c^{\hat{H}_{FZ0}} \widehat{Q}_\alpha(\mathbf{r}_{c,t}) \tag{12}$$

$$\widehat{RES}_{c,\alpha,t}(FZ0) = c^{\hat{H}_{FZ0}} \widehat{ES}_\alpha(\mathbf{r}_{c,t})\,. \tag{13}$$

4 Multiplicative Error Models for Tail-Risk Forecasting

The forecasting procedure for VaR and ES applies dynamic modeling to daily RQ and RES time series. We implement separate MEM specifications, generating distinct forecasts for both risk measures.

Since VaR and ES derive from the lower tail of returns distribution, the MEM specification uses their positive counterparts, $-VaR_t$ and $-ES_t$, with outputs later converted back for interpretation.

Generally, we represent daily series of $\widehat{RQ}_{c,\alpha,t}$ and $\widehat{RES}_{c,\alpha,t}$ as a non-negative random variable X_t within the MEM framework:

$$X_t = \mu_t \epsilon_t\,, \tag{14}$$

where μ_t denotes the conditional mean given past information $\mathcal{F}_{t-1}$ and ϵ_t represents an i.i.d. non-negative error process with unit mean and constant variance.

The conditional mean, μ_t, is specified as:

$$\mu_t = \beta_0 + \beta_1 X_{t-1} + \beta_2 \mu_{t-1}, \tag{15}$$

with positivity constraints $(\beta_0, \beta_1, \beta_2) > 0$.

The error term ϵ_t follows the GF distribution, providing greater flexibility than standard Gamma densities. The probability density function of the GF distribution is:

$$f(x;\underline{\zeta}) = \frac{\gamma x^{\gamma\delta-1}[\phi + (x/\eta)^{\gamma}]^{(-\phi-\delta)}\,\phi^{\phi}}{\eta^{\gamma\delta}\,\mathcal{B}(\delta,\phi)}, \tag{16}$$

where $\underline{\zeta} = (\gamma, \delta, \phi, \eta)'$, $\gamma > 0$, $\delta > 0$, $\phi > 0$ and $\eta > 0$, with $\mathcal{B}(\cdot,\cdot)$ the Beta function.

To ensure unit mean for ϵ_t, we set $\eta = \xi^{-1}$, where:

$$\xi = \phi^{1/\gamma}\left[\Gamma(\delta + 1/\gamma)\Gamma(\phi - 1/\gamma)\right]\left[\Gamma(\delta)\Gamma(\phi)\right]^{-1}.$$

The GF distribution offers remarkable versatility, encompassing several distributions as special cases, including Weibull ($\delta = 1$, $\phi \to \infty$), Generalized Gamma ($\phi \to \infty$), Log-Logistic ($\delta = 1$, $\phi = 1$), and Exponential ($\gamma = \delta = 1$, $\phi \to \infty$).

5 Data and Risk Forecasting Design

Our analysis relies on 1-min interval price data for 27 large-cap NYSE stocks included in the DJIA traded during standard trading hours, 9:30 a.m. to 4:00 p.m., between January 1, 2001, and December 31, 2016. We implement a dynamic forecasting approach with daily parameter calibration using a rolling window scheme, based on calendar time sampling with $c = 390$ observations per day.

For risk assessment, the strictly consistent scoring rule $QL_{\alpha,t}$ is used to evaluate VaR forecasts, while the zero-homogeneous FZ0 score [15,30] and the Strict ES Regression (ESR) backtest [4] are employed for ES evaluation. Finally, the statistical significance of performance differences across competing models is tested using the Model Confidence Set (MCS) procedure of [21].

We estimate RQ and RES at α-risk levels of 1% and 2.5%, while the forecasting framework involves three MEM specifications with different scaling indices: $MEM(H_{0.5})$ with fixed $H = 0.5$, $MEM(H_{QL})$ and $MEM(H_{FZ0})$ with dynamically derived H parameters following Eqs. 7 and 11.

Alternatively, we implement the HAR model of [7] on RQ and RES series. Following [10], we use $HAR(H_{0.5})$ with fixed $H = 0.5$ as our primary reference model, and propose $HAR(H_{QL})$ and $HAR(H_{FZ0})$ alternatives using dynamic H parameter estimation.

6 Forecasting Results

We evaluate the predictive performance of our proposed specifications through an out-of-sample forecasting comparison over the out-of-sample period December 2006–December 2016. To capture the temporal evolution of market dynamics, model parameters are recursively estimated on a daily basis using a rolling window of 1,500 observations. In accordance with Basel Accord guidelines, we assess model performance at the 1% VaR and 2.5% ES risk levels.

Table 1. Summary results for QL at $\alpha = 0.01$ risk level. $\overline{QL}_{rank}$: average rank across the 27 stock; $\overline{QL}$: average QL computed over all 27 stocks; $\%QL_{min}$: percentage of time a model minimizes the QL; $\%QL_{MCS0.75}$: percentage of times a model enters the 75% MCS using the Semi-Quadratic statistic.

	$\alpha = 0.01$			
	$\overline{QL}_{rank}$	$\overline{QL}$	$\%QL_{min}$	$\%QL_{SQ0.75}$
$HAR(H_{0.5})$	2.815	1.048	3.704	33.333
$HAR(H_{QL})$	4.389	1.094	0.000	7.407
$HAR(H_{FZ0})$	5.926	1.135	0.000	0.000
$MEM(H_{0.5})$	1.574	1.039	70.370	88.889
$MEM(H_{QL})$	2.296	1.049	25.926	77.778
$MEM(H_{FZ0})$	4.000	1.091	0.000	3.704

The forecasting performance for VaR is evaluated using consistent loss functions. Table 1 presents the MCS results for the QL at the 1% risk level, along with the average model rankings where lower loss values correspond to superior performance rankings.

The empirical analysis yields several key insights regarding the comparative performance of different models. The MEM specifications demonstrate significantly superior performance relative to their HAR counterparts, as evidenced by lower average QL values and higher inclusion frequencies in the 75% MCS. Specifically, MEM($H_{0.5}$) and MEM(H_{QL}) exhibit higher overall performance showing superior MCS inclusion rates, while HAR($H_{0.5}$) performs less favorably. Notably, MEM($H_{0.5}$) and MEM(H_{QL}) achieve the highest rankings in quantile loss minimization, emphasizing their superior forecasting ability. Conversely, model specifications employing H optimized through the $FZ0$ criterion encounter greater difficulty in achieving MCS inclusion compared to those utilizing H_{QL}-based optimization.

We evaluate ES forecast validity using the Strict ESR backtesting framework. Table 2 shows that fixing $H = 0.5$ proves inadequate for ES forecasting, as evidenced by low Acceptance Rates (ARate) for both HAR($H_{0.5}$) and MEM($H_{0.5}$) models. However, estimating H through consistent loss functions substantially improves performance. In particular, HAR(H_{QL}), HAR(H_{FZ0}), MEM(H_{QL}), and MEM(H_{FZ0}) specifications achieve notably higher acceptance rates. Remarkably, MEM(H_{QL}) is the only model that always passes the Strict ESR backtest at both 1% and 2.5% risk levels, demonstrating superior predictive accuracy. In contrast to the VaR framework, $FZ0$-based specifications also achieve satisfactory ES accuracy.

For joint VaR and ES evaluation, we employ the FZ0 loss function. Table 3 presents MCS results and average model rankings. The analysis confirms that the MEM($H_{0.5}$) and MEM(H_{QL}) models provide superior overall performance. In line with QL-based findings, MEM specifications consistently outperform their HAR counterparts across all evaluation criteria. Finally, models using H esti-

Table 2. ESR bactesting results. Acceptance Rate (ARate): percentage of times models pass the *Strict ESR* backtest of [4] according to the 5% (ARate 5%) and 1% (ARate 1%) significance levels. Risk levels: $\alpha = 0.01, 0.025$.

	$\alpha = 0.01$		$\alpha = 0.025$	
	ARate 5%	ARate 1%	ARate 5%	ARate 1%
$HAR(H_{0.5})$	22.22	29.63	11.11	22.22
$HAR(H_{QL})$	92.59	100.00	92.59	100.00
$HAR(H_{FZ0})$	70.37	100.00	44.44	74.07
$MEM(H_{0.5})$	18.52	33.33	14.81	14.81
$MEM(H_{QL})$	100.00	100.00	100.00	100.00
$MEM(H_{FZ0})$	85.19	96.30	70.37	96.30

Table 3. Summary results for the $FZ0$ loss at $\alpha = 0.025$ risk level. $\overline{FZ0}_{rank}$: average rank across the 27 stock; $\overline{FZ0}$: average FZ0 computed over all 27 stocks; $\%FZ0_{min}$: percentage of time a model minimizes the FZ0; $\%FZ0_{SQ0.75}$: percentage of times a model enters the 75% MCS for the FZ0 using the Semi-Quadratic statistic.

	$\alpha = 0.025$			
	$\overline{FZ0}_{rank}$	$\overline{FZ0}$	$\%FZ0_{min}$	$\%FZ0_{SQ0.75}$
$HAR(H_{0.5})$	3.222	–3.490	0.000	22.222
$HAR(H_{QL})$	4.278	–3.472	0.000	7.407
$HAR(H_{FZ0})$	5.889	–3.431	0.000	0.000
$MEM(H_{0.5})$	1.611	–3.514	66.667	85.185
$MEM(H_{QL})$	1.889	–3.511	33.333	81.481
$MEM(H_{FZ0})$	4.111	–3.474	0.000	7.407

mated via FZ0 show a reduced MCS inclusion rate compared to QL-based alternatives.

7 Conclusions

The paper proposes a novel framework for rescaling intraday quantiles to forecast daily VaR and ES. We develop data-driven scaling factors estimated through consistent loss functions, enhancing flexibility in capturing market heterogeneity. The proposed methodology is further extended introducing the MEM with GF innovations to directly capture the dynamics of quantile-based risk measures. An empirical analysis of 27 DJIA stocks reveals that our adaptive scaling methodology significantly improves the accuracy tail-risk forecasting. In particular, MEM specifications consistently outperform HAR models across several risk metrics. The results also highlight the importance of employing distinct modeling dynamics for VaR and ES to capture different tail-risk behaviors, offering important insights for risk management practitioners and regulatory applications.

Acknowledgments. Giuseppe Storti and Antonio Naimoli acknowledge financial support under the National Recovery and Resilience Plan (NRRP), Mission 4, Component 2, Investment 1.1, Call for tender No. 104 published on 2.2.2022 by the Italian Ministry of University and Research (MUR), funded by the European Union NextGenerationEU Project Title: Methodological and computational issues in large-scale time series models for economics and finance (ID: 20223725WE) CUP D53D 2300610 0006 - Grant Assignment Decree No. n. 967 of 30/06/2023.

References

1. Alessio, E., Carbone, A., Castelli, G., Frappietro, V.: Second-order moving average and scaling of stochastic time series. Eur. Phys. J. B Condens. Matter Complex Syst. **27**(2), 197–200 (2002)
2. Basel Committee on Banking Supervision: Overview of the amendment to the capital accord to incorporate market risk (1996)
3. Baur, D.G., Dimpfl, T., Jung, R.C.: Stock return autocorrelations revisited: A quantile regression approach. J. Empir. Financ. **19**(2), 254–265 (2012)
4. Bayer, S., Dimitriadis, T.: Regression-based expected shortfall backtesting. J. Financ. Econ. **20**(3), 437–471 (2020). https://doi.org/10.1093/jjfinec/nbaa013
5. Cai, C.X., Kim, M., Shin, Y., Zhang, Q.: Farvar: Functional autoregressive value-at-risk*. J. Financ. Econ. **17**(2), 284–337 (2018). https://doi.org/10.1093/jjfinec/nby031
6. Calvet, L., Fisher, A.: Multifractality in asset returns: theory and evidence. Rev. Econ. Stat. **84**(3), 381–406 (2002)
7. Corsi, F.: A simple approximate long-memory model of realized volatility. J. Financ. Economet. **7**(2), 174–196 (2009)
8. Danielsson, J., Zigrand, J.P.: On time-scaling of risk and the square-root-of-time rule. J. Bank. Finance **30**(10), 2701–2713 (2006)
9. Di Matteo, T.: Multi-scaling in finance. Quant. Finance **7**(1), 21–36 (2007)
10. Dimitriadis, T., Halbleib, R.: Realized quantiles. J. Bus. Econ. Stat. **40**(3), 1346–1361 (2022)
11. Embrechts, P.: Selfsimilar Processes. Princeton University Press, Princeton (2002). https://doi.org/10.1515/9781400825103
12. Engle, R.: New frontiers for arch models. J. Appl. Economet. **17**(5), 425–446 (2002)
13. Engle, R.F., Gallo, G.M.: A multiple indicators model for volatility using intra-daily data. J. Econ. **131**(1–2), 3–27 (2006)
14. Engle, R.F., Ghysels, E., Sohn, B.: Stock market volatility and macroeconomic fundamentals. Rev. Econ. Stat. **95**(3), 776–797 (2013)
15. Fissler, T., Ziegel, J.F.: Higher order elicitability and Osband's principle. Ann. Statist. **44**(4), 1680–1707 (2016). https://doi.org/10.1214/16-AOS1439
16. Gerlach, R., Wang, C.: Semi-parametric dynamic asymmetric Laplace models for tail risk forecasting, incorporating realized measures. Int. J. Forecast. **36**(2), 489–506 (2020)
17. Gneiting, T.: Making and evaluating point forecasts. J. Am. Stat. Assoc. **106**(494), 746–762 (2011)
18. Hallam, M., Olmo, J.: Semiparametric density forecasts of daily financial returns from intraday data. J. Finan. Econ. **12**(2), 408–432 (2013). https://doi.org/10.1093/jjfinec/nbt016

19. Hallam, M., Olmo, J.: Forecasting daily return densities from intraday data: A multifractal approach. Int. J. Forecast. **30**(4), 863–881 (2014). https://doi.org/10.1016/j.ijforecast.2014.01.007. https://www.sciencedirect.com/science/article/pii/S016920701400051X
20. Hansen, P.R., Huang, Z., Shek, H.H.: Realized garch: a joint model for returns and realized measures of volatility. J. Appl. Economet. **27**(6), 877–906 (2012)
21. Hansen, P.R., Lunde, A., Nason, J.M.: The model confidence set. Econometrica **79**(2), 453–497 (2011)
22. Hautsch, N.: Assessing the risk of liquidity suppliers on the basis of excess demand intensities. J. Financ. Economet. **1**(2), 189–215 (2003)
23. Koenker, R.: Quantile regression. Cambridge Univ Pr (2005)
24. Koenker, R., Xiao, Z.: Quantile autoregression. J. Am. Stat. Assoc. **101**(475), 980–990 (2006)
25. Lazar, E., Xue, X.: Forecasting risk measures using intraday data in a generalized autoregressive score framework. Int. J. Forecast. **36**(3), 1057–1072 (2020)
26. Mandelbrot, B.: The variation of certain speculative prices. J. Bus. **36**(4), 394–419 (1963)
27. Menkens, O.: Value at risk and self-similarity. In: Numerical methods for finance, pp. 225–254. Chapman and Hall/CRC (2007)
28. Naimoli, A., Gerlach, R., Storti, G.: Improving the accuracy of tail risk forecasting models by combining several realized volatility estimators. Econ. Model. **107**, 105701 (2022)
29. Naimoli, A., Storti, G.: Heterogeneous component multiplicative error models for forecasting trading volumes. Int. J. Forecast. **35**(4), 1332–1355 (2019)
30. Patton, A.J., Ziegel, J.F., Chen, R.: Dynamic semiparametric models for expected shortfall (and value-at-risk). J. Econ. **211**(2), 388 – 413 (2019). https://doi.org/10.1016/j.jeconom.2018.10.008. http://www.sciencedirect.com/science/article/pii/S030440761930048X
31. Peng, C.K., Buldyrev, S.V., Havlin, S., Simons, M., Stanley, H.E., Goldberger, A.L.: Mosaic organization of DNA nucleotides. Phys. Rev. E **49**(2), 1685 (1994)
32. Sánchez-Granero, M., Fernández-Martínez, M., Trinidad-Segovia, J.: Introducing fractal dimension algorithms to calculate the hurst exponent of financial time series. Eur. Phys. J. B **85**, 1–13 (2012)
33. Sun, J., Zhao, X., Xu, C.: Crude oil market autocorrelation: Evidence from multiscale quantile regression analysis. Energy Econ. **98**, 105239 (2021)
34. Taylor, J.W.: Forecasting value at risk and expected shortfall using a semiparametric approach based on the asymmetric Laplace distribution. J. Bus. Econ. Stat. **37**(1), 121–133 (2019)
35. Wang, J.N., Yeh, J.H., Cheng, N.Y.P.: How accurate is the square-root-of-time rule in scaling tail risk: A global study. J. Bank. Finance **35**(5), 1158–1169 (2011)

High-Profile GDPR Fines and Their Financial Impact on Listed Firms: An Exploratory Analysis

Albina Orlando and Serena Pulcini(✉)

Consiglio Nazionale delle Ricerche (CNR), Istituto per le Applicazioni del Calcolo, Napoli, Italy
{albina.orlando,serenapulcini}@cnr.it

Abstract. This study investigates the short-term market impact of General Data Protection Regulation (GDPR) enforcement actions on publicly listed firms using an event study methodology. In the context of the GDPR's stringent compliance framework—enforced since May 25, 2018—monetary sanctions act as adverse information signals with potential implications for firm valuation. Focusing on high-profile enforcement cases involving fines exceeding one million euros, this analysis evaluates whether data privacy violations produce statistically significant abnormal returns. The methodology combines a window-based approach, identifying the most sensitive time intervals, with an event-specific analysis that captures heterogeneity in company reactions. A cluster analysis further highlights divergent response patterns, distinguishing firms more exposed to reputation and regulatory risk from those exhibiting greater market resilience. The results offer empirical insights into investor behavior in the face of data protection breaches.

Keywords: GDPR enforcement · data privacy violations · event study · abnormal returns

1 Introduction

In the digital age, data privacy and data sovereignty have emerged as critical pillars of individual rights and national governance. As personal data become a valuable economic asset, ensuring that individuals retain control over how their information is collected, processed, and stored has become a central concern for policy makers around the world. Data sovereignty, in particular, emphasizes the need for data to be subject to the laws and governance structures of the country where they are collected, reinforcing the idea of digital autonomy and accountability. In this context, the European Union introduced the General Data Protection Regulation (GDPR), which came into effect on May 25, 2018. The GDPR establishes a comprehensive legal framework for data protection, imposing strict obligations on organizations and allowing Data Protection Authorities (DPAs) to issue significant fines for non-compliance, up to €20 million or 4% of global annual turnover. A growing body of recent scientific contributions examine the

M. La Rocca et al. (Eds.): WMAF 2025, *New Perspectives in Mathematical and Statistical Methods for Actuarial Sciences and Finance*, pp. 226–237, 2025.
https://doi.org/10.1007/978-3-032-05551-4_20

application of the GDPR, as well as the categories of infringements perpetrated by the various stakeholders subject to the Regulation across Europe (among the others, [8] and [9]).

Within this broader framework, we focus on regulatory sanctions as adverse signals to investors, with potential implications for the reputation of a company and perceived risk. In particular, this study aims to explore the short-term financial impact of GDPR-related fines on publicly traded companies employing an event study methodology, a statistical and an econometric approach used to evaluate the effect of specific events on outcome variables (for details, see [6]). Our goal is to provide a preliminary evidence on whether data privacy violations translate into measurable market consequences. The primary motivation for this study is the belief that analyzing the impact of past events can support the development of predictive models to estimate the future financial or stock market consequences of new fines. This approach could also allow analysts to treat compliance risk as a more quantifiable factor within ESG (Environmental, Social, and Governance) investment strategies. Among the limited literature on GDPR enforcement and financial markets, the study by Ford et al. [4] appears to be the only one applying the event study methodology to such events. Based on data up to 2019, it employs a standard Market Model (MM) to estimate abnormal returns. Building on the work of Ford et al. [4], this paper focuses on more recent GDPR enforcement actions involving landmark cases (fines exceeding one million euros), thus offering a more up-to-date and targeted perspective. Along with the updated dataset, the study also introduces more refined selection criteria. Furthermore, the analysis adopts two complementary perspectives. The first is a window-based approach, which identifies the most relevant event windows—periods where enforcement actions have the strongest market effect—and highlights the firms most affected during each. The second is a firm- or event-based approach, which examines each company (or individual event, in cases of multiple sanctions for the same firm) to assess overall sensitivity and determine the specific windows where market reactions are most evident. Finally, the results are examined through a cluster analysis to identify patterns in firm responses, distinguishing companies more severely affected by GDPR enforcement actions from those that appear more resilient. The remainder of the paper is organized as follows. Section 2 outlines the methodology, Sect. 3 presents a case study, and Sect. 4 offers concluding remarks.

2 Event Study Methodology Based on the Market Model

Event studies represent a well-established methodology for evaluating how specific events affect a company's stock price and market value, as illustrated by Mackinlay et al. [6] and Huntington-Klein [5]. The core idea is straightforward: compare actual stock returns to expected returns—estimated without accounting for the event—to identify abnormal returns and evaluate the market's reaction. To this aim, two periods, measured in days, are defined: an estimation window, that is the period before the event used to calculate expected returns, and the event window, the time period around the event during which abnormal returns are assessed. As this is an initial exploratory analysis, we rely—while fully aware of its limitations—on the Market Model (hereinafter MM), previously employed by Ford et al. [4]. The MM, formalized by [2], models asset returns directly as a linear function of the market return, expressed as:

$$R_{it} = \alpha_i + \beta_i \cdot R_{mt} + \epsilon_{it} \tag{1}$$

where R_{it} is the return on asset i, and R_{mt} is the corresponding market return. The parameter α_i captures the asset average return that is not explained by market movements, β_i measures the sensitivity of asset i to changes in overall market, and ϵ_{it} is the idiosyncratic error term.

Basing on Eq. 1 , abnormal returns (AR) are computed as the difference between actual returns and model—based expected returns within a defined event window. Formally, for asset i at time t:

$$AR_{it} = R_{it} - E[R_{it}|X_t], \tag{2}$$

Abnormal returns within the event window are summed to obtain the cumulative abnormal return (CAR):

$$CAR_i = \sum_{t=T_2}^{T_3} AR_{it} \tag{3}$$

where T_2 and T_3 denote the start and end of the event window. For multiple events, the cumulative average abnormal return (CAAR) is calculated by averaging the CARs across all events:

$$CAAR = \frac{1}{N}\sum_{i=1}^{N} CAR_i \tag{4}$$

The respective t-test statistics are defined as follows:

$$t_{AR_{it}} = \frac{AR_{it}}{S_{AR_{it}}}, \tag{5}$$

$$t_{CAR_i} = \frac{CAR_i}{\sqrt{(T_3 - T_2 + 1)S^2_{ARi}}}, \tag{6}$$

$$t_{CAAR} = \sqrt{N}\frac{CAAR}{S_{CAAR}} \tag{7}$$

where the standard deviation of AR for asset i is computed as:

$$S_{AR_i} = \sqrt{\frac{1}{M_i - K}\sum_{t=T_0}^{T_1}(AR_{it})^2}, \tag{8}$$

with M_i the number of observations in the estimation window, and K the degrees of freedom of the chosen model. Finally, the standard deviation of CAAR is calculated as:

$$S_{CAAR} = \sqrt{\frac{1}{N-1}\sum_{i=1}^{N}(CAR_i - CAAR)^2} \tag{9}$$

with N total number of events. The three t-tests are conducted as one-sided tests, reflecting the alternative hypothesis that abnormal returns are negative.

3 A Case Study

In the following, we focus on high-profile GDPR fines exceeding one million euros and issued in 2024. As previously noted, since our interest lies in market reactions to these enforcement actions, the analysis is limited to publicly traded companies. The methodology follows the approach outlined in Sect. 2, and the primary data source is the GDPR Enforcement Tracker (hereafter referred to as ET) [3]. The ET dataset compiles fines issued by Data Protection Authorities (DPAs) across the 27 EU member states, as well as the United Kingdom and Norway, since the GDPR came into effect. Each entry includes a unique Enforcement Tracker ID, the country of issuance, the decision date, the fine amount, the name and industry sector of the penalized entity, the relevant GDPR articles, a general classification of the violation, a source link, and a case summary.

3.1 Data Description and Pre-processing

The first step, prior to data collection, is to identify the events of interest that occurred in 2024 using the ET dataset. The records were manually filtered to include only those involving publicly listed companies or their subsidiaries, as well as fines greater than one million euros. This selection process results in a sample of 16 enforcement events from 2024, involving 14 different companies, as shown in Table 1. For each company in the sample, Table 1 presents core details of the GDPR fines examined in the study, including firm identity, enforcement date, fine amount, related GDPR articles, country of origin, and industry sector.

Despite the sample consisting of only 16 events, it includes three of the largest individual fines among the top 10 since the GDPR came into effect: LinkedIn (October 24, whose parent company is Microsoft), Uber (July 22), and Meta (December 17, the second Meta-related event in the sample). Regarding the severity of the violations, the GDPR establishes two levels of infringement. The more serious category allows for fines up to the greater of €20 million or 4% of the company's total worldwide annual turnover from the preceding financial year. Looking at Table 1, the violations committed by the companies in our sample fall into this more severe category. In fact, all companies—except Uber—breached the basic principles of data processing (Articles 5 and 6), as well as provisions related to controllers and processors (including, among others, Articles 25, 26, 28, 32, and 33). Uber is the only company in the sample that violated solely Article 44, which pertains to serious infringements involving the transfer of personal data to a third country or an international organization. Despite the severity of these violations, none of the fines exceeded 1% of annual revenue. Uber came closest (0.85%), followed by Meta (0.20%) and Microsoft (0.14%). Of the 16 fines, 5 come from Spain, accounting for 31.25% of the sample; 3 from Italy and 3 from Ireland, each representing 18.75%; 2 from the Netherlands (12.5%); and one each from France, Sweden, and the Czech Republic (6.25% each).

Table 1. Overview of GDPR fines included in the sample

Parent Company	Fine Date	Fine Value[a]	Violated Articles	Country[b]	Sector[c]
Amazon.com Inc.	January 23	32,000,000	5 (1) c), 6, 12, 13, 32	FR	25
Iberdrola1	February 5	3,500,000	5 (1) f), 32	ES	55
Iberdrola2	February 7	3,000,000	5 (1) f), 32	ES	55
Enel S.p.A.	February 8	79,100,000	5 (1) f), 5 (2), 24 (1), 25, 28, 32	IT	55
UniCredit S.p.A.	February 8	2,800,000	5 (1) f), 32 (1), (2)	IT	40
Caixabank	April 12	1,200,000	6 (1)	ES	40
Gen Digital Inc.	April 15	13,900,000	5, 6 (1) f), 13 (1) c)	CZ	45
Eni S.p.A.[d]	June 6	6,419,631	5 (1) a), d), f),5 (2), 6, 24, 25, 28	IT	10
Avanza Bank Holding AB	June 24	1,300,000	5 (1) f), 32 (1)	SE	40
Uber Technologies, Inc.	July 22	290,000,000	44	NL	45
Telefonica	September 26	1,300,000	5 (1) f), 5 (1) c)	ES	50
Meta1	September 27	91,000,000	5 (1) f),32 (1),33 (1), (5)	IE	50
Microsoft Corporation	October 24	310,000,000	5 (1) a), 6 (1)a), e), f), 13(1) c), 14 (1) c)	IE	45
Netflix, Inc.	November 26	4,750,000	5 (1) a), 12 (1), 13 (1) c), e), f),(2), 15 (1) a), c), d)	NL	50
Assicurazioni Generali S.p.A.	December 10	4,000,000	5 (1) f), 25, 32, 35	ES	40
Meta2	December 17	251,000,000	33 (3), (5), 25 (1), (2)	IE	50

[a]All fine amounts are expressed in euro (€)
[b]Country codes follow the ISO 3166-1 alpha-2 standard
[c]Business sectors are classified according to the Global Industry Classification Standard (GICS)
[d]Eni has also violated Article 130 of the Italian Privacy Code

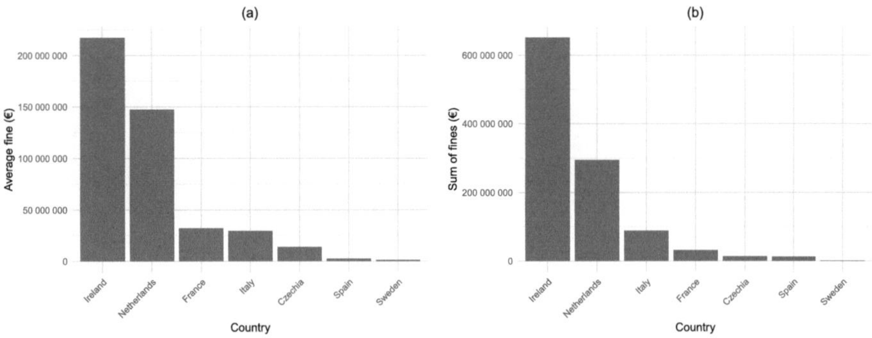

Fig. 1. Comparison of average and total GDPR fine amount per country: (**a**) average fine per country; (**b**): total fine amount per country

Figure 1 presents the analysis of fines by country: panel (a) displays the average fine for each country, while panel (b) shows the total fine amount for each country. As shown in Fig. 1, Ireland ranks first by a wide margin in both total and average fine amounts, followed by the Netherlands. This ranking is influenced by the inclusion of three major fines among the top 10 under the GDPR: two issued by Ireland (LinkedIn and Meta) and one by the Netherlands (Uber). Italy ranks third in terms of total fines, with three fines included in the sample, while France ranks fourth according to panel (b) in Fig. 1. However, as shown in panel (a) of Fig. 1, France holds the third position for average fine amount despite having only one fine in the dataset. Notably, the sole French fine is a €32 million penalty imposed on Amazon. In contrast, among the three Italian fines, the lowest is approximately €2.8 million for Unicredit, while the highest reaches €79.1 million for Enel. This contrast highlights how a single significant fine can elevate a country's average fine amount, whereas Italy's wider distribution of fines results in a higher total but a lower average penalty. The Czech Republic, with a single fine of approximately €13.9 million issued to Gen, ranks fifth. Even though Spain leads in number of fines, it records the second-lowest total amount, indicating that the selected cases, all above one million, are still relatively small in value. In fact, Spain's highest fine is €4 million, imposed on Generali, while the lowest is €1.2 million for Caixabank, reflecting a relatively narrow range of penalties compared to other countries. Sweden completes the ranking, occupying the seventh and last position with a single fine of €1.3 million imposed on Avanza Bank.

The companies in the sample represent a diverse range of sectors. Within this sample, the most affected industries are Communication Services and Financial Services, each accounting for four fines and representing 25% of the total. Following them are the Technology and Utilities sectors, with three fines each, making up 18.75% respectively. The remaining two sectors—Energy and Consumer Cyclical—each appear only once in the dataset, corresponding to 6.25% of the sample. Figure 2 provides a sector-level overview of the fines. Specifically, panel (a) illustrates the average fine per sector, while panel (b) depicts the total fine amount for each. Technology and Communication Services rank first and second, respectively, in both panels. In fact, the Technology sector includes three fines in total, two of which are among the highest GDPR penalties: LinkedIn (€310 million) and Uber (€290 million). Even the lowest fine in this sector is relatively high—€13.9 million—imposed on Avast, a subsidiary of Gen Digital. The Communication Services sector includes three fines as well, ranging from €1.3 million for Telefónica to €251 million for Meta (the second Meta-related case in the sample), one of the highest GDPR fines ever issued. This notable spread reflects a high degree of variability in the fines within the sector. The Utilities

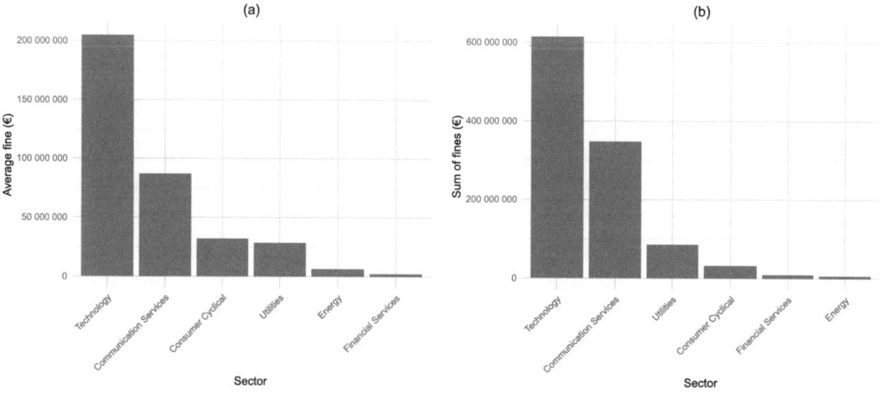

Fig. 2. Comparison of average and total GDPR fines by business sector: (**a**) average fine by sector; (**b**) total amount of fines by sector

sector ranks fourth and third in panel (a) and panel (b) of Fig. 2, respectively, with fines ranging from €3 million (Iberdrola) to €79.1 million (Enel), indicating moderate variability. On the other hand, the Consumer Cyclical sector ranks third in panel (a) and fourth in panel (b) of Fig. 2, with its only fine being the one previously mentioned for Amazon. The last two positions are occupied by the Energy and Financial Services sectors: Energy ranks fifth in average fines and sixth in total fines, while Financial Services ranks sixth in average and fifth in total. Energy has only one fine, imposed on Eni for over €6 million, whereas Financial Services includes four fines, with the highest being the €4 million fine for Generali.

As for the financial information, share prices and market index data were obtained from Yahoo Finance! ([10]), along with firm demographics such as market capitalization, annual revenue, and industry classification. For subsidiaries that are not directly listed, the parent company's ticker was used instead. Where companies were listed on multiple stock exchanges, the ticker from the exchange based in the company's country of origin was used. The selection of the most appropriate market index for each firm was based on index membership, visual trend alignment, and statistical correlation. For U.S. firms, the following indices were selected: the Nasdaq 100 index for Microsoft, Meta, and Amazon; the Nasdaq Composite index for Netflix and Uber Technologies, Inc.; and the Standard & Poor's 500 index for Gen Digital Inc. For companies based in other countries, the most representative stock index was chosen. The sample includes events involving the following parent companies: Amazon, Enel, Unicredit, Caixabank, Generali, Eni, Avanza Bank Holding, Uber, Telefónica, Microsoft, Netflix, Meta (2 events), and Iberdrola (2 events). A comprehensive dataset has been established for each firm, encompassing daily records of the adjusted closing price and the corresponding market index. Data pre-processing began with the conversion of prices into returns, following the common practice of using log returns for daily financial data. This transformation was applied to both firms' closing prices and market index values. With log returns computed, the market model could be estimated directly.

The final step in data pre-processing was to define the estimation and event windows. Eight event window configurations, measured in trading days, were considered: [–2,2], [–1,1], [–1,0], [0,1], [0,2], [0,3], [0,4], and [0,5]. These event windows are kept brief to minimize the risk of overlapping announcements by the firm during the event period, consistent with Neuhierl et al. [7]. In all cases, the estimation window length was set to 120 days, consistent with Andoh-Baidoo et al. [1] and Ford et al. [4], with no gap between the estimation and event windows. Event and estimation windows accounted for country-specific holidays and non-trading days; events after market close or on non-trading days were moved to the next trading day. As a robustness check, each event window was screened for confounding negative events unrelated to the GDPR fines. No such events were identified, supporting the attribution of market reactions to the enforcement announcements.

3.2 Results

The analysis was carried out on 16 GDPR fine events using the market model, across the eight event window configurations previously specified.

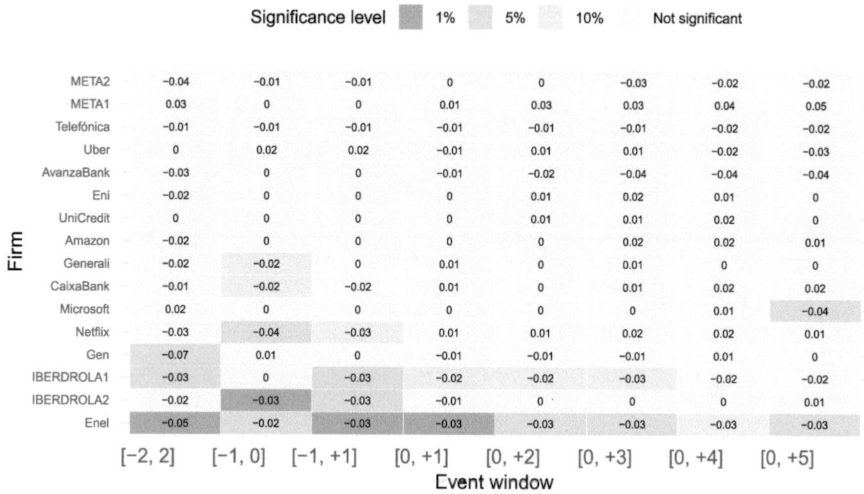

Fig. 3. Heatmap of CARs from the Market Model

Figure 3 presents a heatmap of CARs from the Market Model, where the shading intensity reflects the significance of the corresponding t-statistics (tCARs) across events and event windows. Darker shades indicate higher levels of statistical significance: the darkest cells correspond to the 1% level (one-tailed test), medium gray to the 5% level, light gray to the 10% level, while white cells represent non-significant results. This figure allows for a visual interpretation of the results by examining the impact either by column—showing the CAR values for each event window across all events—or by row, observing how each company's CAR changes with different event window lengths. Looking at the matrix by column, the [–1,0] window exhibits the highest concentration of significant events (five), followed by [–1,1] with four, [–2,2] with three, and most of the remaining windows with two; notably, the [0,4] window shows just one. This suggests that market reactions may partially precede official announcements, possibly due to information leaks or early media coverage. Looking at Fig. 3 by row, Enel clearly emerges as the most negatively affected firm, showing statistically significant CARs across all event windows—three at the 1% level, four at 5%, and one at 10%. Iberdrola also exhibits notable impacts, with the first event significant in several windows and the second in only two. As shown in Table 1, the two Iberdrola events occurred just two days apart, raising concerns of overlapping windows. For this reason, the CAAR offers a more reliable measure, with significantly negative results in windows like [–1, +1] (–0.03800;

t = –12.43) and [–2, +2] (0.02490; t = –3.84). Additional, though less frequent, significant windows are observed for Gen, Netflix, Microsoft, CaixaBank, and Generali. Other firms, such as Telefónica and Avanza, show consistently negative CARs across most windows, although none reach statistical significance. A similar pattern holds for Meta's second event, which also displays negative but non-significant CARs. In contrast, some firms such as UniCredit, Amazon, Eni, and Uber exhibit zero or positive abnormal returns across most windows, indicating a more resilient or favorable market reaction.

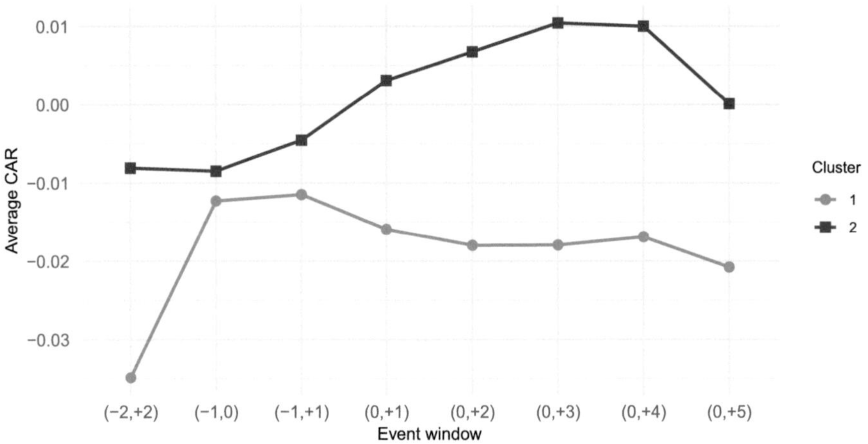

Fig. 4. Average CAR trends over event windows, grouped by cluster membership, illustrating distinct response patterns

To gain clearer insights into the results, a cluster analysis was performed to identify homogeneous groups of firms. The clustering variables were the CARs for most companies, while for Meta and Iberdrola, which have two events each, the CAARs were used. Clustering was performed using the k-means method with $k = 2$, a choice supported by the silhouette index and the limited number of firms. The underlying assumption was that firms could be broadly categorized into those reacting positively and those reacting negatively to GDPR enforcement. Indeed, Fig. 4 shows the average CAR trends across event windows for the two clusters. Cluster 1, depicted with lighter circles, exhibits a consistently negative average CAR, representing the group of firms that react adversely. Conversely, Cluster 2, shown with darker squares, displays a positive trend starting from the [0,1] window, indicating firms with a more favorable market reaction. Figure 5 displays the firms in the revenuemarket capitalization space, with points distinguished by both shape and shading intensity according to cluster membership. The lighter-shaded cluster, associated with negative market reactions, includes firms with predominantly negative CARs such as Enel, Iberdrola, Gen, Avanza Bank, and Telefónica. The darker-shaded cluster represents more resilient firms and comprises Amazon, UniCredit, Generali, CaixaBank, Uber,

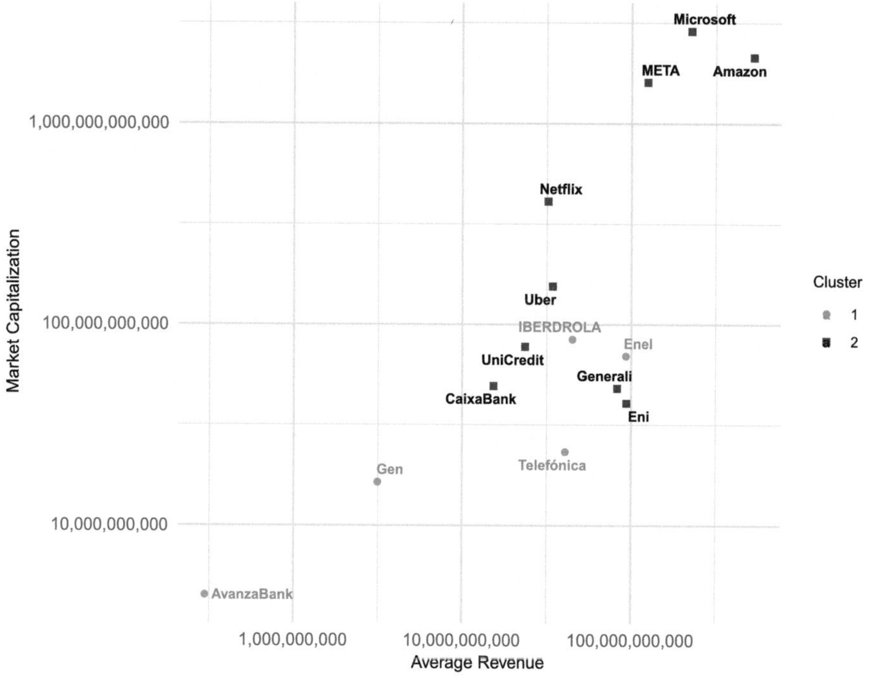

Fig. 5. Firm clusters derived from the market model, displayed in the revenuecapitalization space; axes use base-10 logarithmic scaling to enhance interpretability

Eni, Netflix, Meta, and Microsoft. The composition of the two clusters reflects the same dynamics highlighted in the heatmap (Fig. 3), reinforcing the distinction between firms that responded more negatively and those that appeared more resilient. These findings offer a coherent picture of market reactions to GDPR enforcement, which is further discussed in the concluding section.

4 Conclusive Remarks

Building on and updating previous event study research, this paper investigates how financial markets reacted in the short term to recent GDPR enforcement actions. The proposed analysis contributes to a relatively underexplored area of research. Indeed, up to date we have found only one contribution in this area ([4]). Nevertheless, given the growing importance of data protection and regulatory enforcement, this topic is considered to hold substantial academic relevance and is likely to stimulate further research in the future. Assessing the financial impact of GDPR enforcement, allows investors, regulators, and companies to better understand how privacy violations affect market value. It also can be a useful tool for the integration of data protection risks into ESG and compliance-related investment strategies. Although the event study methodology is extensively employed used in finance and economics, its application to the effects of GDPR fines remains very limited. As in previous studies, we refer to a relatively small sample size due to the need to focus exclusively on publicly listed companies because only for these firms are market reactions observable. This study further refines the scope by including only the most significant cases, defined as those involving fines exceeding one million euros and issued in 2024. To provide context for the financial analysis, we detailed the nature and severity of the violations committed by the companies in the sample, along with their country of origin and industry sector. The majority of the fines were imposed for serious infractions, mainly involving breaches of the core principles of data processing. We employed an event study methodology and estimated abnormal returns using the well-known Market Model, analyzing eight different event windows. While the earlier analysis by Ford et al. [4] identified the strongest effects in slightly longer windows ([0,2] and [0,3]), this updated study finds more pronounced abnormal returns in shorter windows, namely [–1,0] and [–1,1]. This may indicate a quicker market response in recent years, possibly due to greater investor awareness or early information leaks. Within the sample, Enel appears to be the most adversely affected firm, corresponding to the largest fine imposed by the Italian data protection authority. Iberdrola also experienced significant negative impacts, particularly in response to two consecutive fines (see Table 1). By contrast, firms such as Meta, Uber, and Microsoft (through its subsidiary LinkedIn), despite receiving three of the largest GDPR fines to date, did not exhibit notably negative market reactions showing a strong resilience.

This study represents a preliminary step in exploring the topic, and we acknowledge its limitations. The primary constraint lies in the choice of the model used to estimate abnormal returns. The standard Market Model is widely adopted for its simplicity but relies solely on market returns as the explanatory variable, which causes it to neglect potentially important firm- or industry-specific factors. Moreover, as a static model, it assumes stable relationships over time, which may not adequately reflect the dynamic nature of market reactions to regulatory events. Looking ahead, future research could enhance the robustness of the analysis by expanding the sample to include smaller fines and by employing more advanced methodological approaches. Multi-factor models, such as Fama-French, allow for the inclusion of additional risk dimensions, while machine learning techniques offer greater flexibility in capturing non-linear patterns and time-varying relationships typical of financial data. These advancements would contribute to a more comprehensive understanding of how markets respond to regulatory actions.

Acknowledgments. This research was partially supported by the SERICS project (PE00000014) under the MUR National Recovery and Resilience Plan funded by the European Union, NextGenerationEU.

References

1. Andoh-Baidoo, F.K., Amoako-Gyampah, K., Osei-Bryson, K.M.: How internet security breaches harm market value. IEEE Secur. Priv. **8**(1), 36–42 (2010)
2. Black, F., Jensen, M.C., Scholes, M.S.: The capital asset pricing model: some empirical tests. In: Jensen, M.C. (ed.) Studies in the Theory of Capital Markets, pp. 79–121. Praeger Publishers Inc., New York (1972)
3. CMS Law.Tax. GDPR Enforcement Tracker. https://www.enforcementtracker.com/
4. Ford, A., Al-Nemrat, A., Ghorashi, S.A., Davidson, J.: The impact of GDPR infringement fines on the market value of firms. Inf. Comput. Secur. **31**(1), 51–64 (2023). https://doi.org/10.34190/EWS.21.088
5. Huntington-Klein, N.: The Effect: An Introduction to Research Design and Causality, 1st edn, Chap. 17: Event Studies. Chapman and Hall/CRC, Boca Raton (2021). https://doi.org/10.1201/9781003226055
6. MacKinlay, A.C.: Event studies in economics and finance. J. Econ. Literat. **35**(1), 13–39 (1997)
7. Neuhierl, A., Scherbina, A., Schlusche, B.: Market reaction to corporate press releases. J. Finan. Quant. Anal. **48**(4), 1207–1240 (2013)
8. Ruohonen, J., Hjerppe, K.: The GDPR enforcement fines at glance. Inf. Syst. **106** (2022). https://doi.org/10.1016/j.is.2021.101876
9. Wolff, J., Atallah, N.: Early GDPR penalties: analysis of implementation and fines through may 2020. J. Inf. Policy **11**, 63–103 (2021)
10. Yahoo Finance, Historical data. https://finance.yahoo.com/quote. Accessed 25 Feb 2025

Delay-Adjusted Modeling of Cybersecurity Breaches Using INLA: Evidence from State Attorney General Data

Marco Pirra, Sofia Sarubbo(✉), and Fabio Viviano

Department of Economics, Statistics and Finance, University of Calabria, Ponte Bucci Cubo 0C, 87036 Rende, Cosenza, Italy
{marco.pirra,sofia.sarubbo,fabio.viviano}@unical.it

Abstract. This paper presents a statistical framework for analyzing cybersecurity breach data, with a focus on delayed reporting dynamics. Using legally mandated breach notification records from U.S. state attorneys general, we construct a monthly panel of breach occurrences and disclosures for California and Indiana from 2015 to 2025. We implement a Bayesian model with a negative binomial likelihood, incorporating structured temporal, delay-specific, and seasonal effects, and estimate it using Integrated Nested Laplace Approximation (INLA). The model adjusts for reporting lags and enables probabilistic estimation of latent breach incidence. Empirical results reveal significant cross-state differences in breach volume and reporting behavior, underscoring the importance of jurisdiction-specific models. Our findings contribute to the literature on cyber risk forecasting and offer actionable insights for insurers, regulators, and policymakers. The proposed framework supports delay-adjusted risk monitoring and can be extended to additional jurisdictions or enriched with covariates to capture sectoral or organizational heterogeneity.

Keywords: Cyber risk modeling · Reporting delays · Bayesian inference · Integrated Nested Laplace Approximation

1 Introduction and Scope

Understanding and quantifying cyber risk - particularly data breaches - has become a critical priority for researchers, insurers, and policymakers. These breaches often involve highly sensitive information, posing direct financial and reputational threats. More broadly, they pose systemic risks to organizations operating within increasingly digital infrastructures. While much of the early literature has focused on sectors like healthcare due to its data sensitivity and regulatory exposure, recent work has expanded toward cross-sectoral analyses that combine insights from operational risk, actuarial science, and cybersecurity. Empirical studies have employed a variety of statistical methods to capture breach frequency and severity. For instance, hurdle models, e.g., [13], have been applied to account for zero-inflated count data, while breach sizes are frequently

M. La Rocca et al. (Eds.): WMAF 2025, *New Perspectives in Mathematical and Statistical Methods for Actuarial Sciences and Finance*, pp. 238–248, 2025.
https://doi.org/10.1007/978-3-032-05551-4_21

modeled using log-normal or heavy-tailed distributions [3,16]. Dynamic and multivariate extensions, including Markov-modulated processes [7] and state-level frequency-severity models [14], have been developed to reflect the evolving nature of cyber threats and their dependence on organizational or contextual factors [6,11,15]. These models are particularly valuable in settings where reporting practices or risk exposure shift over time, as is common in cybersecurity. Despite progress in modeling techniques, consistent and transparent data remains a core challenge. Widely used breach datasets, such as the Privacy Rights Clearinghouse (PRC), Advisen Cyber Loss, and SAS OpRisk Global Data, aggregate information from diverse sources, including media reports, public disclosures, and state-level notifications. While these sources offer broad coverage, their evolving and sometimes opaque data collection methods complicate longitudinal analyses of breach frequency and severity trends [4,5,10]. In contrast, breach data released by U.S. state attorneys general offer distinct empirical advantages rooted in their regulatory origin. These datasets are collected under legally mandated, state-specific notification laws, ensuring a high degree of legal consistency. Reports are submitted directly by the affected organizations, which helps reduce the selection biases common in media-sourced or third-party-aggregated databases. The reporting process is typically granular and timely, with many jurisdictions providing daily updates that include breach occurrence and disclosure dates, the number of individuals affected, and, in many cases, the original notification letter. Importantly, the legal frameworks governing breach notification have remained relatively stable within each state, which enhances the reliability of longitudinal analyses and supports the detection of temporal patterns in breach activity. These attributes make attorney general datasets particularly valuable for studying reporting delays, an often-overlooked but critical aspect of cyber risk measurement. The lag between breach occurrence and public disclosure can obscure real-time signals, complicating both risk estimation and regulatory response. Moreover, such datasets support the analysis of systemic exposures, such as breaches propagated through shared service providers, which are difficult to isolate in aggregate datasets. In this paper, we contribute to the literature by leveraging state attorneys general datasets to model the timing and reporting of data breaches. Our empirical analysis centers on two representative states, California and Indiana, which offer rich, regularly updated breach records. We develop a Bayesian modeling framework that adjusts for reporting delays and decomposes breach counts into interpretable temporal, seasonal, and delay-related components. This approach builds on methods originally developed in epidemiology and insurance reserving, but adapted here for cyber risk applications. The remainder of the paper is organized as follows. Section 2 introduces the statistical methodology, outlining our Bayesian framework for delay-adjusted estimation using Integrated Nested Laplace Approximation (INLA). Section 3 presents the empirical results for California and Indiana, highlighting predictive performance and latent breach dynamics. Section 4 concludes with a discussion of implications for cyber risk regulation and insurance, and suggests directions for future research.

2 Modeling Framework for Delayed Reporting

Delayed reporting is a well-known challenge in fields such as epidemiological surveillance, insurance reserving, and cyber risk monitoring. To address this, we develop a statistical framework that models observed event counts $n_{t,d}$, representing events that occurred at time t and were reported after a delay of d months. While early approaches in epidemiology, such as those in [2], used a multinomial specification over delays conditioned on the total number of events N_t, we instead adopt a count-based formulation that directly models $n_{t,d}$ using a negative binomial distribution. This formulation accommodates overdispersion and allows for greater flexibility in modeling cyber incident data. Such delay-adjusted approaches are particularly valuable in settings where real-time estimation of total event counts is critical, as in nowcasting or regulatory early-warning systems. Our work builds on these epidemiological foundations and adapts them to cyber breach data disclosed through U.S. state attorneys general portals. These portals provide detailed, time-stamped reporting of breach occurrences and disclosures, enabling robust estimation of latent breach dynamics and delay structures. A closely related tradition in actuarial science is the chain-ladder method, which estimates future claims using triangle structures of $n_{t,d}$. Originally introduced as a distribution-free estimator [9], this approach was later generalized to a linear additive model [12]:

$$\mathbb{E}[n_{t,d}] = \lambda_{t,d} = \mu + \alpha_t + \beta_d, \tag{1}$$

where temporal and delay-specific effects are estimated additively. These models have gained traction in insurance reserving, public health, and other domains involving right-censored or partially observed data. When embedded in a Bayesian framework, such models support full posterior inference and uncertainty quantification, capabilities that are increasingly emphasized under regulatory standards such as Solvency II. We extend this structure by incorporating structured random effects and covariates, enabling flexible yet interpretable modeling of breach occurrence and reporting dynamics. The data-generating process is defined over a reporting triangle indexed by $t \in \{1, \ldots, T\}$ and $d \in \{0, \ldots, D\}$, with the constraint that $t + d \leq T$ to reflect observability. This setup naturally accommodates right-censoring in recent periods. We now detail the probabilistic structure and latent components underlying our model.

2.1 Model Specification

We model the observed counts $n_{t,d}$ using a negative binomial distribution:

$$n_{t,d} \sim \text{NegBin}(\lambda_{t,d}, \phi), \quad \lambda_{t,d} > 0, \quad \phi > 0, \tag{2}$$

where $\lambda_{t,d}$ is the mean and ϕ is a dispersion parameter. As $\phi \to \infty$, the distribution converges to the Poisson. To regularize dispersion in the absence of strong empirical signal, we place a weakly informative prior $\phi \sim \text{Exp}(0.1)$. The log-mean is additively decomposed to capture multiple sources of structural and temporal heterogeneity:

$$\log(\lambda_{t,d}) = \mu + \alpha_t + \beta_d + \gamma_{t,d} + \eta_{w(t)} + X_{t,d}^{\top}\boldsymbol{\delta}, \tag{3}$$

where:

- $\alpha_t \sim \mathcal{N}(\alpha_{t-1}, \sigma_\alpha^2)$ is a first-order random walk capturing smooth evolution in calendar time;
- $\beta_d \sim \mathcal{N}(\beta_{d-1}, \sigma_\beta^2)$ models gradual variation across reporting delays;
- $\gamma_{t,d} \sim \mathcal{N}(\gamma_{t-1,d}, \sigma_\gamma^2)$ introduces interaction dynamics between time and delay, capturing localized irregularities in delay behavior;
- $\eta_{w(t)} \sim \mathcal{N}(2\eta_{w-1} - \eta_{w-2}, \sigma_\eta^2)$ represents seasonal variation using a second-order random walk over calendar months $w(t) = t \bmod 12$;
- $X_{t,d}$ contains covariates such as breach type, organizational size, or policy indicators, with coefficients $\boldsymbol{\delta}$.

All random effects are centered at zero to ensure identifiability of the global intercept μ. Hyperpriors on precision parameters $\tau = 1/\sigma^2$ are specified using penalized complexity priors, which shrink toward smoother trajectories in the absence of strong data support. This formulation balances flexibility and parsimony, enabling the model to capture both persistent structure and regime shifts such as regulatory changes or sudden breach surges.

2.2 Model Implementation

We perform inference using the Integrated Nested Laplace Approximation (INLA), a deterministic and computationally efficient approach designed for latent Gaussian models. Unlike traditional Bayesian approaches such as Markov Chain Monte Carlo (MCMC), INLA approximates posterior marginals through nested Laplace approximations and numerical integration, enabling fast and accurate inference even in high-dimensional settings. We implement the model using the `R-INLA` package, which exploits the sparse structure of precision matrices associated with Gaussian Markov Random Fields (GMRFs), [8]. This structure allows for computational scalability while preserving full Bayesian inference over both fixed effects and structured latent components. Our implementation yields posterior distributions for delay-adjusted event counts, credible intervals for unreported breaches, and smoothed estimates of structured effects. In the empirical analysis, we apply this INLA-based framework to monthly breach reporting data from California and Indiana. The results illustrate the model's ability to recover latent breach trajectories, quantify reporting uncertainty, and adapt to jurisdiction-specific reporting patterns.

3 Empirical Analysis

This section presents an empirical investigation of breach disclosure dynamics using data from U.S. state attorneys general. These records offer timely, structured, and legally mandated breach notifications with detailed information on breach timing and scale. Crucially, they include both the date of occurrence and the date of reporting, enabling a robust analysis of reporting delays, an often-overlooked but vital aspect of cyber risk modeling. In contrast to aggregated datasets sourced from media or third-party aggregators, state attorney general data are submitted directly by affected entities under consistent legal frameworks. This structure enhances reliability, reduces source bias, and ensures consistency across time. For a comprehensive review of the legal and structural characteristics of these datasets, see [1]. We focus on two high-activity jurisdictions, California (CALI) and Indiana (INDI), selected for their data richness and regulatory relevance. Using monthly aggregated breach reports from 2015 through April 2025, we analyze temporal trends in breach activity, examine discrepancies between occurrence and disclosure, and implement a delay-adjusted model to estimate latent breach incidence. The results inform state-level reporting behavior and support nowcasting for cyber risk surveillance.

3.1 Temporal Dynamics of Reported and Occurred Events

To examine temporal trends in cyber incident disclosure, we plot monthly time series of both *occurred* and *reported* breach events for two representative jurisdictions, California (CALI) and Indiana (INDI), covering the period from 2015 through April 2025. Figure 1 shows breach occurrences by disclosure date, while Fig. 2 displays the timeline of corresponding report filings. A pronounced divergence emerges between breach occurrence and its subsequent reporting, particularly in earlier years. Indiana exhibits more volatile and higher-frequency breach activity overall, especially from 2019 onward. California, while recording fewer breaches, shows notable spikes, most visibly in early 2020 and late 2023. The apparent decline in reported events after 2023 likely reflects censoring effects due to ongoing reporting lags, rather than a true reduction in breach frequency. This underscores the need for statistical frameworks that adjust for incomplete reporting near the edge of the triangle. Additionally, the temporal misalignment between peaks in occurred and reported events reinforces the operational importance of nowcasting methods in cyber risk monitoring and regulatory response.

3.2 State-Level Reporting Behavior: CALI Vs INDI

To explore jurisdiction-specific reporting dynamics, Figs. 3 and 4 compare monthly series of occurred versus reported breach events for California and Indiana, respectively. These plots highlight the timing and magnitude of disclosure relative to breach occurrence. In California (Fig. 3), the gap between reported and occurred events varies significantly over time. From 2019 through early 2023, reported counts often exceed occurrences, suggesting lagged but eventually complete disclosure. The presence of sharp spikes, especially in early 2023, points to episodic bulk reporting or backlogged disclosures. This volatility implies operational irregularities in either detection timelines or administrative reporting processes. Indiana (Fig. 4), by contrast, shows a more consistent alignment between

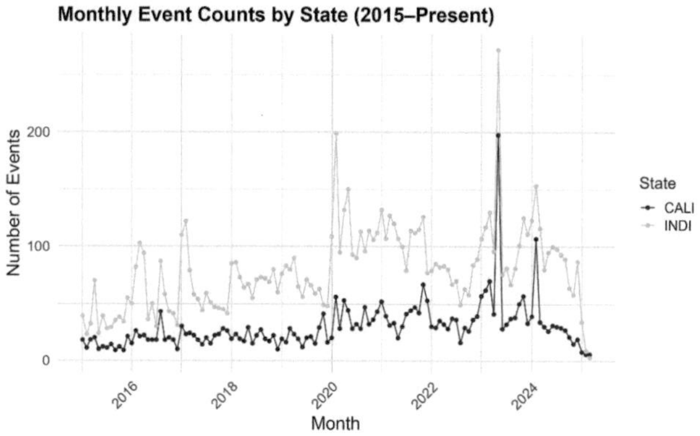

Fig. 1. Monthly Occurred Event Counts by State (CALI vs. INDI, 2015–2025/04)

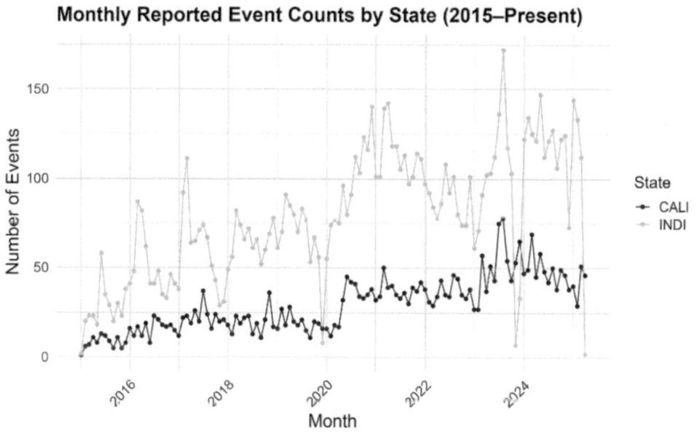

Fig. 2. Monthly Reported Event Counts by State (CALI vs. INDI, 2015–2025/04)

the two series, with moderate reporting delays that are relatively stable post-2020. Although some divergence is visible during the 2019–2021 period, Indiana's temporal structure appears less erratic than California's, with smoother transitions and fewer abrupt surges. These comparisons underscore cross-state differences in breach reporting behavior. California exhibits more volatile and asymmetric disclosure dynamics, while Indiana reflects steadier, more synchronized reporting. This variation highlights the importance of flexible modeling frameworks capable of capturing state-specific reporting regimes.

3.3 Distribution of Reporting Delays: CALI vs INDI

Figure 5 shows the distribution of reporting delays, in months, for California and Indiana. The histogram highlights important structural differences in the timing of breach disclosures across the two jurisdictions. Both distributions are right-skewed, indicating that most breaches are reported within the first few

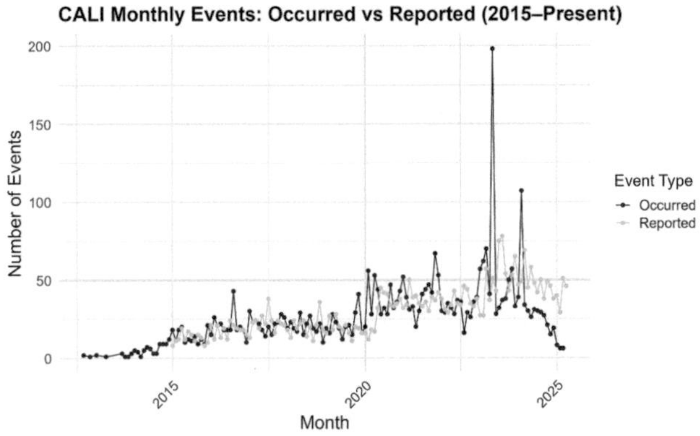

Fig. 3. California Monthly Events: Occurred vs Reported

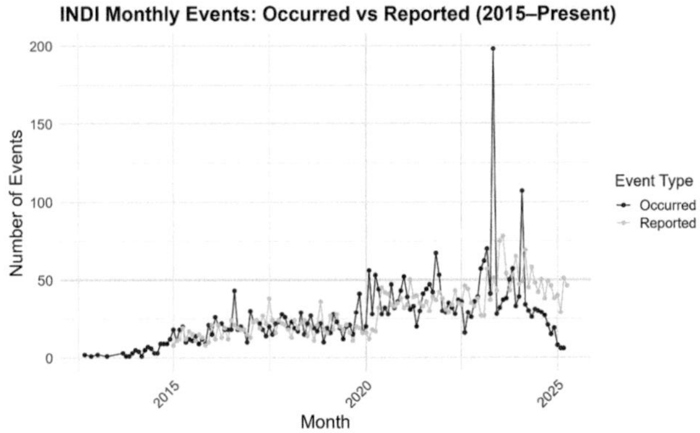

Fig. 4. Indiana Monthly Events: Occurred vs Reported

months following occurrence. Indiana's delay pattern is more tightly concentrated in the 0–3 month range, suggesting prompt reporting in the majority of cases. California, in contrast, exhibits a broader spread and heavier right tail, reflecting a subset of breaches with longer notification lags. Summary statistics confirm this visual pattern: Indiana has a median delay of 3 months, with only 8.8% of breaches reported after 10 months. California's median delay is slightly longer at 4 months, and 16% of breaches exceed a 10-month delay. These findings stress the need to model jurisdiction-specific delay structures explicitly. Ignoring such structural variation can distort breach incidence estimates, particularly in near-real-time estimation tasks such as nowcasting or regulatory surveillance.

3.4 Delay-Adjusted Forecast and Latent Effects: California

To correct for underreporting and forecast breach incidence, we apply the delay-adjusted INLA model to California. Figure 6 presents the resulting predictions, which integrate both observed and latent breaches. The model reveals a moderate

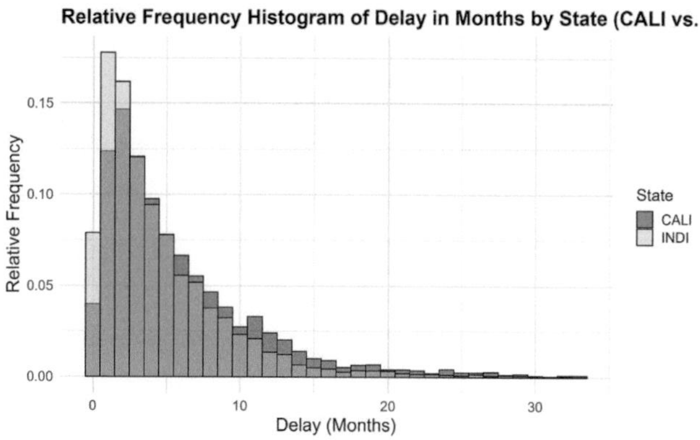

Fig. 5. Relative Frequency Histogram of Delay in Months: California vs Indiana

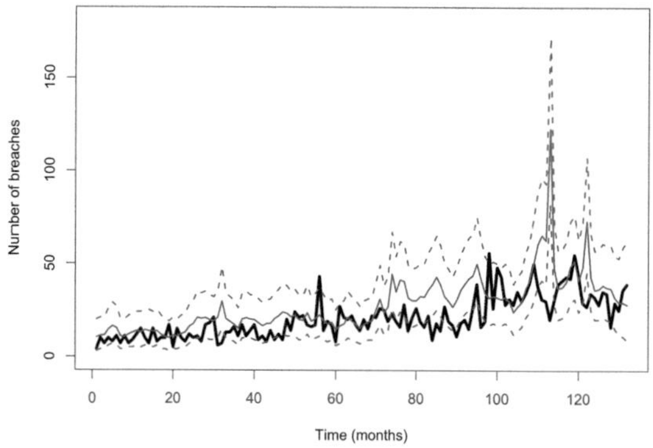

Fig. 6. INLA-based delay-adjusted predictions for monthly breach counts in California. The solid black line shows the observed number of breaches eventually reported, while the solid grey line represents the model's posterior mean estimate accounting for unreported events due to delay. The shaded area indicates the 95% posterior predictive interval, quantifying uncertainty about latent breaches not yet disclosed at the time of analysis.

upward trend in breach activity over time, with uncertainty widening toward the most recent months, consistent with right-censoring effects introduced by reporting delays. Forecast metrics reflect the inherent challenges in nowcasting with partially observed data: the mean Continuous Ranked Probability Score (CRPS) is 9.345, indicating moderate forecast error. Figure 7 decomposes the model into three latent components. The time effect increases steadily from –0.99 to 1.60 (average 95% CI width: 0.75), confirming the long-term rise in breach intensity. The delay effect ranges from 1.47 to –1.66 (CI width: 0.43), revealing a sharp decay in reporting probability with elapsed time. The seasonal effect is centered near zero, with minimal amplitude (±0.08) and narrower uncertainty (CI width: 0.28), indicating low cyclicality in reporting. These results suggest

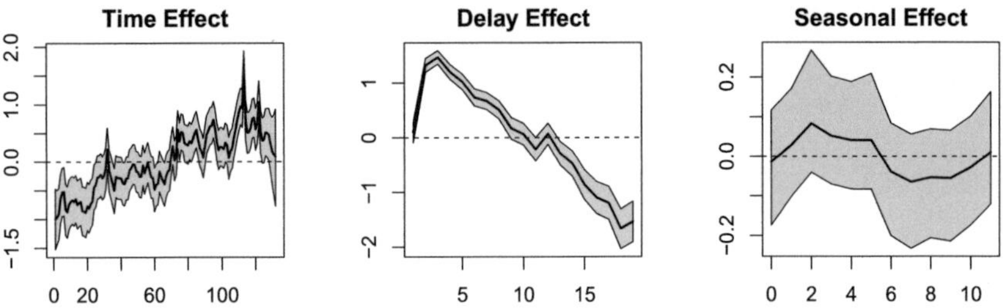

Fig. 7. Estimated Latent Effects: Time, Delay, and Seasonality (California)

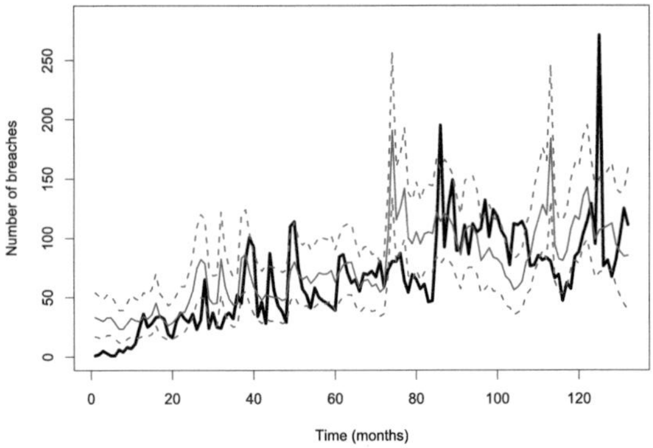

Fig. 8. INLA-based delay-adjusted predictions for monthly breach counts in Indiana. The solid black line shows the observed number of breaches eventually reported, while the solid grey line represents the model's posterior mean estimate accounting for unreported events due to delay. The shaded area indicates the 95% posterior predictive interval, quantifying uncertainty about latent breaches not yet disclosed at the time of analysis.

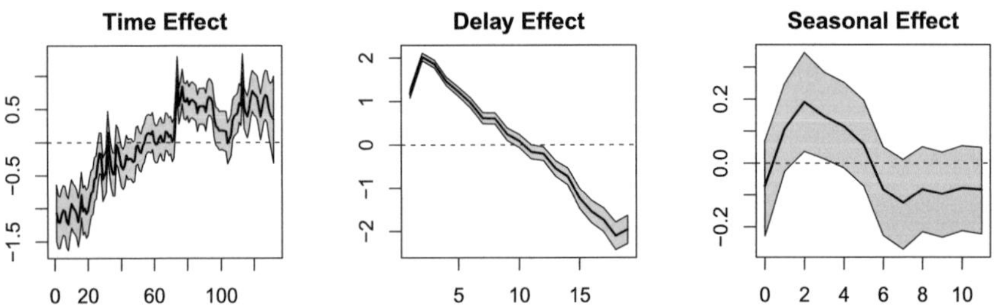

Fig. 9. Estimated Latent Effects: Time, Delay, and Seasonality (Indiana)

California's reporting process is structured, with predictable lags and a gradual rise in breach activity. However, conservative interval construction or additional covariates may be required to improve predictive calibration.

3.5 Delay-Adjusted Forecast and Latent Effects: Indiana

We apply the same modeling framework to Indiana, a jurisdiction characterized by higher breach volume and greater variability in reporting patterns. Figure 8 displays the model's delay-adjusted predictions, which show accelerated growth in breach incidence since 2020. Prediction intervals widen in recent months, reflecting greater uncertainty due to unreported but expected events. Indiana's predictions exhibit higher forecast error, as reflected by a mean CRPS of 30.889: the score reflects greater variability and potential underfitting, likely due to unmodeled heterogeneity in organizational behavior or reporting practices. Figure 9 decomposes the model into latent components. The time effect ranges from –1.24 to 1.04 with an average credible interval width of 0.63, reflecting moderate fluctuations and a less linear upward trend than observed in California. The delay effect decays sharply (–2.10 to 2.02, CI width: 0.34), confirming that most breaches are reported within a short timeframe. The seasonal effect ranges from –0.13 to 0.19 with a CI width of 0.28, indicating mild but slightly stronger within-year cyclicality than in California. Together, these patterns highlight the challenges of modeling breach dynamics in more volatile jurisdictions. While the core INLA framework remains applicable, additional refinements, such as more flexible priors or overdispersion adjustments, may be needed to capture Indiana's burstier breach behavior and improve predictive performance. These contrasts highlight the need for jurisdiction-specific modeling. Differences in volume and delay patterns suggest that standardized approaches may not generalize well.

4 Conclusion and Future Work

This paper presents a delay-adjusted Bayesian modeling framework for cybersecurity breach data, leveraging the Integrated Nested Laplace Approximation (INLA) to quantify and forecast breach incidence in the presence of reporting lags. Applying the model to data from California and Indiana reveals substantial differences in temporal patterns, delay structures, and predictive calibration, differences with direct implications for both cyber insurance pricing and regulatory oversight. For insurers, correcting for underreporting improves the accuracy of loss reserving and risk estimation, particularly in jurisdictions with volatile or burst-prone disclosure behaviors. For regulators, delay-aware models offer early-warning capabilities for systemic risk, such as vendor-driven contagion across organizations. Accurate nowcasting in such settings depends on timely recognition of latent breach dynamics obscured by reporting lags. Several extensions can further improve the framework's utility and adaptability. Expanding the model to include additional jurisdictions could uncover regional heterogeneity and cross-state spillover effects. Incorporating firm-level or breach-type covariates may improve predictive sharpness and interpretability. Hierarchical formulations could account for nested variation across industries or regulatory regimes.

Finally, embedding the framework in a real-time monitoring system would support more proactive governance by enabling dynamic estimation of emerging threats in real time, even when data remain incomplete or delayed.

References

1. Avanzi, B., Tan, X., Taylor, G., Wong, B.: On the evolution of data breach reporting patterns and frequency in the united states: a cross-state analysis. North Am. Actu. J. 1–32 (2025)
2. Bastos, L.S., et al.: A modelling approach for correcting reporting delays in disease surveillance data. Stat. Med. **38**(22), 4363–4377 (2019)
3. Edwards, B., Hofmeyr, S., Forrest, S.: Hype and heavy tails: a closer look at data breaches. J. Cybersecur. **2**(1), 3–14 (2016)
4. Eling, M., Ibragimov, R., Ning, D.: Time dynamics of cyber risk. SSRN 4497621 (2023)
5. Harry, C., Gallagher, N.W.: Categorizing cyber effects. In: The Elgar Companion to Digital Transformation, Artificial Intelligence and Innovation in the Economy, Society and Democracy, pp. 7–31. Edward Elgar Publishing (2023)
6. Hu, K., Levi, R., Yahalom, R., Zerhouni, E.G.: Supply chain characteristics as predictors of cyber risk: a machine-learning assessment. ArXiv arxiv:2210.15785 (2022). https://api.semanticscholar.org/CorpusID:253224006
7. Li, Y., Mamon, R.: Modelling health-data breaches with application to cyber insurance. Comput. Secur. **124**, 102963 (2023). https://doi.org/10.1016/j.cose.2022.102963
8. Lindgren, F., Rue, H.: Bayesian spatial modelling with r-inla. J. Stat. Softw. **63**(19), 1–25 (2015). https://doi.org/10.18637/jss.v063.i19. https://www.jstatsoft.org/index.php/jss/article/view/v063i19
9. Mack, T.: Measuring the variability of chain ladder reserve estimates. ASTIN Bull. **23**(2), 213–225 (1993)
10. Maillart, T., Sornette, D.: Heavy-tailed distribution of cyber-risks. Eur. Phys. J. B **75**, 357–364 (2010)
11. McLeod, A., Dolezel, D.: Cyber-analytics: modeling factors associated with healthcare data breaches. Decis. Supp. Syst. **108**, 57–68 (2018)
12. Renshaw, A.E., Verrall, R.J.: A stochastic model underlying the chain-ladder technique. Br. Actu. J. **4**(4), 903–923 (1998)
13. Sun, H., Xu, M., Zhao, P.: Modeling malicious hacking data breach risks. North American Actuarial Journal **25**, 484–502 (2020). https://api.semanticscholar.org/CorpusID:225634330
14. Sun, H., Xu, M., Zhao, P.: A multivariate frequency-severity framework for healthcare data breaches. Ann. Appl. Stat. **17**(1) (2023). https://doi.org/10.1214/22-aoas1625
15. Sun, M., Lu, Y.: A generalized linear mixed model for data breaches and its application in cyber insurance. SSRN Electron. J. (2022). https://doi.org/10.2139/ssrn.4077521
16. Wheatley, S., Hofmann, A., Sornette, D.: Data breaches in the catastrophe framework & beyond. arXiv Applications (2019). https://api.semanticscholar.org/CorpusID:155100199

Addressing Long-Term Care Risk Through Pension-Linked Insurance in the Italian Context: A Stochastic Approach Using Severance Pay Scheme

Alberto Piscitelli(✉)

Sapienza University of Rome, Rome, Italy
alberto.piscitelli@uniroma1.it

Abstract. This paper addresses the growing challenge of long-term care (LTC) dependency in aging populations by proposing the integration of LTC insurance with pension funds through the allocation of severance pay. A stochastic model is introduced to evaluate the financial sustainability and effectiveness of this approach, using Monte Carlo simulations to analyse the trade-offs between pension income and LTC benefits. The findings emphasize the need for welfare reforms and improved health and financial literacy to ensure broader adoption of LTC solutions, contributing to a more sustainable and equitable management of aging-related risks.

Keywords: Long Term Care · LTC risk · Silver economy · Pension fund · Severance pay scheme · Healthy life expectancy

1 Introduction

In recent years, life expectancy has shown a consistent upward trend; however, the additional years of expected life are not accompanied by a proportional increase in healthy life expectancy [14]. Consequently, there has been an increase in the number of individuals who depend on LTC, also, this trend is more evident among women than men [13].

For older adults, this trend poses particular challenges, where a larger portion of life expectancy is spent in conditions of poor health or dependency changes significantly affect social welfare systems.

Aging populations exert mounting pressure on public healthcare and pension frameworks, exacerbating issues of sustainability. As stated by Fornero (2015) [10], the need for structural reform in welfare systems is evident, as they must adapt to the shifting demographic landscape while addressing the limitations of current models. Alongside financial concerns, a lack of awareness about risks such as long-term care dependency further complicates the implementation of effective solutions.

M. La Rocca et al. (Eds.): WMAF 2025, *New Perspectives in Mathematical and Statistical Methods for Actuarial Sciences and Finance*, pp. 249–260, 2025.
https://doi.org/10.1007/978-3-032-05551-4_22

As emphasized by Brown et al. (2009) [2], the LTC risk is one of the least insured across European countries, including Italy, where the private sector struggles to provide sufficient coverage, and public systems face increasing financial constraints.

To address these issues, a shift toward integrating private insurance solutions with existing pension frameworks has been proposed. This approach leverages products such as LTC insurance integrated with pension plans, offering a dual solution for income security and dependency risks. However, the effectiveness of such solutions hinges on widespread understanding and acceptance among the population. Economic and health literacy emerge as critical factors, enabling individuals to make informed decisions about their financial and health-related futures [1].

In Italy, the growing demand for long-term care in the private sector [4] highlights both the inherent risks associated with aging and the difficulties of the public welfare system to meet the demand. Many elderly individuals lack the financial resources to manage potential disability-related expenses, and these challenges are often exacerbated in socially and economically disadvantaged groups [5].

This underscores the need for a welfare reform that fosters closer collaboration between the public and private sectors. The financial burden of long-term care cannot be sustained solely by public expenditure, necessitating the involvement of private solutions. Among these, LTC insurance products have been identified as essential tools for managing the LTC risk.

Particular emphasis has been placed on pension funds with an LTC option, which is linked to pension schemes. This type of coverage is versatile, addressing both longevity risk and LTC risk.

In this study, a stochastic model is proposed to evaluate this solution, offering a method to analyse the effectiveness of integrating private insurance solutions with existing pension frameworks in mitigating the LTC risk.

The paper is structured as follows: Sect. 2 introduces the theoretical model used in the analysis and provides the model estimates, Sect. 3 presents a policy suggestion and Sect. 4 concludes the study.

2 Methodologies and Estimates

2.1 Framework Description

This study proposes a stochastic framework for evaluating Long-Term Care (LTC) insurance products linked to a pension fund, designed to address the growing challenges posed by aging populations and rising care costs. The analysis focuses on pension-linked LTC coverage, where individuals at retirement can allocate a fraction of their pension benefits to secure financial support in case of LTC dependency.

Two key decision variables guide the model:

1. **Contribution Amounts:** Determined during the working life and to be paid during the accumulation phase.
2. **Benefit Reduction Rate (q)**: The fraction of the pension benefit foregone to exercise the LTC option.

In the numerical application, we assume that contributions are fixed and equal to the Italian severance pay scheme, which evolves according to wage growth. However, in a more general setting, alternative contribution rules, either fixed, income-related, or voluntarily chosen, could be incorporated within the proposed framework.

The model also incorporates demographic and financial factors to estimate the adequacy of the proposed product and can assess its sensitivity to varying parameters.

Using this approach, the model accounts for the various possible trajectories of the pension fund's value and the corresponding LTC benefit levels. To implement the quantitative design, the following key elements are required:

- **Stochastic Process for Interest Rates**: Defining the dynamics of interest rates to represent the term structure of yields over different maturities.
- **Pension Fund Value Formation**: Determining the stochastic evolution of the pension fund in relation to the benchmark returns.
- **Contribution Rule:** Establishing the periodic contribution levels to be paid during the accumulation phase.
- **Stochastic Process for Benchmark Returns**: Modelling the investment benchmark guiding the pension fund's performance.
- **Demographic Assumptions**: Using projected technical demographic bases for the conversion year to account for mortality and morbidity rates.
- **Conversion Factors**: Calculating the conversion of the accumulated pension amount into either a simple annuity or an LTC-linked annuity, specifying the benefit reduction rate (q) allocated to cover LTC risk.

The application of this framework relies on the Monte Carlo simulation approach, enabling the exploration of a wide range of possible scenarios and assessing the robustness of decisions under varying conditions.

The demographic and financial components are modeled independently. Even if the demographic assumptions are derived from stochastic population projections, while the interest rate curve is obtained from a large set of Monte Carlo simulations, both mortality and long-term care probabilities, as well as the term structure of interest rates, are specified deterministically, without any explicit dependence between the two processes. The only stochastic element in the model concerns the accumulation of the pension fund, which is subject to uncertainty through investment returns.

2.2 Interest Rate Process

To compute annuity rates and discount future cash flows, a yield curve from simulated interest rate paths is derived. Specifically, the term structure of interest rates is simulated using the Cox-Ingersoll-Ross (CIR) model proposed by Cox et al. (1985) [6], which captures the dynamics of risk-free returns over time.

The CIR process is defined as:

$$dr_t = \kappa(\theta - r_t)dt + \sigma\sqrt{r_t}dW_t \quad (1)$$

where r_t is the interest rate at time t, κ is the speed of mean reversion, θ is the long-term mean, σ is the volatility, and dW_t is a Wiener process.

For the parameter estimation of the CIR model, the time series of Euribor 1-year rates from 1994 to 2021 was used. The last years of the series were excluded, as they were deemed potentially misleading for the parameter determination due to recent market events. The estimation has been done via OLS; results are shown in Table 1.

The yields curve is determined by simulating 10,000 possible interest rate paths through a Monte Carlo approach, starting from the last observed value in 2024 and extending to 2100. Based on the simulated values, and considering that the rates are continuously compounded, a deterministic yield curve is then obtained by averaging the simulated outcomes, which simplifies the modeling process without compromising the overall structure of the analysis.

Table 1. Parameter Estimation of CIR Model

Parameter	Estimated Value
θ	0.0031
κ	0.0833
σ	0.047

2.3 Pension Fund Process

To describe the evolution of the pension fund, is proposed a model that simulates the growth of the fund over time based on contributions and the performance of the benchmark index. The following assumptions are made for the model:

- The fund is structured as a defined contribution fund based on a capitalization system, and for simplicity, assuming the absence of any transition costs.
- The contribution is variable and increases progressively, being paid annually.
- The fund follows a passive management strategy, needing the definition of a benchmark. Consequently, the fund's value evolves in line with the performance of the chosen index.

The equation that describes the evolution of the fund's value is as follows:

$$V_t = \begin{cases} P_0 & t = 0 \\ P_t + V_{t-1}\left(1 + \tilde{j}_{t-1}\right) & t > 0 \end{cases}. \tag{2}$$

V_t represents the fund's value at time t, which is determined by the sum of the annual contributions paid and the growth of the fund's value. P_t indicates the contribution paid at time t, $\tilde{j}_{t-1}$ is a random variable representing the annual return of the benchmark observed between year t − 1 and year t. Furthermore, assume that the contributions are determined year by year according to the following dynamic:

$$P_t = \begin{cases} P_0 & t = 0 \\ P_{t-1}(1 + y_t) & t > 0 \end{cases}. \tag{3}$$

where P_t represents the contribution at time t, and P_0 is the initial contribution. For each trajectory, the contributions are the same, and the amount to be paid increases annually by a factor y_t, which represents the annual growth rate of the contributions, which can also stay constant.

In addition to modelling the dynamics of interest rates, the benchmark returns for the pension fund are also modelled. Without loss of generality, is assumed that the benchmark used to guide the fund's management is the FTSE MIB index.

To model the benchmark, the assumption is that it follows a Geometric Brownian Motion (GBM) process, a widely supported by academical findings concerning the stock markets (Fama et al., 1965) [9]. Other stochastic process could be taken in account for other kinds of benchmarks.

The GBM process is defined as:

$$dI_t = I_t(\mu dt + \omega dW_t) \tag{4}$$

In this formulation, I_t represents the value of the benchmark at time t, while μ corresponds to the drift term, ω denotes the diffusion term while dW_t is a Wiener process.

The historical data used for the estimation are the annual price levels of the FTSE MIB index, spanning the period from 1998 to 2024. The resulting parameters are as follows (Table 2):

Table 2. GBM parameters estimates

Parameter	Estimated Value
μ	0.0299
ω	0.2215

2.4 Demographic Assumptions

The model adopted is a three-state multi-state model, which considers the transitions between active, LTC-disabled and deceased states. Reactivation from the LTC-disabled state is not modelled, given the chronic nature of long-term care dependency.

The data used to model the demographic assumptions are derived from various life tables published by De Angelis and Di Falco (2016) [7]. The demographic bases are constructed using data from the Istituto Nazionale della Previdenza Sociale (INPS) on individuals receiving disability benefits ("indennità di accompagnamento") between 2002 and 2012.

The model operates in discrete time and is non-homogeneous, allowing for the projection of transition probabilities over time. These projections were obtained using the Lee-Carter model extended by Brouhns et al. (2002) [3].

The projections produced three life tables, named LTC15L, LTC15M, and LTC15H, which include transition probabilities for individuals aged 20 to 95 years. These tables are projected from 2013 to 2043 and differentiated by gender.

Each table corresponds to a distinct scenario, obtained by 1000 simulations:

- **LTC15L (low scenario)**: Derived from the 95th percentile of the simulations.
- **LTC15M (central scenario)**: Represents the average of the simulations.
- **LTC15H (high scenario)**: Derived from the 5th percentile of the simulations.

These tables were combined using weighting factors that reflect the reliability of each table. Given that the data cover the decade 2002–2012 and the estimates were published in 2016, the tables do not incorporate the impacts of the COVID-19 pandemic.

To address this limitation, greater weight was assigned to the high-mortality scenario (LTC15H), which reflects higher mortality rates for both active and non-self-sufficient individuals. The applied weights are 50% for the high scenario, 30% for the central scenario, and 20% for the low scenario, resulting in the following formula for the transition probabilities:

$$p^{ij}(x, 2024) = \sum_{k \in S} \alpha_k p_k^{ij}(x, 2024) \tag{5}$$

where $S = \{L; C; H\}$ is the set of the possible scenarios (L for Low, C for Central and H for High), while α_k is the weight associated with the generic scenario k, so $a = \{20\%; 30\%; 50\%\}$. Furthermore, since the technical bases are gender-specific, a simple 50% weighting for both genders was applied to construct the unisex table.

2.5 Modelling LTC Option

The conversion factor calculation can be modified to consider the LTC option. As previously mentioned, in the case of a pension fund with an LTC option, at the time of retirement, the fund participant can decide whether to cover the LTC risk. Depending on the choice, different annuity conversion factors will apply. The actuarial notation used is the most widespread and commonly adopted in the literature (De Angelis et al., 2016 [7]; Dupourquè et al., 2019 [8]; Habermann et al., 1999 [11]; Levantesi, 2011 [12]). In the case where the risk is not covered, the actuarial equilibrium equation between the accumulated capital S and the pension annuity R will be:

$$S = R\left(\ddot{a}_x^{aa} + a_x^{ai}\right) \tag{6}$$

And the conversion factor to apply, defined as τ, will be:

$$\tau = \frac{1}{\ddot{a}_x^{aa} + a_x^{ai}} \tag{7}$$

where $\ddot{a}_x^{aa}$ represents the present value of an immediate annuity received by an individual who remains self-sufficient, and a_x^{ai} represents the present value of a deferred annuity, which is received by an individual who is active at the time of valuation but will begin receiving payments once they become LTC-dependent.

Using this factor, the pension annuity R is obtained by applying the formula:

$$R = S \cdot \tau \tag{8}$$

This annuity is paid to the participant until their death, regardless of their health status. If the participant opts to cover the LTC risk, they will receive different amount of annuity depending on his health status. The pension annuity received during the active phase is defined as R_a and the pension annuity received during the LTC dependency as R_i. In this case the relation between the two amounts and the accumulated capital S is determined by:

$$S = R_a \ddot{a}_x^{aa} + R_i a_x^{ai} \tag{9}$$

By combining Eq. (4) and Eq. (7), the following expression is obtained:

$$\left(\ddot{a}_x^{aa} + a_x^{ai}\right)R = R_a \ddot{a}_x^{aa} + R_i a_x^{ai} \tag{10}$$

Additionally, $R_a < R < R_i$.

It is necessary to impose an additional condition to solve the equation, as there are two unknown variables. In the proposed framework, the issue is resolved as follows.

Let's assume that at the time of exercising the option, the individual can choose the portion of the annuity R that they would forgo during the active phase to cover the LTC risk. This portion is denoted by q, defined as the Benefit Reduction Rate, with $q \in [0,1]$. In formula:

$$R_a = R(1-q) = S\tau(1-q) = S\tau^a \tag{11}$$

where τ^a is the conversion factor for the active phase, expressed as:

$$\tau^a = \tau(1-q) = \frac{1-q}{\ddot{a}_x^{aa} + a_x^{ai}} \tag{12}$$

Combining Eq. (7) and Eq. (9) leads to:

$$R_i = S\tau^i \tag{13}$$

where τ^i is the conversion factor for the LTC invalid phase, expressed as:

$$\tau^i = \tau\left(1 + q\frac{\ddot{a}_x^{aa}}{a_x^{ai}}\right) \tag{14}$$

3 Applications on LTC Costs

3.1 Policy Suggestion

Based on the socio-demographic context outlined in Sect. 1, a policy proposal is made to use Trattamento di Fine Rapporto (TFR) contributions to support LTC insurance subscription, focusing on pension funds with an LTC option. The TFR, Italy's severance pay scheme, is a deferred compensation accumulated at 6.91% of the employee's annual gross salary, adjusted annually for inflation and a fixed rate.

This framework explores how integrating TFR contributions into LTC coverage can improve accessibility and affordability of long-term care, leveraging TFR's nature as a deferred benefit to address growing LTC financing needs in aging populations.

A key question arises: are average TFR-based contributions sufficient to build a fund that can provide an annuity covering LTC costs?

The following sections apply the evaluation model from Sect. 2 to assess whether, for an average accumulation, an acceptable benefit reduction rate (q) exists that adequately covers LTC risk.

3.2 TRF Dynamics

To evaluate the proposed policy, it is essential to define the vector of contributions, which corresponds to the annual TFR allocations. Determining these contributions requires starting from the individual's gross annual salary. This data is derived from ISTAT, which provides the average hourly wage for age groups (15–29, 30–49, and over 50) from 2014 to 2021.

Hourly wage data have been interpolated to obtain specific values for each age and year of observation, enabling the calculation of two growth rates:

- **Age-related growth rate** (g_x): This reflects the increase in hourly wage due to age increasing. It is calculated as the geometric mean of the annual gross growth rates across all years of observation:

$$g_x = \left[\prod_{t=2014}^{2021} \left(1 + \overline{g}_x(t)\right)\right]^{\frac{1}{8}} - 1 \tag{15}$$

with $\overline{g}_x(t) = \left[\prod_{i=15}^{64}\left(\frac{w_{i+1,t}}{w_{i,t}}\right)\right]^{\frac{1}{50}} - 1$

Where $\overline{g}_x(t)$ represents the average age-related growth rate determined for year t, and $w_{i,t}$ is the gross hourly wage of an individual aged i recorded in year t.

- **Time-related growth rate** (g_t): This reflects the increase in hourly wage over time, independent of age. It is calculated as the geometric mean of annual growth rates across all ages:

$$g_t = \left[\prod_{x=15}^{65} \left(1 + \overline{g}_t(x)\right)\right]^{\frac{1}{51}} - 1 \tag{16}$$

with $\overline{g}_t(x) = \left[\prod_{j=2014}^{2020}\left(\frac{w_{x,j+1}}{w_{x,j}}\right)\right]^{\frac{1}{7}} - 1$

Where $\overline{g}_t(x)$ represents the average time-related growth rate determined for age x, and $w_{j,t}$ is the gross hourly wage of an individual aged x recorded in year j.

From these formulas, the growth rates obtained are $g_x = 0.636$ and $g_t = 0.69\%$.

Using these rates, the hourly wage for subsequent years is projected as:

$$w_{x+1,t+1} = w_{x,t}(1 + g_x)(1 + g_t) = w_{x,t}(1 + g) \tag{17}$$

where $g = 1.33\%$ represents the annual growth rate of gross hourly wages. This rate represents the growth of (TFR) contributions, denoted by y_t in the model presented in the Sect. 2.

3.3 LTC Costs

As previously mentioned, LTC insurance must cover expenses for individuals who become dependent on long-term care. The type of care significantly affects costs, with two common options being home care by a caregiver or nursing home care (Residenze Sanitarie Assistenziali, RSA).

Caregivers provide personal assistance at home, with wages regulated by national labor agreements. According to the CCNL Lavoro Domestico, depending on qualifications and responsibilities, monthly caregiver costs range from about €1,000 to €1,250, excluding benefits like meals and lodging allowances, leading to a gross monthly salary around €1,500.

Nursing homes provide comprehensive care for elderly individuals, with average monthly costs around €3,500—an expense that can impose a substantial financial burden on families.

3.4 Policy Evaluation

Let consider the case of a 46-year-old individual in 2024. Assume that the individual has already accumulated a pension fund through TFR contributions since 2001, when they were 23 years old and entered the workforce. This choice is due to the limitation on demographic data, since they are projected until 2043, year in which 46-years-old of 2024 will be able to retire.

Assuming the individual has worked 40 h per week for 52 weeks per year since the start of their career, their gross annual salary can be calculated through a simple multiplication. The gross hourly wage for a 23-year-old in 2001 is derived as:

$$w_{23,2001} = w_{23.2014}(1 + g_t)^{-13} = 9.01\,€/_h \tag{18}$$

The initial contribution, which grows annually at the rate g, is then calculated as:

$$p_{2001} = w_{23.2001} \cdot 40 \cdot 52 \cdot 6.91\% \tag{19}$$

The evolution of the accumulated fund from 2001, the year of the first contribution, to 2043, the year of retirement, is obtained via simulation. The empirical distribution, based on 10,000 trajectories, yields a lognormal distribution for the accumulated fund at maturity, consistent with the model discussed in the Sect. 2.

Once the distribution is obtained, it is necessary to convert the accumulated fund into an annuity. In case the option is not exercised, the standard conversion coefficient is applied. Conversely, if the option is exercised, the value of q must be determined to apply the formulas presented earlier.

As can be noted by analysing Eq. 11 to Eq. 14, an increase in q results in a larger increase in the LTC benefit annuity (R_i) than the corresponding reduction in the active-phase annuity (R_a). For the evaluation, the relationship between q and the LTC costs is explored.

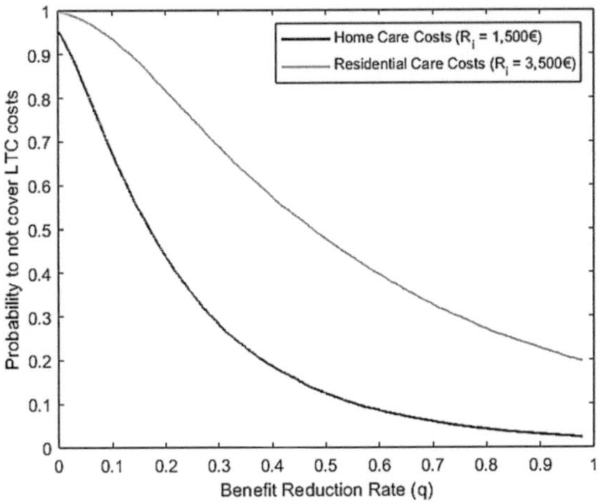

Fig. 1. Relation between the benefit reduction rate and the probability to not cover LTC costs

Depending on the type of care required, LTC costs range from €1,500 to €3,500 per month, as outlined in Sect. 3.3. Using the distribution of the disability annuity, percentiles are analyzed to determine the values of q needed to ensure that the annuity at a given percentile covers these costs.

Figure 1 shows the relationship between q and the probability of not covering LTC expenses. The graph displays the values of q required for the disability annuity at a given percentile to cover either home care costs (€1,500, darker line) or residential care costs (€3,500, lighter line). This relationship can be formally expressed through the following probability measure:

$$\alpha(q) = p\{R_i < \overline{R}_i\} = p\left\{S\tau\left(1 + q\frac{\ddot{a}_x^{aa}}{a_x^{ai}}\right) < \overline{R}_i\right\} = p\left\{S < \frac{\overline{R}_i}{\tau}\frac{a_x^{ai}}{a_x^{ai} + q\ddot{a}_x^{aa}}\right\} \tag{20}$$

where $\alpha \in [0; 1]$ and $q \in [0; 1]$.

Thus, considering the empirical distribution of S obtained through a Monte Carlo approach, the values of q corresponding to any given percentile α can be derived.

Note that for certain percentiles, there may be no value of q satisfying this equation due to the constraint on q. The values of $\overline{R}_i$ considered are €1,500 and €3,500.

More details are shown in Table 3. For instance, by forfeiting 74.39% of the complementary pension annuity, there is only a 5% probability that the disability annuity will fall below €1,500. Similarly, forfeiting 84.40% ensures that the probability of not covering residential care costs (€3,500) is limited to 25%.

Table 3. Relationship between the benefit reduction rate and some levels of the probability to not cover LTC costs.

α	q	
	$\overline{R}_i = 1500$	$\overline{R}_i = 3500$
0.5%	-	-
1%	-	-
5%	0.7439	-
10%	0.5530	-
25%	0.3281	0.8440
50%	0.1669	0.4748

As evident from the results, certain percentile values have no corresponding q, as there are no solutions to the equation under the $q \in [0; 1]$ constraint. Additionally, if the individual forfeits their entire pension annuity ($q = 1$), the probability of failing to cover the costs of a caregiver is 2.1%, while the probability to cover residential care costs is 18.91%.

Setting a more reasonable threshold, such as forfeiting 30% of the pension annuity ($q = 0.3$) results in a 28.12% probability of not covering caregiver costs and a 68.86% probability of not covering residential care costs.

4 Conclusions

From the application of the model to the data, is observed that, by forfeiting a significant portion of the complementary pension annuity, it is possible to achieve a reasonable confidence level to cover at least the cost associated with caregiver assistance.

However, the primary issue lies in the fact that the wage data, and thus the TFR contributions, represent averages across all workers. While could be conclude that this solution could effectively manage the risk of LTC dependency, albeit with considerable sacrifices, this conclusion applies only on average. This means that workers with wages in the lower tail of the distribution, below the average, may not achieve the same outcome.

This issue is particularly significant because lower-wage workers typically have lower levels of education, which correlates negatively with health literacy and preventive healthcare awareness. As a result, individuals with fewer financial resources are not only at higher risk of becoming LTC-dependent but are also more exposed to the risks of long-term care due to limited economic means.

Future developments may focus on enriching the financial and demographic modeling, incorporating career heterogeneity, diversified investment strategies, and more granular longevity and dependency projections.

References

1. Apicella, G., De Giorgi, E.G.: A behavioral gap in survival beliefs. J. Risk Insur., 1–29 (2024)
2. Brown, J.R., Finkelstein, A.: The private market for long-term care insurance in the united states: a review of the evidence. J. Risk Insur. **76**(1), 5–29 (2009). Special Issue on LongTerm Care Insurance and Health Insurance
3. Brouhns, N., DenuitM, V.J.K.: A Poisson log-bilinear regression approach to the construction of projected lifetables. Insur. Math. Econ. **31**(3), 373–393 (2002)
4. Censis: 58° Rapporto sulla situazione sociale del Paese 2024 (2024)
5. Costa, G., et al.: Salute in Italia e livelli di tutela: approfondimenti dalle indagini ISTAT sulla salute (2016)
6. Cox, J.C., Ingersoll, J.E., Ross, S.A., Jr.: A theory of the term structure of interest rates. Econometrica **53**(2), 385–407 (1985)
7. De Angelis, P., Di Falco, L.: Assicurazioni sulla salute: caratteristiche, modelli attuariali e basi tecniche. il Mulino (2016)
8. Dupourquè, E., Planchet, F., Sator, N.: Actuarial Aspects of Long-Term Care, pp. 95–192. Springer (2019)
9. Fama, E.F.: The behavior of stock-market prices. J. Bus. **38**(1), 34–105 (1965)
10. Fornero, E.: "Reform, Inform, Educate": A New Paradigm for Pension Systems' Sustainability. The Future of Welfare in a Global Europe (2015)
11. Haberman, S., Pitacco, E.: Actuarial Models for Disability Insurance. Chapman & Hall/CRC (1999)
12. Levantesi, S.: La valutazione del rischio di non autosufficienza nelle assicurazioni Long Term Care. Seminario, Centro studi Adepp (2011)
13. Ministero della Salute: Il genere come determinante di salute. Lo sviluppo della medicina di genere per garantire equità e appropriatezza della cura. Quaderni del Ministero della Salute, n. 26 aprile 2016 (2016)
14. Ministero della Salute: Relazione sullo Stato Sanitario del Paese 2017–2021 (2022)

Author Index

M. La Rocca et al. (Eds.): WMAF 2025, *New Perspectives in Mathematical and Statistical Methods for Actuarial Sciences and Finance*, p. 261, 2025.
https://doi.org/10.1007/978-3-032-05551-4

MIX
Papier aus verantwortungsvollen Quellen
Paper from responsible sources
FSC® C105338

FSC
www.fsc.org

If you have any concerns about our products,
you can contact us on
ProductSafety@springernature.com

In case Publisher is established outside the EU,
the EU authorized representative is:
Springer Nature Customer Service Center GmbH
Europaplatz 3, 69115 Heidelberg, Germany

Printed by Libri Plureos GmbH
in Hamburg, Germany